IOC Ocean Forum II

In this series:

Coastal zone space: Prelude to conflict?

UNDERSTANDING THE INDIAN OCEAN

Perspectives on oceanography

by T.S.S. Rao and Ray C. Griffiths

Environment and Development

UNESCO Publishing

The contributions to this book by T.S.S. Rao were supported by the US Office of Naval Research Grant No. N0014-92-J-1517, and constitute Contribution No. 8482 from the Woods Hole Oceanographic Institution.

Published in 1998 by the
United Nations Educational, Scientific and Cultural Organization
7, place de Fontenoy, 75352 Paris 07 SP, France

Cover design: Jean-Francis Chériez
Layout and page make-up: Micheline Turner
Printed by Imprimerie Jouve, 53100 Mayenne

ISBN 92-3-103448-0

Printed in France

PREFACE

In response to the wishes of its Member States, UNESCO and its Intergovernmental Oceanographic Commission (IOC) have made commitments to the strengthening of national and regional marine science and technology infrastructures and research programmes on the world ocean and its environment. The dissemination of oceanographic research results and scientific information has taken high priority in the Organization's marine science programmes. One specific avenue for such support is the publication of various forms of oceanographic literature, including material of a historical nature.

The present volume is the third published by the Organization on various historical aspects of marine scientific research in the Indian Ocean, a marine expanse regarded by scientists in the early 1950s as relatively unexplored, oceanographically speaking. The two previous publications dealt with specific expeditions. *Assault on the largest unknown*, by Daniel Behrman (published originally in English in 1981, followed by translations in Arabic, French, Russian and Spanish), explored the human dimensions of the International Indian Ocean Expedition (IIOE, discussed in some detail in Chapter 1 and elsewhere in this book). *Deep-sea challenge*, edited by Anthony L. Rice (published in English in 1986, and subsequently in Arabic), included the account of and ancillary texts concerning the Anglo-Egyptian John Murray Expedition to the Indian Ocean, which took place in 1933-34 on board the Egyptian research vessel *Mabahiss*. In contrast to the previous two publications, the present volume combines historical information with a summary of our oceanographic knowledge of this ocean.

UNESCO, following its 50th anniversary year and in anticipation of the International Year of the Ocean in 1998, declared by the United Nations General Assembly in December 1994, renders homage to the principle of international cooperation in this important field of science. The IOC of UNESCO, three decades after the completion of the IIOE, during which it played a coordinating role, salutes the mammoth ensemble of work carried out by scientists and institutions of many nations in the long march towards a better understanding of this great body of water. The resulting knowledge and information are crucial in piecing together the global scheme of ocean processes and resources, which – enmeshed with those of the land, the air and the cryosphere – greatly influence the environment of the Planet Earth.

The Organization is pleased to recognize the generous support of the following co-sponsors towards the publication of this work: Global Environmental & Ocean Sciences (GEOS) Ltd., France's Institut Français de Recherche Scientifique pour le Développement en Coopération (ORSTOM) and the United States Office of Naval Research (ONR). Also, the dedicated work of the authors and others who assisted in the preparation of the manuscript are commended.

The authors

Professor T.S.S. Rao, formerly with the National Institute of Oceanography (1964-86), in Goa, India, is a marine biologist. From his travels and experience, one can cite his tour as visiting professor at the Instituto Oceanográfico at the Universidad de Oriente, Cumaná, Venezuela (1973-74). Later, he was Dean of the Faculty of Applied Sciences at Goa University (1986-91), where he established the Department of Marine Sciences and Marine Biotechnology. As head of the Indian Ocean Biological Centre at Cochin and Scientific Liaison Officer for the US Program in Biology during the International Indian Ocean Expedition (1962-65), he brings to bear on the book a broad understanding of the oceanography of the Indian Ocean. Besides the treatment of sea-related mythology, as well as early exploration and oceanography in the region, Professor Rao has provided a succinct review of the biology of the Ocean, supported by a brief account of geological/geophysical knowledge of the Indian Ocean basins and continents.

A complementary role was played by Mr. Ray C. Griffiths, also a marine biologist by training and currently a consultant in marine science. Mr. Griffiths brings to this book his diversified background of study and professional experience gathered during a varied career of research, consulting and international service. Following aquatic and fishery research in Canada, in Scripps Institution of Oceanography (California, USA) and in Venezuela and other parts of the Caribbean, Mr Griffiths worked for twenty-five years as an FAO specialist, mostly detached to the IOC. He is thus well versed in international cooperation in marine scientific endeavour.

CONTENTS

ABOUT THIS BOOK

The event that led eventually to the development of the text of this book was a visit to the USA by Professor T.S.S. Rao in 1991. By coincidence, the US National Science Foundation and the US Office of Naval Research (ONR) had, at that time, finalized plans to mount a major study in the Arabian Sea as a US contribution to a new international marine research programme – the Joint Global Ocean Flux Study (JGOFS), co-sponsored by the Scientific Committee on Oceanic Research (SCOR) and the Intergovernmental Oceanographic Commission (IOC of UNESCO) within the International Geosphere-Biosphere Programme. Professor Rao approached Dr. Bernard J. Zahuranec, then Associate Programme Director at ONR, with a request for support for his stay in the USA to write a book on the oceanography of the Indian Ocean as a background for the proposed research programmes in the area. Dr. Zahuranec agreed to support the preparation of such a book, so long as it could be sponsored by an appropriate institution in the USA.

Professor Rao was appointed Guest Investigator at the Woods Hole Oceanographic Institution where he completed the first draft of the manuscript. His work there was supported by a grant from ONR.

Subsequently, the manuscript was submitted to Dr. Gunnar Kullenberg, Executive Secretary of the IOC, following the latter's visit to India, with the suggestion that UNESCO might be willing to publish it. Dr. Kullenberg agreed to sponsor the publication project under the aegis of IOC. However, it was deemed appropriate first to broaden the manuscript's coverage to include the work of other components of the international oceanographic community, notably those of Australia, France and South Africa, particularly concerned with the Indian Ocean. Also, Dr. Kullenberg wished to make adequate reference to the role of IOC in future oceanographic work in the Indian Ocean region. To this end, he invited Mr. Ray Griffiths, with the agreement of Professor Rao, to incorporate the new material and generally to ensure conformity with IOC requirements for such a publication, and thus to become the junior author.

Within the constraints of a historical and commemorative approach in which the IIOE – devised by the SCOR (member of the International Council of Scientific Unions) and co-ordinated by the IOC – is revealed as the gateway to the modern oceanography of this great ocean, the authors have tried, by style and approach, to appeal to a wide readership. This engenders the inevitable risk of disappointing, on the one hand, the comparatively new breed of 'meteo-oceanographers', with their complex models of ocean-atmosphere coupling, based mainly on remotely sensed data, and, on the other hand, the informed layman who may find the details and complexities of even classical oceanography and marine biology not always easy to absorb.

Nevertheless, the specialist and the layman will, it is hoped, appreciate the fact that the basic findings in the different disciplines are closely interlinked and influence one another, thus vindicating the hypotheses of Redfield and Lovelock (among others), according to which the physical and chemical conditions of the surface of the earth, the atmosphere and the oceans have been and are actively determined by the presence of life itself and the need of such life to ensure its own survival and evolution. This elegant idea behind the ancient Greeks' conception of *Gaia*, passionately espoused by Lovelock and others, also lies behind the Hindus' *Lakshmi*. The authors have kept this idea in mind in reviewing the knowledge and understanding of the oceanography of the Indian Ocean.

In publishing this volume, therefore, the Organization does not claim to be producing a complete treatment of the subject; rather, an effort has been made to publish a synthesis and a building block for a more complete edifice. Furthermore, it was the intent here to maintain the book's essentially historical emphasis and to keep it to a modest size, thus enabling it to be published in the form of a commemorative volume, coming on the heels of the thirtieth anniversary of the completion of the IIOE as well as the fiftieth anniversary of UNESCO itself. Opportunely, it will appear in print at about the time when the global marine community begins celebrating 1998 – the International Year of the Ocean. In view of these special circumstances, the colour-illustrated section, with texts and images from the sponsors, was included. Due to our summary presentation, the many new avenues of research, especially modelling of the Indian Ocean and its interaction with the atmosphere, are only lightly touched upon, not least because these avenues are opening up at a considerable rate. We nevertheless consider that this book will be of value to a broad range of readers as a good synopsis of this Ocean's oceanography, and that it will afford a view on various recent national and international efforts.

GARY WRIGHT, *Editor*
Coastal and Marine Science Publications
UNESCO

ACKNOWLEDGEMENTS

The authors, individually or jointly, as the case may be (see previous section), wish to express their sincere thanks to: Dr. Bernard J. Zahuranec, Associate Programme Director, US Office of Naval Research (ONR) and to Dr. John H. Ryther, Woods Hole Oceanographic Institution (WHOI), for their support from the start, in processing the ONR grant, and during the preparation of the original version of the manuscript; likewise, Dr. Robert B. Gagosian, Director of WHOI, and to Dr. Joel C. Goldman, Chairman of the WHOI Biology Department; to the staff of the WHOI Marine Biology Laboratory Library, particularly Judy Ashmore, Peg Costa, Colleen Hurter, Heidi Nelson and Patricia Pratson, for their invaluable help with documentation; to Judy Kleindinst, Biology Department Administrator, for all administrative help; to Jane McLaughlin, for early language editing; and to Ethel Lafave, for typing the original manuscript.

Special thanks are also given to Dr. George Grice, for reviewing the manuscript at various stages; to Prof. Eugene LaFond and Mrs. LaFond, for the first (historical) chapter; to Prof. K.O. Emery, for the second (geological) chapter; to Dr. Bruce Warren, for the third (oceanographic) chapter; to Dr. David McGill, for the fourth chapter (on nutrients); and to Dr. Howard Sanders, for the fifth (biological) chapter.

The authors wish to thank Dr. Gunnar Kullenberg, Executive Secretary of the Intergovernmental Oceanographic Commission (of UNESCO) for having placed his confidence in them for the restructuring of the original manuscript and for approving the publication of the revised version.

The authors wish also to thank a number of scientists who kindly provided references and/or bibliographies on various aspects of the book's subject that were not originally covered: Professor Henri Lacombe, Laboratoire d'Océanographie Physique, Musée d'Histoire Naturelle, Paris, France; Michèle Fieux and Pascale Delecluse, Laboratoire d'Océanographie Dynamique et du Climat, Université Pierre et Marie Curie, Paris, France; Alain Poisson and Nicolas Metzl, Laboratoire de Physique et Chimie Marines, Université Pierre et Marie Curie, Paris, France; Jean-René Donguy, ORSTOM, Plouzané, France; J. Stuart Godfrey, Division of Oceanography, CSIRO, Hobart, Tasmania, Australia; Marten Gründlingh, Division of Earth, Marine and Atmospheric Science and Technology, Stellenbosch, South Africa; and Professor J.R.E. Lutjeharms, Department of Oceanography, University of Cape Town, Rondebosch, South Africa.

The authors wish to acknowledge the encouragement in the present endeavour kindly given by Dr. B.N. Desai and Dr. E. Desa, respectively former and present Director of the National Institute of Oceanography, Goa, India.

The authors share with the publisher their gratitude to those organizations that have generously supported this publication, morally and financially: Global Environmental & Ocean Sciences (GEOS) Ltd., the Institut français de recherche scientifique pour le développement en coopération (ORSTOM, France), the US Office of Naval Research (ONR, USA). Without ONR's major support for the basic research and preparation, the original manuscript would not have been possible.

The senior author wishes particularly to thank his wife Padma, and his daughter Seetha Murthy, for their moral support, encouragement and assistance. The junior author can do no less for his wife, Jeannie Dombret-Griffiths, and his daughter, Lili.

Text illustration credits

The kind permission to reproduce certain figures, given by the copyright holders of the original sources, free of charge, is gratefully acknowledged. The relevant information, with the respective figure numbers, is given below; the corresponding sources are indicated (by name of author(s) and year) in the legends of the figures concerned and are included in the list of references (pp. 175-185).

British Museum of Natural History, London, United Kingdom: Fig. 4.1.
Butterworth-Heinemann, Oxford, UK: Figs. 2.7, 3.4a, 3.4b, 3.9, 3.11, 3.12a, 3.12b, 3.14, 3.18a, 3.18b.
David Higham Associates Ltd., London (for the Hakluyt Society): Fig. 1.4a, 1.4b, 1.4c, 1.4d.
Elsevier Science, Amsterdam, The Netherlands: Fig. 2.8.
Elsevier Science Ltd., Kidlington (Oxford), United Kingdom: Figs. 2.9, 3.16, 4.6a, 4.6b, 4.6c, 4.6d.
Geological Society of America, Boulder, CO, USA: Figs. 2.5, 2.11, 2.12.
Indian Journal of Marine Science: Fig. 4.4.
John Wiley & Sons, New York, NY, USA: Figs. 2.2, 2.4.
Kluwer Academic Publishers, Dordrecht, The Netherlands: Fig. 2.6.
Plenum Publishing Corporation, New York, NY, USA: Fig. 2.10.
South African Journal of Science, Pretoria, Republic of South Africa: Fig. 3.17.
SPES (for Oceanologica Acta), Gauthier-Villiers Editeur, Paris, France: Fig. 3.3.
Springer Verlag, Heidelberg, Germany: Fig. 4.7, 4.8a, 4.8b, 4.8c, 4.8d.
The Open University, UK, as author and publisher: Fig. 2.3.
Vivekananda Rock Memorial and Vivekananda Kendra: Figs. 1.2, 1.3.
Zoologisk Museum, Copenhagen, Denmark: Fig. 4.2a, 4.2b, 4.5.

ACRONYMS AND ABBREVIATIONS

AABW	Antarctic Bottom Water
AAIW	Antarctic Intermediate Water
AAMW	Australasian Mediterranean Water
AASW	Antarctic Surface Water
ACC	Antarctic Circumpolar Current
ACMRR	Advisory Committee on Marine Resources Research (FAO)
AGOR	Advisory Group on Ocean Research (WMO)
AL	Amoebocyte lysate
AOU	Apparent Oxygen Utilization
APT	Automatic Picture Transmission
ASW	Arabian Sea Water
BBW	Bay of Bengal Water
CCD	Carbonate Compensation Depth
CDW	Circumpolar Deep Water
CMFRI	Central Marine Fisheries Research Institute (India)
CNES	Centre National d'Etudes Spatiales (France)
CPR	Continuous Plankton Recorder
CSIR	Council of Scientific and Industrial Research (India)
CSIRO	Commonwealth Scientific and Industrial Research Organization (Australia)
DBCP	Drifting Buoy Co-operation Panel (IOC-WMO)
DDT	Dichloro-diphenyl-trichloroethane
DSDP	Deep Sea Drilling Project
DSDS	Deep Sea Drilling Ship
EAC	East Arabian Current
EACC	East African Coastal Current (=ZC)
ECC	Equatorial Countercurrent
EEZ	Exclusive Economic Zone
EMC	East Madagascar Current
ENSO	El Niño-Southern Oscillation
FAO	Food and Agriculture Organization of the United Nations
FGGE	First GARP Global Experiment
FV	Fishing Vessel
GARP	Global Atmospheric Research Programme (WMO)
GCOS	Global Climate Observing System
GEOS	Global Environmental and Oceanic Services Ltd.
GEOSECS	Geochemical Ocean Sections Study
GIPME	Global Investigation of Pollution in the Marine Environment (IOC-UNEP)
GLOBEC	Global Ocean Ecosystems Dynamics (SCOR-IOC-ICES-PICES)
GLOSS	Global Sea-Level Observing System (IOC)
GOEZS	Global Ocean Euphotic Zone Study (ICSU/IGBP)
GOOS	Global Ocean Observing System (IOC-WMO-UNEP)
HMIS	His (Her) Majesty's Indian Ship (UK)
HMS	His (Her) Majesty's Ship (UK)
ICES	International Council for the Exploration of the Sea
ICSU	International Council of Scientific Unions
ICW	Indian Central Water
IDOE	International Decade of Ocean Exploration
IEW	Indian Equatorial Water

IGBP	International Geosphere-Biosphere Programme (ICSU)
IGOSS	Integrated Global Ocean Services System (IOC-WMO)
IGY	International Geophysical Year
IIOE	International Indian Ocean Expedition
IIW	Indonesian Intermediate Water
IMC	International Meteorological Centre
INCOR	Indian National Committee on Oceanic Research
INDEX	Indian Ocean Experiment
INS	Indian Navy Ship
IOBC	Indian Ocean Biological Centre
IOC	Intergovernmental Oceanographic Commission (of UNESCO)
IOCSP	Indian Ocean Climate Studies Panel (WCRP)
IODE	International Oceanographic Data and Information Exchange (IOC)
IODW	Indian Ocean Deep Water
IOSN	Indian Ocean Standard Net
IUW	Indonesian Upper Water
JGOFS	Joint Global Ocean Flux Study (SCOR-IOC; ICSU/IGBP)
JOIDES	Joint Oceanographic Institutions for Deep Earth Sampling
LME	Large Marine Ecosystem
LOICZ	Land-Ocean Interaction in the Coastal Zone (ICSU/IGBP)
MBL	Marine Biological Laboratory (WHOI)
MONEX	Monsoon Experiment
MYBP	million years before the present
NADW	North Atlantic Deep Water
NASA	National Aeronautics and Space Administration (USA)
NEC	North Equatorial Current
NEM	Northeast Monsoon
NIO	National Institute of Oceanography (Goa, India; Karachi, Pakistan)
NIOSW	North Indian Ocean Surface Water
NOAA	National Oceanic and Atmospheric Administration (USA)
OCOS	Ocean Climate Observing System
ODP	Ocean Drilling Programme
ONR	Office of Naval Research (USA)
OOSDP	Ocean Observing System Development Panel (WCRP)
ORSTOM	Institut Français de Recherche Scientifique pour le Développement en Coopération
OSLR	Ocean Science in Relation to Living Resources (IOC)
OTEC	Ocean Thermal Energy Conversion
PF	(Antarctic) Polar Front
PGW	Persian Gulf Water
pH	measure of alkalinity/acidity
PICES	North Pacific Marine Science Organization
RIDGE	Ridge Inter-Disciplinary Global Experiment
RRS	Royal Research Ship (UK)
RSPGIW	Red Sea - Persian Gulf Intermediate Water
RSW	Red Sea Water
RV	Research Vessel
SAF	Sub-Antarctic Front
SC	Somali Current
SCOR	Scientific Committee on Oceanic Research (ICSU)
SEC	South Equatorial Current
SECC	South Equatorial Countercurrent
SICW	South Indian Ocean Central Water
SINODE	Surface Indian Ocean Dynamics Experiment (France)
SJC	South Java Current
SSS	Sea-surface salinity
SST	Sea-surface temperature
STF	Sub-tropical Front (Convergence)
SWM	Southwest monsoon
TEMA	Training, Education and Mutual Assistance in the Marine Sciences (IOC)
TOGA	Tropical Ocean and Global Atmosphere Study (WCRP)
TOPEX	Ocean Topography Experiment
T/S	Temperature/Salinity (diagram)
UN	United Nations
UNCLOS	United Nations Conference on the Law of the Sea
UNEP	United Nations Environment Programme
UNESCO	United Nations Education, Scientific and Cultural Organization
WCRP	World Climate Research Programme (ICSU-WMO-IOC)
WGB	Weddell Gyre Boundary
WHO	World Health Organization
WHOI	Woods Hole Oceanographic Institution (USA)
WMO	World Meteorological Organization
WOCE	World Ocean Circulation Experiment (WCRP)
WWW	World Weather Watch (WMO)
XBT	expendable bathythermograph
ZC	Zanzibar Current (=EACC)

INTRODUCTION

The accumulation of scientific information on the Indian Ocean can be considered to have gone through three phases.

The first phase covers a long period of history, from Vedic times (2500-1500BC) up to the end of the 15th century when Vasco da Gama, in 1497-98, rounding the Cape of Good Hope, discovered the sea route to India. During this period, knowledge of the sea and sea routes, harbour locations, favourable winds etc. was mainly in the hands of the Greeks, Egyptians and Arabs, in respect of the Arabian Sea, and by the Indians, in respect of the Bay of Bengal and the East Indies. The writings of the Arab navigator, Ahmad Ibn Magid of the 15th century AD and the *Periplus of the Erythrean Sea* by an unknown author (1st-2nd century AD) were considered to be very valuable documents for marine navigation and commerce.

In the second phase, starting from the 15th century AD up to the time of the International Indian Ocean Expedition (IIOE, 1959-65), the vast expansion of the European colonial powers made it essential for them to acquire detailed knowledge of sea routes, depths, currents and tides for the use of their naval ships and merchant marines. In pursuit of such knowledge, many scientific expeditions were launched, not only into the Indian Ocean, but also throughout the world ocean. The first of these European expeditions, the Danish *Arabia Felix* Expedition (1761-67) to Arabia is considered as being the first scientific survey to be undertaken in the Indian Ocean. Several others followed, organized by the Danes again, the British, the Austrians, the Germans, the Swedes and the Dutch. The British *Challenger* Expedition (1872-76), although perhaps the most famous, bypassed the northern Indian Ocean.

Even so, by the middle of the 20th century, the oceanography, the monsoons and their effects, the productivity and the fisheries of the Indian Ocean were still little known, when compared with the information available on the Atlantic and Pacific Oceans. To remedy this situation, the IIOE was organized and was carried out with scientists of 23 countries and 40 survey ships from thirteen of those countries participating. As a result, vast amounts of data were collected and analysed and more than a thousand research papers and some unique atlases on plankton, hydrography and geology of the Indian Ocean were published. In these efforts, except for India and Pakistan, most of the research was done by European and American scientists. In this second phase, the contributions of the Indian Ocean countries were marginal.

The third phase commences after the IIOE, when the discoveries made during the Expedition attracted scientists, particularly from America, Australia, France, Germany, India, Pakistan, the UK and the USSR, to organize special oceanographic cruises to study and understand the role of the reversing monsoons in the Arabian Sea and elsewhere. As a result, many unknown or unsuspected features of the Indian Ocean came to light, which have become the objects of specific study, not only from research vessels, but also from remote-sensing systems – notably satellites and drifting buoys – in recent years.

Following the IIOE, India's involvement in oceanographic research in the Indian Ocean was enhanced by the establishment of the National Institute of Oceanography, in Goa, and by the acquisition of many large research vessels which were manned by scientists of the Institute and of the Geological Survey of India, Calcutta, the Central Marine Fisheries Research Institute, Cochin, and the Indian Navy which had a fleet of hydrographic survey vessels. Other countries in the area have also either enhanced their existing facilities or established new marine research centres, mainly concerned with fisheries. However, the publication of research results from most of these countries is sporadic and the scientific papers are not readily available for review, except those of Indian and Pakistani researchers, whose publications are found mostly in their respective national journals.

The material in this book is taken or adapted from a large number of research papers published in various journals, collated IIOE reprints issued by UNESCO, expedition reports and proceedings of symposia held in India and elsewhere. The references to various authors are used mainly to underpin the themes of the subjects covered herein, rather than to expose the numerous details the references provide. There are thousands of research papers and reports, 90 to 95 percent of which mostly deal with the biology of the coastal seas; these contributions have likewise come mainly from Indian scientists and institutions.

The first chapter of this book gives a brief history of oceanography (in a broad sense) of the Indian Ocean, starting in ancient times, when Egyptians, Arabs and Greeks ruled the commercial routes in and out of the Arabian Sea ports, and when navigation required precise knowledge of astronomy, tides and currents, and of the depth of the sea, particularly in coastal areas and harbour approaches. This knowledge was only slowly acquired by the pilots of the ships. It was only during the 16th century and thereafter that scientific survey of this ocean was undertaken. The accumulation of scientific knowledge was also slow, until the time of the IIOE which marked a major turning point. Not only was it the first major international exercise of its kind, but, because of it, the conduct of oceanographic research finally broke out of its national shackles. The addition to knowledge, once the data of the Expedition were digested and published, was dramatic, though raising as many questions as were answered, if not more. The level of co-operation between scientists from many countries who were interested in the Indian Ocean also rose considerably, so that, although national oceanographic programmes still exist, they are no longer conducted in the comparative isolation, and indeed competition, that prevailed before. This brief history therefore ends with the completion of the IIOE; the later international oceanographic programmes are mentioned as appropriate in the context of the related subjects. After tracing the genesis of the IIOE, the chapter ends with a description of the main benefits and results.

Chapter 2 describes the geological origin and evolution of the Indian Ocean. Whereas the ocean water was formed during the first billion years after the formation of Earth, nearly 4.5 billion years ago, the floor of the ocean is merely 150-190 million years old at most and is in constant renewal at the mid-ocean ridges and in constant destruction at the edges of the crustal plate on which each such ridge occurs. The Indian Ocean did not exist 100 million years ago; as the crustal plate carrying India broke away from the huge primeval continent of Gondwanaland, in the southern hemisphere, and drifted north, the Indian Ocean began to form in its wake. The mechanism of continental drift (and the underlying plate tectonics) by sea-floor spreading was discovered only about 30 years ago (although the idea of continental drift was advanced, in one way or another, particularly by biologists and palaeontologists, over a hundred years ago and crystallized in Wegener's theory in 1924). The chapter briefly describes the main tectonic features of the Indian Ocean: the ridges, basins and continental margins, as well as sedimentation and mineral deposits. The world's largest sedimentary fan extends from the head of the Bay of Bengal down to about 20°S.

Chapter 3 deals with the physical environment. The Indian Ocean is a natural laboratory where Nature plays with its monsoons, forcing and reversing ocean currents. Because the Indian Ocean has no connection with northern temperate and polar regions, the heat exchange between this ocean and the other major oceans (Atlantic, Pacific and Southern or Antarctic) and with the overlying atmosphere is still an unsolved problem. The deep-water circulation appears to be sluggish, judging by the development and persistence of a very thick, oxygen-poor layer all over the northern Indian Ocean, with significant consequences for the biota and for fisheries. A brief description of the monsoons, the currents, water masses and upwelling is given in this chapter.

Chapter 4 is on nutrients. The concentration and availability of the nutrients, particularly nitrates, phosphates and silicates, primarily determine the productivity of the Indian Ocean. The northern Indian Ocean surface waters are poor in nutrients and their availability to the phytoplankton is transient, depending on the seasonal changes in the current patterns and upwelling systems. The development of a strong hydrochemical/hydrographical front at about 10°S, which acts as a barrier to free exchange of water and fauna between the northern and southern Indian Ocean, is a unique feature of the Indian Ocean; the dissolved-oxygen distribution acts similarly.

The biology of the Indian Ocean is the subject of Chapter 5 which highlights the gradual development of our knowledge of the natural history and distribution of the marine organisms constituting the three major types of sea life: plankton, nekton and benthos. Alcock's observations of the Indian Ocean fauna in his classical book *A Naturalist in the Indian Seas*, published in 1902, are briefly summarized. It is interesting to note the types of observations biologists made and reported in the early part of this century.

The biological studies reported here describe the distribution of the fauna and flora in relation to the prevailing monsoon. Although the primary productivity is fairly high in the northern Indian Ocean, it appears not to be translated into rich fishery resources. This is attributed to the lack of upwelling and mixing over wide areas of the Indian Ocean, the development and persistence of a thick and extensive oxygen-minimum layer, particularly in the Arabian Sea, and the sluggish deep-water circulation. All of these factors are inter-related.

It is believed that there are approximately 180,000 species in the world ocean, constituting a rich biodiversity, but the share of the Indian Ocean in this is not known in detail. Systematics are being neglected much to the disadvantage of the scientific community which requires specific identification of organisms for the purposes of proper management of the biological resources of this ocean, as well as for many global programmes.

Although the Indian Ocean occupies nearly 20 percent of the global marine area, it does not produce proportionately as much fish as the Atlantic and the Pacific Oceans. Its contribution has stagnated at around 6% of the total world catch of fish. The *per caput* consumption of fish for the Indian Ocean area is 3.3kg, whereas, for the rest of the world, it is 19.4kg. Since there are numerous published papers, books and reports of the Food and Agriculture Organization of the United Nations (FAO) dealing with the Indian Ocean fisheries and its various aspects, this chapter only briefly touches on fisheries.

Chapter 6 is an attempt to suggest a possible future for oceanographic research and development in the Indian Ocean. However, experience (e.g., the FAO(ACMMR)-SCOR-WMO(AGOR) report on *Global Ocean Research*, in 1969, often referred to as the Ponza Report, the US NRC's *The Continuing Conquest: Large-Scale Ocean Science for the Future*, in 1979, and *Oceanography in the Next Decade*, in 1992, and the IOC's *Ocean Science for the Year 2000*, in 1984) shows that forecasting the future of oceanographic research is something of a lottery; about half of the proposed research subjects never see the light of day – for many reasons, good or bad. The remainder that do see the light of day are also often appreciably modified by the evolution of relevant knowledge and understanding. With this experience in mind, we make several suggestions in the light of the foregoing chapters and current trends in oceanographic research. The chapter also focusses on the development of ocean-observing systems; this is not only because of the IOC's central responsibility in the development of such systems but also because of the need to discover unknown oceanic phenomena and processes and to measure the variability of known processes and phenomena, in time (e.g., monsoon variation from year to year) and in space (e.g., displacement of the principal currents, the extent of the oxygen-minimum layer), so as drastically to improve our predictive capabilities.

The ocean depths are no longer inaccessible to human probes. This was made possible in the last thirty years or so by the development of extraordinary and versatile acoustic and electronic instrumentation mounted on new kinds of equipment, such as submersibles and remotely operated vehicles, deep-sea drilling platforms, underwater TV cameras, automated drifting or moored buoys (at the surface and at depth), satellites etc., making it possible to explore the ocean's surface, its water column and its deep sea floor without hindrance.

The new knowledge acquired through their use revealed such processes as sea-floor spreading and the related mid-ocean ridges, with their thermal vents and the associated fauna, and such resources as sea-bed mineral deposits and polymetallic (so-called 'manganese') nodules; they also provided quasi-synoptic pictures of the ocean currents, the distribution of photosynthetic pigments, the overlying atmospheric features, such as cloud cover, sea-surface winds and waves, sea-surface temperature and sea level. The relative ages of the ocean floor and the ocean water could also be better measured than hitherto.

Although the IIOE and subsequent national and international efforts have answered some basic questions such as the effects of the monsoons on the surface circulation, on the biology and chemistry, sea-floor spreading in the Indian Ocean, thermal vents and hot brines in the Red Sea, marine pollution etc., oceanographers are still coming to grips with many of the subjects mentioned in this chapter. Many global programmes have been launched to study ocean currents, photosynthetic pigments, sedimentation rates etc., for which international collaboration is essential. Excellence in the basic sciences is equally important, since oceanography is an amalgamation of such basic sciences as biology, chemistry, geology and physics.

Even so, the gap between the developed and developing countries regarding the ability to conduct oceanographic research is ever widening. Under the new UN Convention on the Law of the Sea, every coastal State has acquired (or has the right to acquire), insofar as is possible, full sovereignty over a stretch of the sea extending from the coastal baseline out to 200 nautical miles and, in some special cases, even up to 350 nautical miles. This area of the ocean is defined as the Exclusive Economic Zone (EEZ) of the coastal State. However, of all the countries bordering the Indian Ocean, only Australia, India, Pakistan and South Africa are in a position to survey and exploit their EEZ resources; the others are not, not only because of a lack of expertise and equipment, but also perhaps because of a lack of knowledge of the ocean and its resources.

In the last twenty years, in the widespread belief that all the major oceanic phenomena or processes have been identified, scientific interest has turned away from basic descriptive oceanography and towards assessing the variability of such phenomena and processes, in time (on interseasonal, interannual, decadal and longer time scales) and in space (on metric and kilometric space scales). This interest has looked backwards, to geological and palaeological records in Nature and to archives in long-established research institutions that have never had, till now with the advent of modern computers, the resources to exploit all the accumulated data, and forwards through the development of imaginative models of natural systems depending, at first, on existing data and, later, on new data by which to refine the models and bring them asymptotically closer to reality, perhaps eventually to supplant direct observations of all but a very limited number of parameters common to the natural system and the model thereof.

At the same time, as decriptive oceanography has faded into the background of marine science, the relationship between the ocean and the overlying atmosphere has come to the fore. Although the interface is only a few hundred metres thick (say from the depth of the thermocline up to 200m above the sea surface), it has become a happy hunting ground for oceanographers and meteorologists alike, at first in an atmosphere of general ignorance of each other's subject and later in a shared interest in acquiring the essential data. At times, the interest in data acquisition largely outstripped the interest in understanding the processes or even in delimiting them. Meteorologists measured oceanic sea-surface parameters, oceanographers measured atmospheric sea-surface parameters, and a hybrid discipline, still unnamed, took root. The pursuit of data, largely by remote sensing from satellites, but also from drifting buoys, hence largely detached from the object of main interest – the air-sea interface – ran the risk of being pursued without a null hypothesis (the initial proposal to be tested), but the niche was conveniently filled by the ocean-atmosphere modellers; not, we think, a very convincing name for this hybrid race of marine scientists.

Nevertheless, the Indian Ocean, so rich in specific, even unique, phenomena or processes, of which the monsoons are the best known, provided an early pole of attraction for such modellers – from oceanographers who had only a secondary, if not evanescent, interest in the weather and climate of the Indian subcontinent and beyond, to meteorologists who had a similarly vague interest in ocean currents. That situation has now largely changed and the various efforts to monitor such phenomena or processes are briefly described.

Since the approach followed in this book is historical, and since it is a historian's tenet not to include in history human activities that are on-going and still exciting the interest of journalists, we have taken the uncomfortable decision not to get our feet wet in the domain of interaction between the sea and air above it. To have done so would have greatly extended this book in space and in time, and this at a time when the general understanding of this interaction is still, in many important respects, far from having crystallized (i.e., become 'history'). There is still a significant imbalance in our estimates of the Earth's heat budget even on the large scale. There are comparable imbalances or 'holes' in the carbon dioxide-carbon budget, for example, and still only vague ideas about lesser, but potentially significant, budgets such as that of methane, of which the sea, and not least the Arabian Sea, may be an important source.

We have also taken a similar decision in respect of marine pollution, not only because its level in the Indian Ocean is still far from alarming, but because the means of measuring it on an ocean-basin scale do not exist in this region. There is also some indication, from the difficulties still encountered elsewhere in the intercalibration of some pollution measurements, that it is still impossible effectively to monitor some forms of marine pollution over any extensive ocean area. Yet, we do not discount the potential importance of marine pollution in the well-being of ocean resources, ecosystems, flora and fauna, exploited or not.

Nevertheless, when discussing the future of oceanography of the Indian Ocean, and particularly the role of the IOC, we do not hesitate to mention the relevant IOC activities in respect of ocean-atmosphere coupling and marine pollution research and monitoring.

CHAPTER 1

A BRIEF HISTORY OF OCEANOGRAPHIC STUDIES IN THE INDIAN OCEAN

The International Indian Ocean Expedition (1959-1965), which is described later in more detail, was a major turning point in the oceanography of that ocean. Not only was it the first major international exercise of its kind but, because of it, the conduct of oceanographic research finally broke out of its national shackles. The addition to knowledge, once the data of the Expedition were digested and published, was dramatic, though raising as many questions as it answered, or even more. The level of co-operation between scientists from many countries who were interested in the Indian Ocean also rose considerably, so that, although national oceanographic programmes still exist, they are no longer conducted in the relative isolation and, indeed, competition that prevailed before. This brief history therefore ends with the completion of the IIOE; the later international oceanographic programmes are mentioned only in the context of the relevant subjects.

Figure 1.1
The Vedic story of the Samudra Manthan (churning of the ocean): Deva and Asura churned the ocean to obtain the nectar of immortality. As a result, many jewels emerged, including Sri Lakshmi. The sea was invoked to do good to the people. (From the epic Srinad Bhagavatham, 1st-2nd century A.D.)

Early Indian/Arabic concepts of the oceans

Lakshmi and the Samudra Manthan

Lakshmi, the Goddess of Riches, frequently appears in the Vedic literature of the Hindus as the daughter of the King of the Seas, a thoughtful association of marine resources with the daily invocation to the goddess. Lakshmi also happens to be the consort of Lord Vishnu, one of the trinities of the Hindu pantheon, the provider and protector of all mankind.

There is a fascinating story associated with the birth of *Lakshmi* entitled *Samudra Manthan*, the churning of the oceans. Apparently, the Aryans in Vedic times believed that the oceans contained all sorts of resources and in fact one of the names given to the sea is *Ratna Garbha*, meaning repository of precious stones.

According to the story, related in the epic *Bhagvatham*, gods and demons prayed to the Lord Vishnu to help them obtain the elixir of life (*amrut*) which, by drinking it, would render them immortal. Lord Vishnu suggested that gods and demons together should churn the oceans to obtain the elixir. Thereupon, the two warring groups went to the ocean and began to churn it, using a mountain as a stirrer, and a large snake by the name of Vasuki as a rope (Figure 1.1). Lord Vishnu himself took an incarnation as a turtle and kept the mountain from sinking. As they began to churn the ocean, many things came out of it. The first to come up was Sri Lakshmi followed by *halahala*, a highly potent poison. Then came Kamadhenu, a celestial cow which could grant anything asked by the people, followed by a horse, an elephant, diamonds, a flowering plant (*parijata*), dancing girls, a medical expert named Dhanvantari kalpavriksha, a tree useful to all mankind, the moon, a conch and finally the elixir or nectar. At this point, the gods and demons began to fight to partake of the elixir, but Lord Vishnu managed to give the elixir only to the gods, thus depriving the demons of immortality.

The point of interest to us is that the Vedic Aryans believed that the oceans had many things in them useful to humankind, but they made no effort to be precise nor did they leave any written or pictorial record of obtaining resources such as fish, plants, weeds or other substances from the ocean. In some of the temples built during the 10th-16th centuries (AD), there are stone sculptures of oceanic resources, such as fish, giving clear evidence that the people were aware of their existence.

The Indian Ocean, even more than the Atlantic and the Pacific Oceans, has played an important role in commerce and the adaptive radiation of many people living along the shores of the coastal countries during past millennia. That India was a repository of fabulous wealth and spices drove many European, Mediterranean and Arab countries to explore and establish trade routes across the unknown Indian Ocean from the very beginning of recorded history. Panikkar and Srinivasan (1972) have given an account of the early concepts of oceanographic phenomena of the Indian Ocean deduced from Greek and Arab writings and from Indian sources, such as literary records (Sanskrit and Tamil) and archaeological findings (excavations, numismatics and paintings).

Knowledge of winds, waves and tides – early Indian concepts

The Winds: Some knowledge of winds, waves and tides comes from the Vedic literature of the Hindus. The information is clothed in ritualistic activities, such as prayer to gods of war and rain. The central myth of the monsoons is that Vritra, a demon, would prevent the breaking of the monsoon over India and Indra would be invoked to kill Vritra so that the monsoon would come with attendant rains. Other deities, as Vayu, Vata, Rudra and Parjanya, were also appeased so as to bring rain. The fight between Indra and Vritra was described as a cosmic battle in which lightning and thunderbolts were freely used by Indra to subdue or kill Vritra.

In the epic *Mahabharata*, there is a detailed description of eight types of wind; namely, *pravaha*, *avaha*, *uttarvaha*, *samvaha*, *valahagha*, *vivaha*, *parivaha* and *parrovaha*. It is not clear exactly what is meant by these winds. Four types of wind blowing from different directions are described in the Sangam literature of the Tamils. These winds are *kontal* (east wind), *kotal* (west wind), *vatai* (north wind) and *tenral* (south wind).

The seasonal winds which regularly occur and blow over the Indian subcontinent are the monsoons, a term derived from the Arabic word *mausim*, meaning seasonal winds. The south-west monsoon (SWM) prevails from June to September and the north-east monsoon (NEM), from October to March. Generally, there are calm weather periods between the two monsoons. Archaeological excavations, dating to the period 2500-1700BC, indicate that Indus Valley civilizations, Mohenjo-daro and Harappa, experienced the monsoons. However, in the *Vedas*, which were written much later (*ca.* 1500-1200BC), there is no clear mention of the monsoons. In the *Rg Veda*, there is a reference to *maruts* which are rain-bearing winds, which are subdivided into eastern *maruts* and western *maruts*, probably referring to the NE and SW monsoons. These winds are described as carrying big and inexhaustible clouds which they spread in the sky to bring rain. In the later Vedic texts, the SE or NW monsoon is referred to as *sahlavada* and the Buddhist texts call the same phenomena *kalamegha* and *varshavalshaga*.

Until we come to the writings of the Greeks and Arabs, particularly in the 14th and 15th centuries BC, we are not certain about the uses made of the regularly blowing monsoons for navigation. However, we have conjunctural interpretations by modern scholars; to quote Panikkar and Srinivasan (1972), 'that the expedition of the Queen Hatshepsut (*ca.* 1495BC), to the land of Punt and the circumnavigation of Africa by the Phoenicians under Necho (610-594BC) were accomplished with the help of the monsoons. Herodotus, who has recorded the exploration of the Indus by Scylax of Caryanda, sent by Darius (*ca.* 510BC) does not mention the use of the NE monsoon by Scylax for his return trip. By the end of the second century BC, Eudoxus of Cyzicus, under the direction of Ptolemy II, made two trips to India. The first voyage was made with the guidance of a shipwrecked Hindu sailor who divulged to the Greeks the secrets of the open-sea route to India. Since then, the Arabs, who were acting as trade intermediaries between west and east, could no longer keep the knowledge of the monsoons as a "trade secret" '.

During the reign of Claudius (41-56AD), a Greek mariner called Hippalus found a short-cut route to the Indian ports of Damirica (Tamil Country) and Barygaza (modern Broach). This discovery considerably reduced the travel time across the Indian Ocean, between Mediterranean or Arabian ports and the Indian coast, to about 40 days, and gave a tremendous boost to trade between Rome and India. New navigation routes from southern Arabia to Sigerus (Taigarh), from Bab-el-Mendeb to Muziris (Crangamore) on the Malabar coast, were also discovered.

The Waves: In Sanskrit, the waves are known as *urmi*. In the Tamil Sangam literature they are called *tivai*, *alai*, *punavi* and *otham*.

That the waves are caused by the wind was known to the early Hindus. In the *Rg Veda*, one of the hymns says that 'they who toss the clouds across the surging sea with maruts come hither'. The Vedic book *Satapatha Brahmana* informs us 'the way of waters is the wind, for when he (the wind) blows hither and thither, then the water flows'. In Tamil sources, it is said that waves attack the shores; they also cause accretion of sand dunes. The Sanskrit poet Kalidasa (*ca.* 4th century AD), known for his keen perception of nature, remarks in his drama *Raghuvamsha* that, at the mouths of the River Tamraparni, the shoals of conch shells with their heads transfixed at the jutting points, being dashed at once by the force of the billows against the reef of corals, glide away with great difficulty.

Nearchus, the trusted general of Alexander, had to delay his departure from the mouths of the Indus because of waves. The experience of the unknown author of the *Periplus of the Erythrean Sea*, in the Gulf of Baraca (Cutch), is relevant in this context. He says that those who are drawn into the Gulf are lost; for the waves are high and very violent, and the sea is tumultuous and foul, and has eddies and rushing whirlpools.

The Tides: Knowledge of the tides is vital for coastal navigation. Ancient writings in Sanskrit indicate the Indians' awareness of the ebb and flow of the tides. The Harappans made use of the tides for berthing ships effectively; for this they constructed a dock at Lothal (in the present Ahmadabad district of the State of Gujarat) which, to this day, remains an engineering marvel.

Subsequently, in the *Sama Veda*, there is evidence of a linkage between the tides and the phases of the moon. In *Ramayana* also, the author, Valmiki, states that tides are due to the influence of the moon. In *Matsya purana*, one of the eighteen *puranas* of the Hindus, the information on the tides is more clear. To quote from Panikkar and Srinivasan (*op. cit.*) 'when the moon is in the east, the sea begins to swell. The sea becomes low when the moon wanes. When it swells it does so with its own water (and not with additional waters) and when it subsides, it's swelling is lost in its own waters... the sea rises and falls according to phases of the moon'. The tidal range mentioned in the *puranas* is fairly accurate for the Gulf of Cambay. There the sea rises and falls according to the phases of the moon and 510 angulus (about 9.75 metres) is the measure of its rise and fall on Parva days.

As already mentioned, the dockyard built at Lothal during the period 2350-2000BC (Figure 1.2) is clear evidence of the precise knowledge the Harappans had about the tides and their range at Lothal. The dock, trapezoidal in form, measures 240 metres north-south and 23 metres east-west. Kiln-fired bricks were used to construct the walls which were 4.5 metres high. The dock inlet was 12 metres wide and the basin was very close to the river. However, in the year 2000BC, the river mouth silted up owing to floods, and the course of the river was changed so much that a 2km-long channel had to be dug to connect the Lothal Dock to the river. A spillway with a 1.5m-thick wall was built and, to allow the ships to enter, wooden shutters were used to block the water from flowing out at the ebb tide. Technical details given by Rao (1970) indicate that the minimum water depth was about 2 metres at low tide, the maximum being 3.5 metres at high tide.

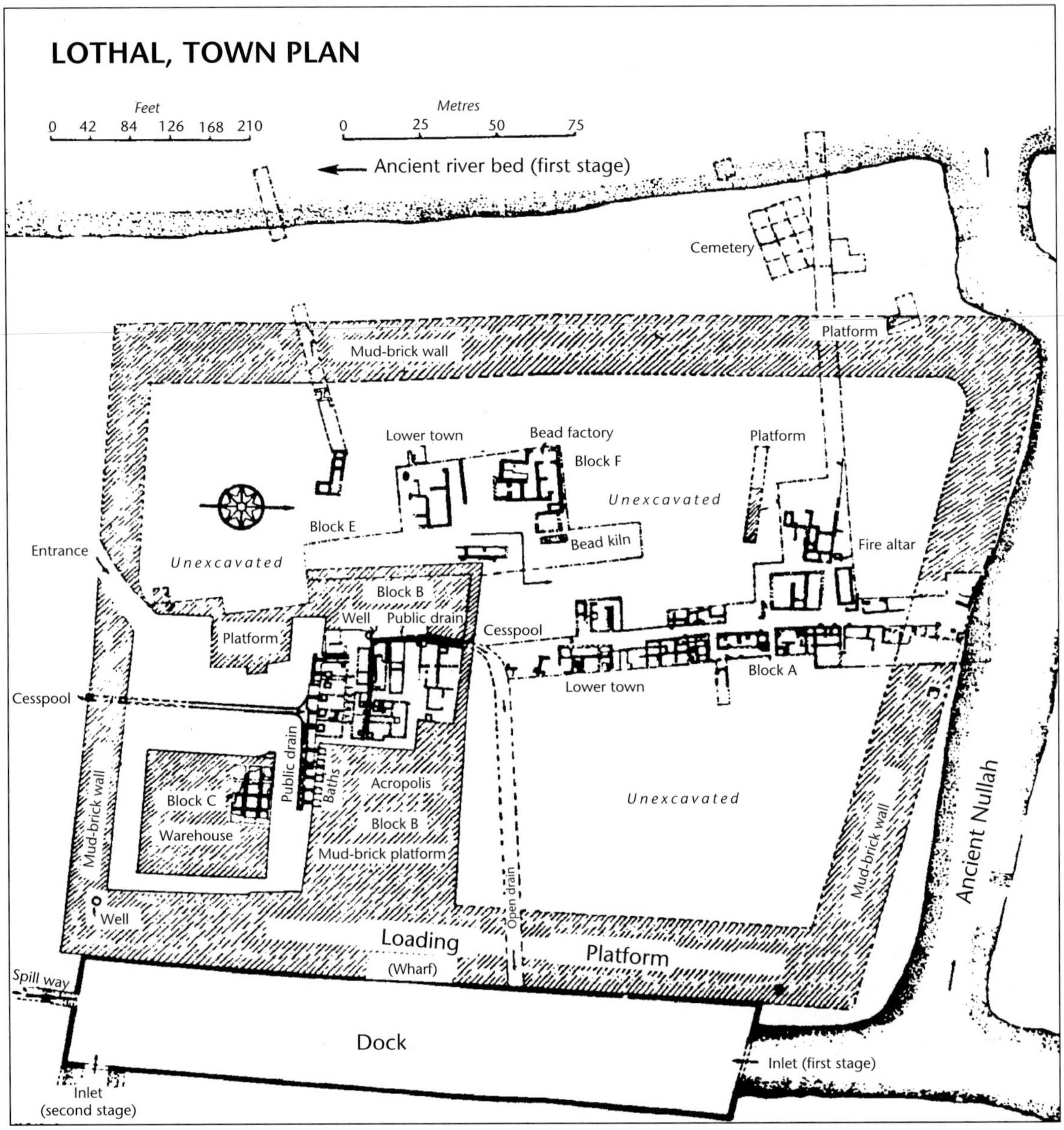

Figure 1.2
Lothal town plan showing the location of the dock. (After S.R. Rao, 1970)

Boats with a capacity of 60-65 tons and a length of 20-25 metres were able to enter and leave the Lothal dock safely. In the archaeological investigations, three perforated stone anchors and a Persian Gulf seal have been found in Lothal. An Indus type of seal has been found in Mesopotamia and this is dated to 2950-2200BC. From this, it appears that marine commerce between the west coast of India and ports in the Red Sea and the Mediterranean was prevalent in those days of the Bronze Age.

Markenday and Srivastava (1980) have given an historical sketch of the development of knowledge leading to early concepts of physical oceanography of the Indian seas, from the beginning of the Vedic period up to British colonial times. There is evidence that trade at the Lothal dock declined around 1500 BC, followed by the decline or emigration of the Harappans.

The Vedic period is dated from 1500BC to 600BC. The important ports mentioned during this period are on the Kathaiwar coast; namely, Prabasa (Somanath) and Dwaraka, the home of Krishna; the others are Sopara (Sopara near Bassein) and Collieana located near Bombay.

During the 7th century BC, Lagadha wrote *Jyotisvedanga*, perhaps the first Indian treatise on astronomy. According to this, the year had 366 days and three seasons. During the Jain and Buddhist periods (from 600BC to 300BC), two astronomical texts were written, one based on the sun (*Surya punnati*) and the other on the moon (*Chandra punnati*), and in *Vrhatksketrasamana* two chapters are devoted to the description of the Lavan and Kalodha Oceans, particularly their dimensions, abyss, tidal action, islands and depths of water.

In *Jatakamala*, written by Arya Sura, there is a description of a maritime code based on some knowledge of oceanography. The navigation in those days was dependent on astronomy, flora and fauna, colour of the sea, birds, rocks, and so forth. The red-water phenomenon was also described. In the court of Menaden or Milinda (in Pali), a Greek king who ruled northwest India during 200BC, caused a book, *Milindapanho*, to be written in a question and answer form between him and an Indian intellectual by the name of Nagesna, wherein the qualities of a ship's navigator, a pilot, an anchor, a mast, and seamanship in general are described. It is also known that the great King Ashoka (302-232BC) maintained trade relations with Sri Lanka and the western kingdoms of Syria, Egypt, Macedonia and Epirus. This could have been possible only when dependable oceanographic and navigational knowledge was available to the sailors of these countries.

During the Mauryan and Kushan period, roughly between the years 322BC and 320AD, significant development in our understanding of the physical aspects of this planet and of techniques of navigation, including the use of star sights, took place. Kautilya's *Arthashastra* furnishes valuable information on the administration of ports and navies; it also lists rules and regulations of navigation.

Most important evidence for the activities of Indians in the Indian Ocean area comes from an unknown author who wrote the *Periplus of the Erythrean Sea* some time during the 1st and 2nd centuries AD. It seems that Indians had already established colonies in western Asia near the upper Euphrates River and on Socotra Island.

In the *Periplus*, the Goddess of the Sea, Sikotari-Mata, who is invoked in Gujarat, is also mentioned. A number of Indian ports active in maritime commerce are also cited in this book. They are Hastabra (Hastab on the Kathaiwar coast), Collieana (Kalyan, near Bombay) and Muziris (near Cranganore). The Tamil classic, *Chilppathikauam*, mentions some of the active ports on the east coast of India at that time. They are Poompahar (Kaveripattinam, in Tamilnadu), Poduke or Paduka (Arikamedu, near Pondicherry) and Kaimarpar (Konark, in Orissa). Excavations at Kaveripattinam have exposed a brick jetty datable to 300BC; at Paduca, Roman pottery belonging to the 1st century AD has been located; at Dhanyakarka, near Amravati, in the Guntur district of Andhra Pradesh, an embankment with a wharf and a navigational channel dated to 200BC. At Tamlik or Tamralipti, at the mouths of the Ganges, an estuarine port dating to the 2nd or 3rd century AD has been unearthed.

Shipping in ancient India

Information on shipbuilding technology is scarce in Indian literature. However, in a book entitled *Yuktikalpatra*, written by King Bhoja of Dhara (4th century AD), details on the types of wood to be used in shipbuilding, dimensions, decorations, mast size and cabins are presented. The use of iron nails was prohibited because it was believed that their presence in the ship would attract magnetic corals or reefs, exposing ships to danger; this notion of magnetic reefs persisted until the 14th century AD.

Figure 1.3a
A boat decorated with fanciful animals, from Sanchi.

Figure 1.3b
A boat carrying ascetics.

At a symposium on India's contribution to world thought and culture, Rao (1970) gave a comprehensive paper on shipping in ancient India, covering a period from around 2500BC to 600AD. According to Vedic texts, particularly in the *Rg Veda* and *Satpatha Brahmana*, ancient inhabitants of India undertook naval expeditions and the god Varuna is credited with the knowledge of the sea routes followed by the ships. According to Rao, three things are clear from Vedic texts. Firstly, ships sailed from Indian ports to foreign countries for trade. Secondly, the ships had multiple oars. And thirdly, many sea routes were known to the Vedic Aryans. However, we have very little to go by in the absence of precise evidence for the period 2500-500BC. In later years, from 500BC to 600AD, there are many references to Indian shipping and maritime trade. Quite a number of ports and docks have been identified as active centres of sea-borne trade. Kautilya's *Arthashastra* and Bhuddhist *Jataka* stories provide valuable information on the administration of the navy and ports and other maritime activities. A Buddhist text known as *Simhalavadan* narrates that Prince Vijaya and his seven hundred followers banished by Simhabahu, King of Bengal, set sail from the mouth of the Ganga (Ganges) and came to Ceylon on the day of the Nirvana of Buddha. There is also mention of the voyage of the merchants from Varanasi and Rajagarh down the Ganga and then westwards to the ports on the seaboard of Sophir in the Gulf of Kutch. From these narratives, it is evident that Indians built sufficiently large ships (Figure 1.3 a, b, c, d). According to Panini (5th century BC), timbers from trees like simpsapa (*Delbergia sisoo*), amra (*Mangifera indica*), somali (*Bombax malabaricum*) and khadira (*Acacia catechu*) were used for ship building. It is believed that Alexander constructed a fleet of ships from pine, cedar and other trees cut from the territory of King Poros. India appears to have supplied timber suitable for shipbuilding to Mesopotamia in the time of King Nebuchadnezzar (604-562BC).

The paintings at Ajanta and the sculptures found at Barhut, Sanchi and Borobudur give some idea of ship construction. The hull and keel were constructed by means of a tenon-and-groove system by joining planks edge to edge, as seen in the Ajanta caves. The use of perforated stone anchors and the introduction of twist drills at Lothal clearly indicate that it was a shipbuilding centre. Rope was preferred to iron nails for joining planks, according to the author of *Yukti-Kalpataru* (11th century AD). Although iron was widely used in India by 400BC, its technology was developed as early as 700BC. Another technique of joining planks was the fish-joinery, whereby wedges of wood were inserted in the joints of the planks, as can be seen in the sculptural representa-

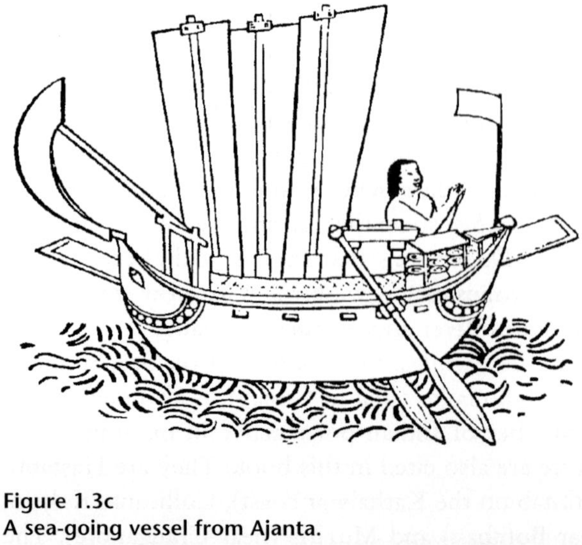

Figure 1.3c
A sea-going vessel from Ajanta.

Figure 1.3d
A multi-masted Indian ship sailing for Java.

tions of a boat carrying three ascetics, on the eastern gateway of Stupa No. 1 at Sanchi.

Milindapanha gives some account of the masts of the ships. On the top of the mast there was provision for the 'look-out man' to sit and the number of masts varied from 1 to 3. If a ship had only one mast, it was fixed at the centre, but two-masted ships, represented on the Andhra coins, had the masts equally spaced so as to balance the ship properly. Three or more masts are seen in the Ajanta murals and Borobudur sculptures, which also show square sails, which were prevalent in those days (5-6th centuries AD). The prow was also often decorated with fanciful animal figures. It is also evident from the Ajanta paintings that a large number of persons and even elephants could be carried in multi-oared ships. It is also believed that long before Hippalus discovered the monsoon winds, Indian sailors possessed a sound knowledge of the periodicity and regularity of the winds in the Indian Ocean. This is evident from the fact that Harappan ships made regular voyages to the East African coast (Socotra), Egypt, the Bahrain Islands and the Persian Gulf ports for the exchange of luxury articles and commercial products. The Indus cities, Lothal being notable among them, traded and exported to western countries, mostly gemstones, beads, bangles, inlays made of shell, different types of timber and clothing.

The Siddhantas

Around the 6th century BC, the Aryans began to write books on astronomy known as *Siddhantas*, numbering 5 in all. The first one, the *Surya Siddhanta*, was written by Lata and contains 14 chapters (Table 1.1).

According to this book, the Earth is a sphere, and 5 planets, with their ascending and descending modes, are indicated, as is the fact that not all the orbits of these planets describe perfect circles. In the second, *Paithamaha Siddhanta*, dating to the 1st century AD, a method is given to locate moving celestial bodies relative to a fixed point and to express angular distances in degrees and minutes. It uses the 12 *rasi* (Zodiac) signs, dividing the sky into 12 parts, each of 30°, thus giving a precise method for the determination of angular distances. The third, *Paulis Siddhanta*, includes methods for determining the exact length of the day and predicting eclipses. Similarly, the fourth, the *Vasishta Siddhanta* (written at some time between the 2nd and 6th centuries AD) gives lunar-solar cycles and a method for evaluating the length of the year. The fifth, *Romaka Siddhanta*, deals with arithmetic, algebra, and the motion of the planets, *inter alia*.

These techniques agree with the methods used by Ptolemy and Hipparchos. These *Siddhantas*, in view of their precise contributions to astronomy, formed the basis for accurate navigation in the unknown seas at that time.

It was during the Gupta period that Indian creative thinking received great support and acclamation. Aryabhata wrote a comprehensive astronomical treatise (5th-6th century AD) introducing the

Table 1.1 The chapters of the Surya Siddhanta

Chapter No.	Subject of discussion
1	Measurement of time
2	Sine tables*
3	Meridians, cardinal points, equinoxes and solstices
4 & 5	Eclipses of moon and sun
6	Graphical projection of eclipses
7	Planetary motions
8	Inclination of the nakshatras (constellations) to the elliptic
9	Helical rising and setting of stars
10	Relative motions of sun and moon
11	Evil conjunctions
12	Cosmography
13	Elementary astronomical instruments
14	Computations of the calendar

* *Earliest known sine tables*

principles of epicycles and the rotation of the Earth. In the 6th or early 7th century AD, another Indian writer, Brahmagupta, described the mean and true planetary motions and conjunctions of planets. In *Mahabhaskariya*, written by Bhaskara I, the longitudes of planets and their relation to time, place and directions are recorded. During the period covering the Rajput dynasties (650-1200AD) in northern India and those of the Tamils in the south, Dhanapala's *Tilakamanjari* gives a good account of the ships and shipping industry and, in a Tamil book, *Permapermarupadai*, mention is made of important naval battles of the period. Raja, in 983AD, led a successful expedition against Sri Lanka. Another Tamil king, Rajendran, conquered Kadarma (Khedah) on the Malay peninsula. During the same period, Bhaskara II compiled the *Romaka Siddhanta*, mentioned above. All this information is basic to oceanography.

The Arabic-Muslim period

For almost 5000 years, commencing from the time of the Mohenjo-daro and Harappan civilizations, dated to 3500-1500BC, until the discovery of a route to India and the Far East via the Cape of Good Hope, the Arabs ruled the great seaways from the Mediterranean to the ports of South Asia and China. Aleem (1968a, 1968b and 1980) has made a detailed study of the history of Arab navigation, based on Arabic literature. We have already mentioned that Queen Hatshepsut of Egypt is credited with organizing the first maritime expedition in 1495BC to explore the southern Red Sea and Somali coast. For many centuries before the advent of Islam (before 622AD), Arab pilots from Oman, Yemen and southern Arabia circumnavigated the Indian Ocean, then known as the Great Sea, and often subnamed parts of it, depending on the coastal country, such as the Sea of India, Sea of Lawry (Arabian Sea), Sea of Faris (Persian Gulf), Sea of Harkand (Bay of Bengal) and so on. The Arabs mainly traded with China, India, East Africa and the intervening islands. The luxurious homes of Arab merchants living in Sri Lanka have been mentioned by a Chinese traveller, Fa Hain, in 414AD. Ibn al-Fakih (also known as al-Hamadami) wrote a book on geography entitled *Kitab of Buldan* (Book of Countries) which contains a discussion of the tides in the China Sea and the river at Canton where there is fresh water and the sea rises and falls twice in the day and night. Fakih also included some causes for the tides, such as the story of the angel who dips his finger in the sea to cause it to rise and, when the finger is removed, causes the sea level to fall; in another story, it is a whale which sucks up the sea to cause it to ebb and, when it exhales, causes the sea to flow.

Another Arab explorer and writer, al-Mokaddesi (985AD), wrote a book entitled *Aksan al-Takaseem* which includes a chapter on seas and rivers. Here he has given a description of tides in the Chinese Sea (Persian Gulf) where it has fluxes (high tides) in the middle of the lunar month and towards its beginnings and ends twice in the day and night. The flow pushes the water of Basra up the river thus filling up its tributaries and irrigating the fields. During the ebb (tide), the water is low. Again, in the 13th century, an Arab scholar, al-Dimiski (Shams ed din Abou Abdullah Mohammad), from Damascus, wrote a remarkable book entitled *On the Wonders of the Land and Sea*. The book was finished in 1325AD. In it there is a detailed and accurate description of the semidiurnal tides at Shat-el-Arab. Medieval Arabs associated currents in the Indian Ocean with winds, and tides with the lunar cycle.

Arab knowledge of sea routes to India and the Far East was sound and accurate. In the book *Voyages of Sulayman the Merchant*, written by Ibn Wahab, there is a vivid description of the maritime route to India and China. Arabian and Persian ships covered this in four steps, as follows:

1. From Siraf and Muscat to Klim (present day Quilon, in southern India).
2. From Klim to Lingbalus Island (Nicobar) in the Sea of Harkand (Bay of Bengal).
3. From Lingbalus to Sunderfulat (an island on the coast of Vietnam) via Kalebar (an island on the coast of Malaysia), then across the Malacca Sound to the island of Tioman (on the east coast of Malaysia, north of Singapore) and then to Kadrang (near Saigon).
4. From Sunderfulat to Canton (Guangzhou) across the South China Sea.

Ships usually commenced their voyages from Muscat in Oman with the NE monsoon during November or December, stayed in summer in Canton and sailed back with NE winds in the autumn.

In the open ocean, navigation in the desired direction would be impossible without knowledge of the stars and their movement. Arabs are credited with the construction of the first astrolabe and they improved the working of the magnetic compass earlier discovered by Indians and Chinese. It was Ibn Magid (1475AD) who improved the magnetic compass by mounting the needle on gimbals and adapting it to a box together with an azimuthal windrose, divided into 32 rhumbs, representing the rise and set of 16 stars. Use of this compass in the Indian Ocean preceded its use in the Mediterranean, according to Aleem (1968a). Another important contribution was the lateen sail. This sail was easy to manœuver in high winds. These sails were later adopted by European powers for their regular use.

The science of astro-navigation was a major Arab contribution aiding the development of sea routes between the east and the west, particularly during the 14-15th century onwards. However, from the 16th century, the Arabs' nautical knowledge lagged behind that of Portugal and Spain. This has been attributed by Aleem (1980) to the support European navigators received from their countries, in contrast to the Arabs who worked as individuals with nobody to back them or their efforts.

The foregoing account of the contributions made by the Arabs would not be complete without paying tribute to two navigators and writers from the past. One of them is the unknown author of the book entitled *Periplus of the Erythrean Sea* from the 1st-2nd centuries AD (Anon., 1st-2nd centuries AD; Huntingford, 1980). The other is Ibn Magid of the 15th century.

The latest translation of the *Periplus of the Erythrean Sea* is by Huntingford (1980). There are several editions of the original Greek, as well as English, German and French translations. In spite of this, we do not know exactly when and where it was first published, nor do we know the identity of the author. He could have been a Greek or an Egyptian well versed in Greek and settled in Alexandria. Based on available evidence, the *Periplus* may have been published at some time between 95 and 130AD.

Periplus simply means circumnavigation. The name Erythrean Sea refers to the Indian Ocean and its branches, the Red Sea and the Persian Gulf. There is some reference to the red colour of the Red Sea being attributed to the weeds or to the reflection of the sun's rays falling on the hills of Arabia. The importance of this work lies in the accurate information given about the ports, harbours, roadsteads (anchorages) and marts (trading counters), together with their imports and exports, besides navigational aids to entering harbours, recognition of approaching storms and the approach of land. In this respect, the *Periplus* turns out to be an early example of a trade directory and an Admiralty Handbook combined in one.

The *Periplus* contains 66 chapters, of which 18 relate to Africa (1-18), 18 to Arabia (19-36), 27 to India (37-63) and the last 3 to China and the unknown lands beyond (east of) India. Being a keen observer, the author of the *Periplus* gives a very accurate account of tidal conditions along the coasts he visited. For example, in Chapters 45 and 46, he describes the tidal bore experienced at Barugaza (the modern Broach) near the Mahi River entrance. To quote: 'at Barugaza the alternation, i.e. between the ebb and the flow, is much greater, so that all of a sudden the sea bottom can be seen and parts of the land dry where a little before (ships) were sailing; and when the tide comes in from the sea, the water of the river is forced back more strongly than the normal flow, for many stades[1]. For this reason, the approach and departure of ships is dangerous for the inexperienced who are coming into the port for the first time. Because of the violent movement (of the water) when the tide is rising it cannot be withstood, and anchors do not hold, ships are caught by its force and are turned sideways by the violence of the current, driven to the shoals and wrecked.'

The *Periplus* indicates that, as late as the 1st century AD, it was possible to go by boat from the Mediterranean ports through the Nile to the Red Sea through a system of canals. However, clear evidence for this is lacking. The first attempt to build a canal to the Red Sea is attributed to Necho II, in the 5th century BC. Since this did not progress well from its starting point a little above Bubastis in Egypt, the Persians under Darius Hystapes, in the 6th century BC, started further extension of the canal up to the Bitter Lakes. Around 276BC, Ptolemy II Philadelphos reconstructed the canal all the way to Arsinoe, a village at the southern end of the Bitter Lakes. Even then, the connection to the Red Sea was not completed until 106AD, when the Emperor Trajan cut a new canal starting from Babylon(ia) opposite Memphis, taking the canal parallel with the Nile through Heliopolis to Thou, where it joined the earlier canal and then turned east through Heliopolis

[1] *Authors' note: stadia understood; 1 stadium equals approximately 186m.*

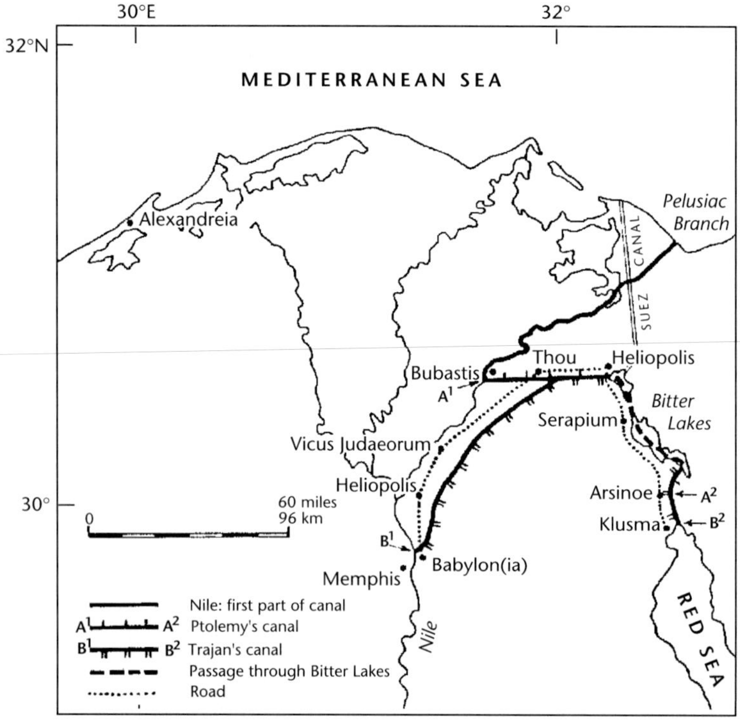

Figure 1.4
A map (facing p. 79 in the Periplus of the Erythrean Sea) showing the Arsinoe Canal. (From Huntingford, 1980)

and Serapium, Bitter Lakes, Arsinoe, with a cut into Klusma (Suez) in the Red Sea (Figure 1.4).

During the past three millennia, people living in the Mediterranean and Red Sea regions have been active in establishing a ship-going canal between them for uninterrupted navigation into the Indian Ocean. The canal connecting the Nile with the Red Sea was a forerunner to the modern Suez Canal which was opened to traffic in 1869. It may also be stated that voyages described in the *Periplus* commenced mostly from some 300 miles south of Klusma, at Muos Hormos on the Red Sea and opposite to Koptos located on the Nile, with which it was connected by a well made road.

It is quite clear that shipping and commerce over the Indian Ocean routes greatly benefitted from the *Periplus of the Erythrean Sea.*

Ahmad Ibn Magid, the Arab navigator of the 15th century

Aleem (1968a) has given a fascinating account of the important contributions made by Ahmad Ibn Magid to our knowledge of the Indian Ocean routes and related oceanographic information (Aleem, 1967). In his paper, Aleem mentions that Magid was the pilot who led Vasco da Gama to the Malabar Coast of India. However, in a later paper, Aleem (1980) suggests that the pilot who assisted Vasco da Gama was a Muslim pilot from Gujarat, as suggested by Portuguese writers. Nevertheless, this should not distract us from paying tribute to Magid for the scholarship and sound knowledge of the Indian Ocean described in his writings.

It would appear that Magid was born between 1433 and 1430AD, witnessed the beginning of the 16th century, and lived mostly in Oman. He spent over 50 years of his life on the sea and had extensive knowledge of astronomical and geographical aspects, as revealed in about 40 of his works of which some 30 are known. His major work is in a prose form entitled: *Kitab al Fawayed fi Osul Ilm et Bahr wal Kawaid* (meaning Book of Useful Instructions and Principles of the Science of the Sea). This book includes nautical instruction, celestial navigation, meteorological and oceanographic observations and descriptions of coasts and islands. Another major work, written in a poetry form, is entitled: *Hawiyat al Iktisa fi Osul Ilm al Bihar* (meaning Abridged Exposure of the Principles of the Science of the Sea). This book deals with signs for astronomical observations, particularly the rise and set of stars. Magid's conception of 'Regional Waters' is also interesting. To quote 'but the sea belongs to no one of these peoples (Chinese, Indians, Persians, Zingi, etc.). Once the coasts disappear from your sight, you have nothing to guide you except your recognition of stars to which you adjust your course'.

When all this knowledge of the Arabs passed on to the Portuguese and the Spanish, the importance of Arab pilots in the voyages across the Indian Ocean waned and the stage was set for the Portuguese to sail round the Cape of Good Hope and reach India in the year 1498.

Subsequent to the Portugese landing on the Malabar Coast of India, other European powers followed and thus began the colonial expansion of the British, Dutch, French, Spanish and others in Africa and Asia. And by the 17th century, practically all countries around the Indian Ocean rim became colonies of the various European powers. Successful and safe navigation between Europe, Africa and Asia required precise knowledge of the Indian Ocean and thus began a series of ocean expeditions sponsored by the colonial powers.

Some important expeditions sponsored by European powers

While the British were active in India with their marine surveys, the Danes and the Dutch made significant efforts in exploring, particularly, the east Indian waters, where they had commercial (the products of the spice islands) and political interests. The Danish contribution is summarized in an excellent book written by Torben Wolff (1967).

The Arabia Felix Expedition (1761-67)

The Danish Arabia Felix Expedition (1761-67) is, perhaps, the first oceanographic expedition in the history of marine science and in the Indian Ocean area. The launching of this scientific venture, with the support of King Frederik V, created a great sensation throughout Europe in those days. The ship *Grønland*, belonging to the Danish navy, sailed from Copenhagen on 7 January 1761. It took almost two years to reach Yemen, then called Arabia Felix or Happy Arabia, which proved to be hell on earth. During this cruise, the ship's crew and scientists faced obstinate authorities, unfriendly populations, an unbearable climate and, worst of all, conflict and hostility among the expedition members. Before the expedition could reach Bombay, five out of the six members of the scientific party had died; the only surviving member, Castén Niebuhr, reached Copenhagen after a lapse of seven years. In spite of all these overwhelming difficulties, scientific observations were systematically made and collections preserved. All the credit goes to Niebuhr, who made sure that all the scientific collections reached Copenhagen by a land route through China! The expedition reports were published, and they cover a large number of fauna and flora collected during the voyage (Wolff, 1968).

Peter Forskal, the naturalist on board, was a student of Linnaeus and made the first known systematic study of the pelagic fauna. He described no less than 151 species of fish, paying particular attention to numerical ratios; e.g., number of fin rays, tooth arrangements and morphological characters. He also described the pelagic amphipod *Pheronima sedentaria* which is normally found in its own gelatinous cover. Unfortunately, Forskal died from malaria during the expedition at the age of 31; thus a brilliant young marine biologist was lost to science. However, Forskal's collections are still preserved in the Copenhagen museums and are important even today.

The first Galathea *Deep-Sea Expedition (1845-47)*

Continuing their tradition of involvement in marine exploration, the Danish government launched the first *Galathea* Deep-Sea Expedition round the world in 1845. According to Wolff (1968), the Royal Danish Academy of Science and Letters received a communication from King Christian VIII on 14 May 1845, intimating that they had decided to send the corvette *Galathea* to the East Indies, and in particular to the Nicobar Islands, over which they had sovereignty, to undertake scientific investigations of the natural products of this group of islands. Another objective was to transfer the Indian settlements of Tranquebar and Frederiksnagore to the British East India Company. In a very short time, the Academy selected the naturalists to take part, and the ship started on its circumnavigational cruise on 24 June 1845 under the command of Steéne Anderson Bille. The *Galathea* was a small ship, only 4.57m beam, 42.5m long, drawing 400 tons last burden, mounted with 36 cannon and carrying 231 men. The ship must have been thoroughly cramped for work, with supplies choking all the available deck space. The biggest obstacle was the lack of mechanical power and steel ropes, making it extremely difficult to operate manually the scientific equipment using thick hemp ropes. After visiting Tranquebar, the *Galathea* reached Nicobar in January 1846. All 19 islands were investigated and careful reports were written on their geography, geology, animal life and vegetation, hydrography of their waters, the character, religion and mode of life of the natives as well as sanitary and commercial conditions. The expedition cost Denmark 462,036 rigsdalers and 72 skillings, and what were the results? Not much! The ship returned to Copenhagen on the evening of 23 August 1847 and King Christian wanted the results to be published expeditiously. Unfortunately, the King died within a few months of the return of the ship, and many members of the scientific party were dispersed and took up other work. The greatest loss was that thousands of herbarium specimens remain unworked and so remain even to this day. However, most noteworthy is Forchhammer's epoch-making publication *On the Constituents of Sea Water and Their Distribution in the Sea* wherein he lists the essential chemical constituents of sea water. His most important discovery was that the salinity[2] of the open oceans varied far less than had been previously thought and the quantitative ratio between the various salts was always the same. To obtain these important results, Forchhammer obtained samples of sea water not only from the *Galathea*, but also from the *Erebus* and the *Terror* navigating under the leadership of Sir James Ross.

The Carlsberg Foundation's Round-the-World Oceanographic Expedition (1928-30)

From the time the first *Galathea* round-the-world expedition returned successfully in 1847, until the launching of the Carlsberg expedition in 1928, the Danes were quite active in building up knowledge of marine organisms and hydrography, particularly in Greenland, Iceland and other Arctic areas, besides solving the great mystery surrounding the breeding grounds of the European eel (*Anguilla anguilla*). In May 1904, Johannes Schmidt, on a cruise on board the *Thor*, collected, for the first time, at a station in the Atlantic and west of the Faroes, the first eel larva, then taxonomically designated as *Leptocephalus brevirostris*, measuring 6.35cm long. Encouraged by this discovery, Schmidt followed relentlessly the survey for this larva in the Ionian Sea, the Mediterranean and the North Atlantic Ocean and finally he solved the problem by collecting numerous and different-sized larvae (nearly 9,000) in the Sargasso Sea on board the fishing schooner *Margarethe* and, later on, the *Dana*. As soon as the problem of this European eel was solved, Schmidt was impatient to discover the breeding areas of the freshwater eels from the Pacific and Indian Oceans. To this end among others, with a generous grant from the Carlsberg Foundation and the support of the Danish government, the *Dana* started its round-the-world cruise on 14 June 1928. Among the scientists were Johannes Schmidt, the hydrographer Helge Thomsen, the zoologist Anton Bruun, who later became the leader of the second *Galathea* Deep-Sea Expedition, and the plankton specialist Steeman Nielsen.

Dana entered the east Indian waters on 29 March 1929, after working across the Pacific, and was immediately successful in locating the spawning grounds of the freshwater eels living in the rivers of the Indo-Malaysian area. The breeding ground is located in the Java Sea and the area stretches parallel and close to the west coast of Sumatra and Java and where the depth is around 500m. They caught 1237 larvae representing all 4 species of freshwater eels of the area. The *Dana* also made a bathymetric survey across the Indian Ocean between Ceylon (now Sri Lanka) and the Seychelles and mapped a great submarine ridge running in the NW-SE direction. In recognition of the support given by the Foundation, the ridge was named Carlsberg Ridge. In every way, the expedition was a great success, and the collections were equally divided among the three oceans. The *Dana*'s collections still stand out as some of the most significant contributions ever made to our knowledge of the three oceans. Whereas the *Challenger* found only one specimen of the cephalopod *Spirula*, the *Dana* found 192 specimens. Similarly, the *Challenger* had caught only one specimen of the deep-sea crustacean *Ceratolepis*, whereas the *Dana* found 174. With these kinds of valuable collections, the Danish people and government were further inspired to undertake yet another circumnavigational deep-sea expedition on board the *Galathea* in 1950-52.

---

[2] In classical oceanography, salinity was expressed as parts per thousand, ‰. Following a specific UNESCO-SCOR study of the definition and measurement of salinity by modern techniques, the Practical Salinity Unit has since been adopted: other things being equal, a given salinity expressed in PSUs is virtually, but not precisely, equivalent to the value determined by earlier procedures. The PSU is dimensionless. Herein, all salinities are expressed simply as a number regardless of the analytical method/procedure used to obtain it. This does not change the sense of the text, however.

The second Galathea *Deep-Sea Expedition (1950-52)*

At first, the intention behind the new expedition was to commemorate the centenary of the first *Galathea* Deep-Sea Expedition (1845-47), but, owing to World War II and its aftermath, the Danish government could not launch it in the centenary year of 1945. However, the desire of many scientists to continue the deep-sea investigations could not be ignored and, accordingly, a Danish Expedition Fund was created for which an appeal went out for contributions from overseas Danes. From the funds received, the organizers bought cigarettes from the USA, honey from Australia and spices from East India – all imported duty free – and sold them in the market for a large profit. The organizers said that these transactions were completely legal! The expedition, thus funded, was launched on 15 October 1950, and was devoted to studying mainly the deep-sea fauna from the deepest trenches in the world oceans. The King and Queen of Denmark came on board the expedition vessel and wished the crew all success. The *Galathea* was well equipped to study the deep oceans. It had perhaps the world's largest deep-sea winch, which carried a 12,160m tapering steel cable with a thickness of 22 mm at the surface end and 9 mm at the other. The ship was 80.8m long, 10.3m abeam, 1600 tons dry weight and was powered by two steam turbines giving the vessel a speed of 12 knots. The ship was commanded by Captain Sv Greve, and the scientific party was led by Anton Bruun. In the Indian Ocean, the ship visited Mombasa, Seychelles, Tranquebar, Serampore near Calcutta, and finally the Nicobar Islands before reaching the Pacific Ocean trenches for deep-sea work. The results of the *Galathea* Expedition proved beyond any doubt that animals inhabit the greatest depths of the ocean, where the temperatures are icy and pressure is enormous, 1000 atmospheres or more. For example, in the Philippine Trench, 10,153m deep, there were many sea cucumbers, shrimps, lobsters, crabs, etc. The expedition was able to demonstrate the occurrence of about 115 species from depths greater than 5776m. Bruun called these deep-water denizens the hadal fauna.

The expedition returned safely to Denmark on 29 July 1952 with glory for the wonderful and successful accomplishment of its objectives. Thus, we find in Denmark a most enlightening chapter in the history of marine investigations in which the Indian Ocean was included.

The Novara *Expedition (1857-59)*

Austria, not to be left behind in ocean research, organized a world-wide expedition, on board the *Novara*, an Austrian frigate commissioned to circumnavigate the globe under the order of the Imperial Government of Austria; the *Novara* sailed from the port of Trieste on 30 April 1857, the year of the Sepoy Mutiny in India. The commander of the ship was Commodore Wollerstorff-Urbair. The narrative of the ship's world-wide cruise is given by Karl Scherzer (1861), a scientific member of the expedition, at the invitation of H.I.H. the Archduke Ferdinand Maximilian of Austria. Alexander Von Humboldt provided the expedition with physical and geognostic suggestions to be followed in various seas and climes.

For the Indian Ocean area, Humboldt suggested studies of the minor volcanoes stretching across the Sunda group of islands, from Sumatra, northwest through the Nicobar and the Andaman Islands, and then through the volcanoes of Barren Island, Narcondam and Cheduba, all located in the eastern part of the Bay of Bengal. Similarly, off the west coast of India, there is a string of three archipelagos, extending from 14°N to 8°S – the Laccadives, the Maldives and the Chagos. Among these, the Laccadives and Maldives are known to consist solely of coral and are therefore true atolls. It was suggested that the *Novara* should examine whether these islands would answer the hypothesis of subsidence rather than an elevated region.

Humboldt was also keen that observations on terrestrial magnetism be made between Ceylon and the Bight of Biafra in view of the fact that nothing was known from that area. The ship was also ordered to make observations on natural phenomena such as zodiacal light and to collect animals and plants. Accordingly, when the *Novara* returned to Trieste in 1859, it brought in an extremely rich collection of about 23,700 specimens: 440 minerals, 300 reptiles, 1500 birds, 1400 amphibians, 1330 fish, 9000 insects, 8900 molluscs and crustacea, 300 bird's eggs and nests, along with numerous skeletons and herbarium specimens.

Scherzer also commented upon the customs and cultures of the lands and people the ship visited, and gave the list of stores carried by the ship before departure, which included, among other things, 50,965 pounds of biscuits, 3,510 gallons of red Istrian wine and 3,165 gallons of rum! The total expenditure for this expedition was £58,000 sterling.

The Valdivia *Expedition (1888-89)*

While the *Siboga* Expedition (1889-90) was active in the East Indies seas under the Dutch flag, Germany organized one of the most important of the circumnavigating expeditions, on board the *Valdivia* (1888-89), to investigate the physical and biological conditions of the Atlantic and Indian Oceans, penetrating even to the Antarctic areas, weather and ice permitting. The scientific reports were issued in a series of magnificent memoirs edited by Chun, who was also the leader of the expedition. These reports, together with the reports of the John Murray Expedition (see page 26, below), form a sound basis of knowledge of the oceanography of the Indian Ocean.

The Siboga *Expedition (1889-90)*
and the Snellius *Expedition (1929-30)*

The Dutch government organized two important deep-sea expeditions. The *Siboga* Expedition in 1889-90 and the *Snellius* Expedition in 1929-30, to the Dutch East Indies to investigate the vast number of islands and the deep channels separating them. The *Siboga* Expedition, under the leadership of Max Weber, made an extraordinarily detailed study of the depth distribution of the fauna and flora of the East Indies seas. Its scientific reports ran into several volumes and include the first large bathymetric charts of the archipelago, drawn by G. F. Tydeman, the Naval Chief of the Expedition. The *Snellius* Expedition, under the leadership of P. M. Van Riel, investigated the depths in the area of the East Indies archipelago in greater detail and took nearly 35,000 soundings. Besides depths, the *Snellius* charts also give the direction of the flow of the bottom water and give an idea of the part which the Pacific and Indian Oceans and the China Sea play in the renewal of the abyssal layers in the isolated basins.

The areas investigated by the *Siboga* and the *Snellius* cover the interface between the Pacific and the Indian Oceans and therefore contribute to our understanding of the communication between these two ocean systems.

The Albatross *Expedition (1947-48)*

Much later, Sweden organized a circumnavigational expedition on board the *Albatross* in the years 1947-48. The reports of the Swedish Deep Sea Expedition have been published in 10 volumes under the leadership of Hans Patterson (1957) who was also the scientific leader. Some of the more important reports are: *The Results of Deep Sea Coring*, by Borje Kullenberg; *Optical Studies of Ocean Water*, by N.G. Jerlov; *Sediment Cores from the Mediterranean and Red Sea*, by Eric Ollanson; *The Distribution of Radium in Deep Sea Cores*, by Viktor St. Kroll; *Diatoms from Equatorial Indian Ocean Cores*, by R.W. Kolbe; and *The Geochemistry of Deep Sea Sediments*, by Stuve Landergren.

While narrating the voyage of the *Albatross* across the Indian Ocean, Hans Patterson makes two interesting observations, not particularly connected with oceanography. The first deals with the double coconut palm growing on the island of Praslin (Seychelles group). The palm tree, *Lodoice seychellarium*, bears giant double coconuts, each weighing up to 13kg. The contents are known to have medicinal value, including an antidote to poisoning. The second observation concerned the building of a dam across Bab-el-Mandeb, a proposal with a dreamlike quality. It is known that Red Sea water is saltier than water in most other areas of the world oceans and is subject to a high rate of evaporation. Without the counteracting effect of the Arabian Sea water flowing through the straits of Bab-el-Mandeb, the Red Sea would become a dead sea in a few centuries to come. Patterson suggests that, if a dam could be raised over the sill, whose depth is about 100 metres, and cut off the connection with the Arabian Sea (Gulf of Aden), in a few hundred years, one would expect that the Red Sea level would fall by several hundred metres and, at that point of time, turbines built into the dam could generate millions of watts of electricity by allowing the Gulf of Aden water to flow in to turn the turbines!

The British contribution

Early voyages and the Bombay
Marine (1601-1830)

For the first hundred years that the British were in India, beginning in the early part of the 17th century, their stay was solely directed toward consolidating their political power and the establishment of a worldwide empire. There was no organized effort by them to establish a scientific culture, much less any research programmes in their colonies. However, hydrographic surveys of the Indian coastal areas were begun in 1832 by the Royal Indian Navy. The

Marine Survey Department of the Indian Navy was established in 1873, at which time the famous *Challenger* Expedition had already completed two years of its circumnavigating cruise.

Markham (1878) did an extremely valuable service to the scientific community at large when he brought together in one volume the entire history of the British Indian marine surveys. That history began on 2 May 1601 when James Lancaster's fleet of four ships sailed for India from Torbay. The East India Company got its charter from the British Government in 1600 and, from then onwards, it sent at least one or two survey vessels every year for the next 20 years. Lancaster and Middleton, the commanders of the first two East India fleets, visited the East Indian islands and returned home loaded with spices. However, Captain Keeling, the leader of the third voyage, in 1607, arrived at Surat, landed a Mr. Finch to establish a factory, and sent Hawkins, second in command, to Agra to obtain trade concessions and just dealings for the British from the Mogul rulers. Thus began the story of the British Empire in India, which, after 347 years, ended on 15 August 1947.

The early English sailors accumulated lots of navigational experience in the form of charts, sailing directions and customs and all these were condensed and reported by a remarkable Arctic explorer, John Davis, of Limehouse, who himself made five voyages to India and published his *Rules for Our East India Navigations*. Meanwhile, the first chairman of East India Company Directors, Sir Thomas Smith, appointed Mr. Richard Hakluyt, Archdeacon of Westminster, as Historiographer of the East Indies, and commissioned Mr. Edward Wright to perfect the charts that were in the custody of the East India Company. Then followed a series of excellent English surveyors who are the unsung heroes of the British Empire. Over the next hundred years, they built up a detailed knowledge of the depths of the coastal areas and navigational charts for the Indian seas which are vital for defence as well as commerce.

What follows is a partial list of some of the officers and their areas of study:

Captain John Ritchie (1770-1790) – Coasts of Bay of Bengal and Outlets of the Ganges
Captain Huddart (1783-1790) – Malabar Coast
Captain John McCluer (1783-1791) – Whole West Coast of India from Cape Comorin to Diu
Lieutenant A. Blair (1777-1795) – Andaman Islands, Kathiawar Coast and Salcette
Captain Michael Topping (1790-1794) – Godavari Mouths
Captain Daniel Ross (1806-1820) – Straits of Malacca
Lieutenant Haines (1820-1830) – Persian Gulf and Bombay Harbour

All of these officers were working under the command of the Bombay Marine. Recognizing the importance of marine surveys, the company created a post of Marine Surveyor General at Calcutta and appointed Captain Court to the post, which he held until his death in 1823. He had under him a brilliant surveyor by the name of James Horsburgh, who published the first edition of the *East Indian Directory* in 1808. Based on his surveys, the company published 14 charts dealing with parts of the Atlantic and Indian Oceans. In appreciation of this work, the company granted him 100 guineas.

In 1832, the Bombay Marine was converted into the Royal Indian Navy by the wish and command of King William IV, and Sir Charles Malcolm was appointed its first Commander-in-Chief.

Work of the Royal Indian Navy (1832-1862)

In 1829, the Indian Government decided to open the sea route via Egypt to India through the Red Sea, and, as Governor of Bombay and the first Commander-in-Chief of the Indian Navy, Sir Charles Malcolm organized the Red Sea Survey which was of immediate necessity. The officers were handpicked and consisted of Captain Moresby to survey the northern half of the Red Sea, from Suez to Jeddah, and Captain Elwon, from Jeddah to Bab-el-Mandeb, on board the *Palinurus* and the *Banares*, respectively. This excellent survey was carried out from 1830 to 1834 and subsequently Moresby surveyed the Maldives and Chagos archipelagos in 1834-38. The hydrographic survey of the Gulf of Mannar, Palk Strait, Pamban channel and the Ceylon coast were taken up by Lieutenants Powell and Ethersey. At the same time, a survey of the Indus River approaches was commenced by Lieutenant Wood, who is also credited with being the first European, after Marco Polo, to have travelled to *Bam-i-Duniya* or *Roof of the World* and discovered the source of the River Oxus. For this splendid achievement, Lieutenant Wood received the Gold Medal of the Royal Geographical Society in 1892.

Colaba Observatory at Bombay was established in 1823 to study astronomy and meteorology and Mr. Curnin was appointed as the Company's first astronomer. He could not do much because the instruments supplied were faulty. Only when new instruments were supplied – they were not unpacked for five years – and used, was a regular register of the observations made and maintained, from September 1841. The office of the draughtsman of the Indian Navy was earlier located in the naval dockyard at Bombay, but most of the records kept there in a loft were eaten away by white ants and cockroaches, and, consequently, the office was shifted to Colaba Observatory.

Owing to wars and other disturbances, the post of the Marine Surveyor General was abolished in 1890. However, Captain Fell continued his work of charting the Andhra coast on board the brig *Krishna*. Captain Ward succeeded Fell as commander and, between 1851 and 1859, he surveyed the Mutlah river, Preparis North channel, the Bassein and Rangoon rivers, Malacca Strait and Penang. Finally, Lt. Heathcote succeeded Captain Ward and surveyed the Bay of Bengal and the Hugli River approaches in 1856-62.

In 1838, Captain Carless surveyed the African coast from Ras Hafun to Ras Gulwaini, followed in 1848 by Lt. Cruttenden who visited and reported on the Somali coast. On the Bengal side, Captain Lloyd, the Marine Surveyor General, commissioned the survey of the Hugli River from its mouth to Calcutta, co-ordinating his observations with the baseline measured by Colonel Everest on the Barrackpore Road in 1833. In 1840, he completed the survey of the surface of the Sunderbans from Chittagong to Hijih. Lloyd also described the existence of the 'Swatch of No Ground', very much like the Indus Canyon. These are deep chasms cut by the Indus and Ganges Rivers across the continental shelf where the depth at some places exceeds 250 fathoms (456m).

In addition to the survey work in the Indian Seas, the Indian Navy made a detailed survey of the Euphrates and Tigris Rivers for which the surveyors spent nearly 21 years (1841-62). For this alone, the Royal Indian Navy ranks among the foremost contributors to our geographical knowledge of these then uncharted areas.

Coming back to the Indian coast, Lt. Dundas Taylor conducted an excellent survey, on a trigonometrical basis, of the coasts of South Konkan, Canara, and Malabar. He also surveyed the harbour at Karwar, the anchorages at Beypore, Cochin and Corangi Bay near Kakinada. After his return to England, Taylor published the sailing directions for the west coast of India and this was issued by the Admiralty in 1866. Taylor also ensured the publication of a chart showing winds and currents in the Arabian Sea during the SW monsoon and a similar chart was prepared by Lt. Heathcote for the Bay of Bengal. Earlier, in 1856, Lt. Fergusson, the Draughtsman of the Navy, also prepared three sets of charts, each set containing a chart for every month in the year, showing winds and currents of the Red Sea, Persian Gulf and Indian and Chinese seas.

Credit for drawing our attention to the study of the Law of the Storms goes to Capt. Piddington who was then Foreign Secretary to the Agricultural Society of India and President of the Marine Court of Enquiry at Calcutta. He obtained extracts from numerous logs preserved in the India House, dealing with hurricanes (cyclones) from 1780 to 1841. He accumulated proof that these great storms are circular, they turn from right to left north of the equator and are progressive. He also traced the tracks along which they moved and their rates. He published 23 memoirs on cyclones in the Journal of the Asiatic Society of Bengal and wrote a book entitled *A Horn Book of the Law of Storms*, in 1845, which went through many editions.

Another interesting contribution relates to a suggestion made by Colonel Fraser, who was responsible for the erection of lighthouses along the Burma coast, that monthly wreck charts should be issued by the Navy. Accordingly, the first wreck chart of India for 1864-65 was issued in 1865. A similar chart showing the wrecks off the Bombay harbour was issued covering the years 1826-1866.

It is sad to record that, in 1861, the Royal Indian Navy ceased to exist and all the records belonging to the Bombay Marine and the Navy were entirely destroyed. However, in the eyes of geographers (to quote Markham) 'the widespread and lasting utility of the excellent surveys made by the officers of the Indian Navy hold an equally important place' in the Indian Ocean history.

The Marine Survey Department of India (1873-77)

For some reason, from 1861, the marine surveys of coastal India were practically abandoned. The Secretary of State proposed that, in the future, all surveys should be conducted by vessels selected by the

Bombay government and paid by the Government of India. Admiral Washington, who was then the hydrographer, did not like the idea. However, in August 1861, the Government of India was directed to complete those surveys which were incomplete and henceforth new marine surveys would be carried out by the Royal Navy, all expenses to be met by the Imperial Government in London. It was concluded by all marine surveyors of the Indian seas that the revision of old work and the production of correct charts were absolutely essential for the safety of the increasing trade and commerce in the Indian Ocean. Since the duties of the British hydrographer were worldwide, it may not have been possible for him to devote the attention required to create necessary priority and support for the surveys in Indian Ocean areas. In view of this and with strong representation from the Geographical Department of the India Office in London, the Government of India relented and requested that Captain A. D. Taylor, of the former Royal Indian Navy, be sent to Calcutta to assist in devising suitable means for restoring the efficiency of the Indian Marine Survey Department, which had ceased to exist in 1861. Accordingly, Taylor arrived in Calcutta on 22 December 1873 to take up his new duties. Meanwhile, survey work on the Makran coast was taken up on the recommendation of Captain Robinson, Superintendent, Marine at Bombay. During the great Orissa famine, assistance to local people could not be given for want of dependable survey charts. To remedy this situation, a survey was undertaken by Mr. H. A. Harris in 1869-70.

In 1873, Taylor completed the new edition of the first part of Horsburgh's *East India Directory*, and this book went through several editions. Meanwhile, an epochal change was brought about by the opening of the Suez Canal in 1869 for trade and commerce between European and the south and southeast Asian countries. In this timely book, Taylor gives complete directions for the Red Sea, including comments on the trade winds, monsoons, currents, land and sea breezes, storms and cyclones, variations and deviations of compass and tides.

Taylor also submitted to the Government of India a detailed review of the then existing charts, their differences, and an estimate of their intrinsic value, with a recommendation that there be appointed a Superintendent of Marine Surveys with sufficient funding and staff (not to exceed £20,000 per year) and jurisdiction extending from the Pakehan estuary at the southern extremity of Tennasarim to Sunmiyani Bay on the western limit of the Sind along the Indian (now Pakistan) coastline. He also recommended the commissioning of a small group of hydrographic vessels. These proposals were accepted by the Government of India and, as a result, Taylor was appointed as the first Superintendent of Marine Surveys.

Taylor obtained the services of six surveyors from the Royal Navy; he was also given a Chief Civil Assistant, and for this post, Mr. R.C. Carrington from the Naval Hydrographic Office was appointed, with duties related to the preparation of charts. At that time, the surveys that were to be undertaken on a priority basis included the Cuttak coast, the great Megna flats off the Ganges and Brahmaputra Rivers, the Cocos, Andaman and Nicobar Islands, the Gulf of Cambay, the Chittagong coast, the Burmese coast, and further examination of the Marugui archipelago. The vessels chosen for the Department were the *Clyde* (a steamer), the *Constance* (a schooner), the *Guide* (a brig) and the *Lady Lawrence* (a schooner). By reorganizing the Marine Survey Department on sound lines and for continuing the surveys in an exemplary manner, Dundas Taylor can rightly be called the Father of Indian Hydrography.

Writing on the inception and significance of the *Challenger* Expedition (1872-76), Sir Maurice Yonge (1972) comments that, apart from the publication of 50 monumental volumes, whose contents include description of thousands of new organic species, and distribution of sediments from the deep sea, the Expedition also awakened many maritime countries to the need for greater efforts to study the uncharted oceans. As a result, a number of oceanographic and marine biological institutions came into existence, among them being the Scripps Institution of Oceanography, in La Jolla, California (1912), the Woods Hole Oceanographic Institution, in Woods Hole, Massachusetts (1930), and the National Institute of Oceanography, in Surrey, England (1949[3]). Yonge (1972) further comments, 'It is worthy of note that the expense to which the British Government was put for the running of the Expedition and the publication of the results was more than covered by the money received from the exploitation of the phosphate deposits on Christmas Island (Indian Ocean), evidence for the presence of which was detected by Murray...'.

Although the *Challenger* did not enter the northern portion of the Indian Ocean for investigation,

[3] *Became Institute of Oceanographic Sciences in 1973.*

presumably expecting that the Indian Government would fill this gap, the ship made some stations in the Indian Ocean sector of the Antarctic Ocean and in the vicinity of Christmas Island, before entering the South China Sea and the Pacific Ocean. A summary of the scientific results of the *Challenger* Expedition was given by Murray (1895).

As already stated, the Marine Survey Department, though started in 1873, was keenly aware of its responsibilities, but had no means to implement them, till the *Investigator* was launched. During the period 1832-1862, when the Marine Survey of India was abolished, valuable surveys were undertaken by the officers of the Indian Navy, as reported by Sir Clement Markham (1878) in his memoir on the Indian surveys published by the Secretary of State for India. The Indian Naval Surveyors covered a large area, from the upper reaches of the Tigris and Euphrates Rivers to the ruins of Babylon and Nineveh and the coral reefs of the Seychelles in the southern Indian Ocean. Among these explorers should be mentioned John Davis who made several trips to India round the Cape of Good Hope, and whose adventures in the Arctic to find a northwest passage to the Far East are well known (Markham, 1880). This great navigator's last adventure proved fatal, for he was treacherously murdered by Malay pirates off the coast of Sumatra and his bones were turned to coral in the Singapore Straits. Another famous Arctic explorer whom destiny brought to the Indian seas, only to die, was William Baffin, whose memory is retained in the permanent names of Baffin Island and Baffin Bay, in northern Canada. Sir Clement Markham calls him the 'first Indian Marine Surveyor'. Baffin made two voyages to India, on the first of which he drew some charts of the eastern seas, which earned him a commendation and an award from his employers, but his second voyage ended tragically when he was killed by the Portuguese at the entrance to the Persian Gulf.

HMIS Investigator – Alcock and Sewell (1881-1933)

In our efforts to understand the oceanography and marine biology of the Indian seas, the contributions made by the 'surgeon naturalists' of His Majesty's Royal Indian Survey Ships are inestimable. Among them, Alcock (1902) wrote an excellent book, perhaps the only one so far written on the Indian seas, entitled *A Naturalist in the Indian Seas*, or *Four Years with the Royal Indian Marine Survey Ship HMIS Investigator*; he has also given an account of the earlier history of marine surveys in Indian waters.

Alcock's own involvement in the surveys of the *Investigator* began in 1888 and ended after four years, in 1892. The ship's main responsibility was depth charting of the coastal approaches to the Indian subcontinent, and during this time, if an opportunity arose, the ship could do some purely scientific work to gain knowledge of the hydrography of the local sea basins, their depths and temperatures, and of the deposits forming in their abysses and of the life that inhabits them.

HMIS *Investigator*, a wooden paddle-steamer built in a Bombay dockyard, with a displacement of 581 tons, was launched in 1881 and was specially designed for hydrographic survey and provided with modern equipment. This included a Sir William Thomson Sounding Machine with a Baillie sounding rod, deep-sea thermometers, and an American reversible trawl. The results of the survey by the *Investigator* were recorded in the annual Administration Reports of the Marine Survey of India. However, her zoological observations have been published in various journals, but chiefly in the *Journal of the Asiatic Society of Bengal*, since the year 1885, and in the *Annals and Magazine of Natural History*, from 1889. In his book *A Naturalist in the Indian Seas*, Alcock vividly describes the results of his four years of survey; they are arranged in 14 chapters with a bibliography and a complete list of all the *Investigator* publications till then. The book is replete with his personal experiences and observations on the customs of the native populations and contains descriptions of deep-sea life in the Bay of Bengal and Arabian and Andaman Seas. Of great interest is his description of the robber crab, *Birgus latro* (Figure 5.2b, chapter 5) of the oriental jungle islands, and of the tree climbing and air breathing (amphibious) gobiid fish *Boleophthalmus boddaertii* (Figure 5.2a, chapter 5) of the Indian creeks and estuarine mudflats. The robber crab is a giant hermit crab and is described as being a foot long, 5-6 pounds (≈2.3-2.7kg) in weight, nearly 8 inches (≈20cm) in breadth, and with the span of its longest legs over 2 feet (≈80cm)! This crab is found all over the Indo-Pacific region and is reported to eat coconuts.

The contributions of Alcock and his contemporaries to our knowledge of the marine fauna of the Indian seas are classic. Alcock has described the *Actinaria* (sea anemones), deep-sea corals, echinoderms, new decapods, stomatopods and amphipods. Among others who worked on the *Investigator* material, mention may be made of: Carpenter's contribution on the mean temperature of the deep waters of the Bay of Bengal; Oldham's report on the topography of the Arabian Sea in the neighbourhood of the Laccadives; Edgar Smith's report on deep-sea mollusca; and Wood-Mason's work on deep-sea Schizopoda.

R. B. Seymour Sewell joined the *Investigator* as a surgeon naturalist, very much on the heels of his illustrious predecessor Alcock who, as indicated earlier, made a significant contribution to our knowledge of the deep-sea animals in the Indian seas. Sewell went a step further to make an in-depth study of all relevant oceanographic parameters, such as salinity, temperature and water movements, to facilitate understanding of their relationships with marine organisms.

Sewell's contributions to geographic and oceanographic research in Indian waters during the period 1925-38 form an important milestone in the history of oceanographic developments in the Indian Ocean. In the *Memoirs of the Royal Asiatic Society of Bengal*, Volume 9, Sewell (1925-38) reported on the following subjects:

- The geography of the Andaman Basin.
- A study of the nature of the deep-sea bed and of the deep-sea deposits of the Andaman Sea and Bay of Bengal.
- Maritime meteorology of the Indian seas.
- The temperature and salinity of the coastal waters of the Andaman Sea.
- The temperature and salinity of the surface waters of the Bay of Bengal and Andaman Sea, with references to the Laccadive Sea.
- The temperature and salinity of the deeper waters of the Bay of Bengal and Andaman Sea.
- The topography and bottom deposits of the Laccadive Sea.
- Studies on coral and coral formations in Indian waters.
- Copepoda of the Indian seas.

In his introduction to these papers, Sewell says, 'one result of my work has been to convince me that, so far as this branch (marine biology) of oceanography is concerned, the value of such expeditions as those of *Challenger*, *Valdivia*, *Siboga* etc. has steadily diminished. Each succeeding expedition has added less and less to the sum total of our knowledge and what seems to be urgently required nowadays is an intensive study of comparatively small areas over a considerable period of time.'

Immediately after the *Challenger* Expedition, many western European and North-American countries launched major oceanographic expeditions, including some to investigate specific areas not covered by the *Challenger*. As already mentioned, the *Siboga* and the *Valdivia* mainly explored the East Indian seas, and the *Sealark* was launched to study the vast area between India and Madagascar.

The Percy Sladen Trust Expedition (1905-09)

The Percy Sladen Trust Expedition to the Indian Ocean in 1905 on board the *Sealark* can be considered an important British contribution. The Expedition was lead by Stanley Gardiner, who also worked on the madrepore corals collected during the Expedition from the various island reefs. Gardiner (1907) paid a tribute to the memory of Percy Sladen, an eminent biologist deeply interested in the echinoderms, by naming the Expedition after him. The Expedition's objective was to find faunal and floral evidence for land connections between Africa and India through island chains. The results of the Expedition have been published in a series of reports appearing in the transactions of the Linnaean Society, London, from September 1907. The equipment used for the collection of data and material on board the *Sealark* deserve some comments. The Lucas Sounding Machine was used for measuring depths. This had a large drum carrying piano wire of a length of 5000 fathoms (9,120m) and the *Sealark* had two such machines. The thermometers used were the minimum and maximum type manufactured by Cary, Negretti and Zambra. For dredging, a steel rope, about 2000 fathoms (3,648m) long, having a one-inch circumference, 114 strands in six cords, a pattern regularly made by Messrs. Bullivant & Co., was used and its breaking strain was estimated to be 4 tons (≈5000kg). Currents were measured by an Ekman Current Meter, and Fowler's Closing Nets were used for sampling plankton and small nekton.

Among the 21 reports published between 1907 and 1909, those on the Stomatopoda of the western Indian Ocean, by Borradaile, the marine fishes, by Tate Regan, the madrepore corals, by Gardiner, the polychaetes, by Potts, and the marine algae, by the Gepps, may be mentioned.

The John Murray Expedition (1933-34)

In the annals of oceanographic history, the John Murray Expedition is considered an outstanding example of technology transfer in the field of oceanography from a well developed country, in this case, England, to a poor and developing country, in this case, Egypt, both of which jointly launched this Expedition to the Indian Ocean. From the medieval Arabs piloting the small vessels into the Indian Ocean ports in the early part of the last millennium to the launching of an expedition by Egypt to study the Indian Ocean was a giant step.

This Expedition also stands out in another respect. During the voyages of the *Challenger*, large phosphate deposits were discovered on some of the Indian and Pacific Ocean islands. These deposits were reported to have brought riches to John Murray who bequeathed a part of these funds to furthering ocean research, and the same funds were utilized for launching the John Murray Expedition in 1933 under the leadership of Seymour Sewell, who, as already stated, had made significant contributions to our knowledge of oceanography from the *Investigator* cruises in the Arabian Sea and Bay of Bengal. Under his leadership, the *Mabahiss* cruised mainly in the Arabian Sea, between Colombo and Mombasa, in the Gulf of Aden and Oman and parts of the Red Sea.

The *Mabahiss* was specially built for coastguard and fishery research work and was well equipped with modern collecting gear for all kinds of oceanographic work. The ship measured 41.9m in length, with a beam of 7.14m, a displacement tonnage of 618, and coal-fired steam engines giving a speed of 11 knots. Among the equipment may be mentioned a trawl winch with 4000 fathoms (7,296m) of wire, a hydrographic winch with two drums, carrying 3000-fathom (5,472m) and 1000-fathom (1,829m) lines, respectively, an Acadia

Table 1.2 John Murray Expedition (1933-34) - list of scientific reports

- Introduction and list of stations (R. B. Seymour Sewell)
- Topography, with an Appendix on magnetic observations (Farquharson)
- Accounts of Addu, Horsburgh and Goifurfehendu Atolls (R.B. Seymour Sewell)
- Values of gravity in the Maldive and Laccadive Islands (Glennie)
- Meteorological observations (James Paton)
- Chemical and physical investigations (E. F. Thompson and H. Cary Gilson)
- The general hydrography of the Red Sea (E. F. Thompson)
- The exchange of water between the Red Sea and the Gulf of Aden over the sill (E. F. Thompson)
- pH concentration in the northwestern Indian Ocean (A. F. Mahamed)
- Cirripedia (H. G. Stubbings)
- Halobates (A. D. Imms)
- Nemertia (I. E. G. Wheeler)
- Cirripedia (Austin H. Clark)
- Pennatulacea (Sydney J. Hickson)
- Amphipoda (K. H. Barnard)
- Scyphomedusae (G. Stiasny)
- Polychaeta (C. C. A. Munro)
- Asteroidea (T. T. Macan)
- Phyllirhoidae (H. G. Stubbings)
- Pteropoda (H. G. Stubbings)
- Crustacea: Penaeidae (M. M. Ramadan)
- Opisthobranchia (N. B. Eales)
- Astacura and Palinura (M. M. Ramadan)
- Pycnogonida (W. T. Calman)
- Fishes (J. R. Norman)
- Copepoda: Harpacticoida (R. B. Seymour Sewell)
- Sipunculids and echiurids (A. C. Stephen)
- Paguridae and Coenobitidae (E. F. Thompson)
- Flabellid and turbinoid corals (J. Stanley)
- Euphausiacea and Mysidacea (W. M. Tattersall)
- Ascothoracica (Cirripedia) (K. A. Pyefinch)
- Echinoidea (T. H. Mortensen)
- Ophiuroidea (H. L. Clark)
- Stomatopoda (B. Chopra)
- Crustacea: Caridea (W. T. Calman)
- Madreporaria, excluding Flabellidae and Turbionoidae (J.S. Gardiner and P. Waugh)
- Stomatoid larvae (G. E. H. Faxon)
- Gorgonacea with notes on 2 species of Pennatulacea (S. J. Hickson)
- Ostracoda (Graham Cannon)
- Freeswimming planktonic copepods (R. B. Seymour Sewell)
- Stylasteridae (Hjalmar Broch)
- Epibionts and parasites of the planktonic Copepoda (R. B. Seymour Sewell)
- Pelagic tunicates (R. B. Seymour Sewell)
- Scaphopoda (N. H. Ludrook)
- Crustacea: Diromiacea (Isabella Gordon)
- Marine Algae (Linda M. Newton)
- Cephalochordata (J. E. Webb)
- Sessile Tunicata (P. Kott)
- Sponges (M. Burton)
- Brachiopoda (H. M. Muir-Wood)
- Crustacea: Penaeidae (N.M. Tirmizi)
- Crustacea: (Chirostylidae) (N.M. Tirmizi)
- A review of cephalopod family Sepiidae (W. Adam and W. J. Rees)
- Crustacea: Galatheiidae (N.M. Tirmizi)
- Deep-sea Bivalvia (Knudsen)

type echo-sounding machine, a modified Petersen grab, Ekman reversing water bottles, deep-sea thermometers, and trawl and plankton nets.

The Expedition reports published by the British Museum during the period 1936-40 contain a wealth of scientific information. The titles of the reports and the names of the contributing scientists are given in Table 1.2.

All these reports are included in 11 volumes and cover mainly the fauna of the western Indian Ocean. Information on physical oceanography, phytoplankton and marine geology is conspicuously absent. However, the vast amount of information available in the expedition reports forms a solid background for further biological studies.

The John Murray Expedition (1933-34) marks an important milestone in the contributions made by the British in the Indian Ocean area. At this juncture, many littoral nations here were in a political turmoil. Liberation from the British yoke was in sight for many countries, notably India, Pakistan, Burma, Ceylon, Egypt, Sudan, Kenya and Rhodesia. These countries were preparing to chalk out their destinies according to their own national conceptions and aspirations, and with India becoming an independent nation in August 1947, the Indian Ocean littoral was set for epochal changes both in politics and commerce, but not so much in science and, more particularly, in fisheries and oceanography. Within a few years of India's gaining independence, practically all the countries bordering the Indian Ocean attained their freedom from the colonial powers. While democracy took firm root in India, all other countries, though becoming independent, adopted different types of political structure spread over a wide spectrum between democracy and absolute dictatorship. The development of marine science in these countries, therefore, depended on political patronage for financial support and in most of the littoral countries this was not fully forthcoming.

In narrating the development of oceanographic research and development in the Indian Ocean, it should be mentioned that, up to the time of launching of the International Indian Ocean Expedition (1959-65), none of the littoral countries had research vessels, except South Africa and Australia. However, many of the countries had Fisheries Departments which were keen to survey and exploit fishery resources in their region. In this respect, India led the countries in its fishery development and related marine biological research.

In the absence of published information available from the various Indian Ocean countries, the authors were constrained to report mostly the development of marine research in India in the foregoing account. Even today, the major research effort in the Indian Ocean is contributed by India, but other countries, as Pakistan and the Gulf countries, East Africa and Thailand, are gradually increasing their marine research facilities and marine surveys.

The French contribution

The French contribution, especially between the end of the second World War and the IIOE, was based on the annual traverses of the French polar support vessels servicing the French Southern and Antarctic Territories (Terres Australes et Antarctiques Françaises) – the Crozet, Kerguelen, St. Paul and Amsterdam Islands, and Terre Adélie in Antarctica, all lying, however, south of the Indian Ocean as defined herein.

The most notable of these support ships was the *Commandant Jean Charcot*. The port of departure for its voyages was Djibouti and the track normally passed east of Madagascar, hence across the Somali basin and across the very important passage (oceanographically speaking, at least) between the northern tip of Madagasar (Cape Amber/Cap d'Ambre) and the Farquhar Islands (of the Seychelles group) to the northeast.

Initially, starting in 1948, basic oceanographic and meteorological observations were made regularly en route, notably the temperature and salinity at the sea surface and in the upper layer, air temperature, wind speed and direction, relative humidity, and the vessel's drift (as a measure of major currents). Most of the results obtained were reported by Tchernia (1949, 1951a, b, c), Le Floch (1951), Lacombe (1951), and Eyries and Menaché (1953), in the *Bulletin du Comité d'Océanographie et d'Etudes des Côtes*.

Later, starting in 1956, the scope of the French oceanographic studies in the western Indian Ocean widened to include more sophisticated methods (e.g., the geomagneto-electro-kinetograph, GEK) to measure surface currents underway (Martin, 1956a, b; Guibout and Lizeray, 1959) and other areas, notably the Mozambique Channel (Menaché, 1955, 1958; Martin 1956c), Gulf of Aden, Gulf of Oman and the Persian Gulf (Gougenheim, 1961a), and North Australian Basin (Tchernia and Lizeray, 1960). Other vessels were deployed for oceanographic purposes, as the *Lapérouse* (Tchernia, 1957a; Gougenheim, 1961b), the *Commandant Robert Giraud* (Menaché, 1958). Needless to say, the scope of the studies also allowed a wider or deeper assessment of the region from the oceanographic standpoint (Tchernia, 1957b; Tchernia *et al.* 1958; Tchernia and Lizeray, 1960).

Pioneers in Indian oceanography

Early Indian contributions

During the period 1930-50, Indians did pioneering work on the marine biology of coastal areas, particularly off the Madras coast. The littoral fauna and flora of the Krusadai Islands in the Gulf of Mannar have been worked out in some detail, as has the fauna and flora of the Madras beaches, by the scientific staff of the Madras Museum and professors and students from the Madras University Colleges. Contributions made by them are of a pioneering nature and form the basis of our knowledge in marine biology of the coastal areas of India. As already indicated, European contributions to the study of the Indian Ocean were mainly through expedition reports in which there was, unfortunately, very little Indian participation, except for Chopra and Tirmizi in the John Murray Expedition reports. It may also be stated here that practically no country around the Indian Ocean coastline, except for India, South Africa, and Australia, made any significant scientific contribution to the development of marine science. Even in the case of India, marine biological research commenced, perhaps, in the early part of this century, thanks to the scientists and teachers of the Madras Museum or the Madras University. It was these people who formed the nucleus for the establishment of the Marine Fisheries and Oceanographic Departments of the various central and state government organizations and universities in the years that followed, particularly after India achieved independence in August 1947.

It would, therefore, be in order to mention the names of those pioneering workers at the Madras Museum and Madras University who made very valuable contributions to the development of the marine sciences in India (Table 1.3).

These publications cover the period from 1927 to 1959. The latter was the year when the preliminary cruises under the International Indian Ocean Expedition were launched by various countries. This year also marked the departure of Professor Eugene LaFond, who had completed two terms of appointment as a Fullbright Scholar at the Andhra University and established a group of well trained scientists to work in the marine sciences, both coastal and offshore. Further details will be presented later.

Table 1.3 List of the Madras Museum publications

Author	Subject
F. H. Gravely (1927)	Hydrozoa, Alcyonaria, Scleractinae, Nemertinea, Chaetopoda, Isopoda, Polyzoa, Decapoda, Stomatopoda, Echinodermata, Urochorda
K. Ramunni Menon (1927)	Zoantharia
R. Winkworth (1927)	Mollusca
B. Sundararaj (1927)	Cirripeda, Peguridea, Pycnogonida
P. Fauvel (1930)	Polychaeta
M. G. K. Menon (1930)	Scyphomedusae
M. G. K. Menon (1932)	Hydromedusae of the Madras coast
Sydney J. Hickson (1931)	Alcyonaria of the Gulf of Mannar
C. C. John (1933)	Sagitta of the Madras coast
M. Burton (1937)	Porifera
M. Krishna Menon (1937-40)	Decapod larvae of the Madras plankton
F. H. Gravely (1941)	Shell and other remains found on the Madras beaches
C. P. Ganamuthu (1943)	Foraminifera of the Krusadan Islands
R. Velappan Nair (1949)	Thalacea of the Madras plankton
S. Thomas Satyamurthy (1952)	Amphiura, Scaphopoda, Pelecypoda and Cephalopoda
A. Daniel (1956)	Cirripedia of the Madras coast
K. Nagappan Nair (1959)	Amphipoda of the Madras coast
N. Daniel (1956)	Cirripedia of the Madras coast
K. Nagappan Nair (1959)	Amphipoda of the Madras coast

In 1947, the central government set up the Council of Scientific and Industrial Research (CSIR) responsible for planning and establishing a chain of national laboratories to be located in different parts of the country. First in this series was the National Physical Laboratory in New Delhi. A University Grants Commission was also created to oversee and finance the ailing Indian universities and to set up new universities. Most important from the viewpoint of oceanography and fisheries of the Indian Ocean, a central Marine Fisheries Research Institute was established at Mandapam Camp, near Rameswaram in southern India. Dr. H.S. Rao was its first Chief Scientific Officer, to be succeeded by Dr. N.K. Panikkar, who later played a vital role in the development of marine science in India, culminating in the establishment of the National Institute of Oceanography at Goa in 1966.

Panikkar was a self-made man. He was born into a Nair family in Kerala and was educated in Trivandrum and later at the Presidency College, Madras, where he worked under Prof. R. Gopala Aiyar, Head of the Department of Zoology.

Panikkar's work on osmoregulation in estuarine prawns attracted much attention in biology circles at that time. Later, he went to England for higher studies, on the award of a prestigious scholarship during the war time. On his return sailing, his ship was torpedoed, but he, along with many others, survived the mishap. On return, he joined the Zoology Department of the Kerala University for a year and later was appointed Chief Scientific Officer of the Central Marine Fisheries Research Institute at Mandapam Camp in 1947. The location of the Institute at Mandapam was considered ideal for marine fisheries research, since there was a string of small islands (Krusadai, Shingle and others) with fringing coral reefs, the Palk Strait and Gulf of Mannar nearby and Sri Lanka and the deep Indian Ocean beyond.

Fortunately for Panikkar, there was a group of scientists who had already done some commendable work on the faunal collections of the Madras Museum and some of these scientists joined the new institute. What was lacking, however, were the trained experts in physical, chemical and geological oceanography, a situation which continued until the arrival of Professor LaFond on the Indian scene, in 1952 as a Fulbright scholar in physical oceanography.

Meanwhile, Panikkar, after establishing the Central Marine Fisheries Research Institute on a firm footing, moved to New Delhi as Fisheries Development Advisor to the Ministry of Food and Agriculture and was responsible for creating the Central Institute of Fisheries Technology, the Central Institute of Fisheries Engineering and Nautical Training at Cochin, the Fisheries Survey of India and the Central Institute of Fisheries Education, both at Bombay. All these institutions organized regional centres at different locations along the Indian coast to attend to local problems. He also formed the Marine Biological Association of India and started two new journals, namely, the *Journal of the Marine Biology Association of India* and the *Indian Journal of Fisheries*. For some years, a bulletin was also issued from the Central Marine Fisheries Research Institute (CMFRI).

To generate manpower for servicing these institutions, many coastal universities, such as Andhra, Madras, Annamalai, Trivandrum and Bombay, organized postgraduate courses leading to a Master's Degree, as well as a diploma in marine biology and fisheries.

While Panikkar was at work in New Delhi, important developments were taking place at Andhra, Madras, Annamalai, Trivandrum/Cochin and Bombay Universities.

Contributions of the Andhra University and others

Andhra University: Located at Waltair and at the foot of the Eastern Ghats and a few kilometres from the natural harbour of Visakhapatnam, Andhra University commands a beautiful panoramic view of the Bay of Bengal. In 1947, the University hired the services of Professor R. Gopala Aiyer, who had then retired from the Presidency College of the Madras University, where he had pioneered marine biological work on the Madras coastal waters and the Cooum estuary. Gopala Aiyer was a very strict orthodox Brahmin, an excellent teacher and very meticulous in habits and had trained a band of students in marine biology at the zoological laboratories of the Madras University for nearly 20 years, establishing a reputation for excellent research work. During his five-year tenure at Andhra University (1947-52), the Department of Zoology developed into an important research centre for marine biological studies.

Meanwhile, Prof. V.S. Krishna, the Vice Chancellor of the Andhra University, appointed

Dr. C. Mahadevan as Professor and Head of the Department of Geology and Professor S. R. Savour as Head of the Department of Meteorology. Prof. Krishna, an economist, was well known in the country as a progressive thinker and educationist who later became the Chairman of the University Grants Commission. Anticipating that India must go in for ocean research and development in a big way, so as to exploit its vast fishery and mineral resources, Krishna requested the US Embassy in New Delhi to obtain the services of an American oceanographer to develop ocean science at the Andhra University. This request was readily agreed to and, as a result, Eugene C. LaFond arrived in India with his wife and two young children in September 1952. He also brought with him basic oceanographic equipment such as Nansen bottles, bathythermographs, thermometers, plankton nets, snappers and corers. More importantly, his vast experience in oceanographic research was an asset for training the staff at Andhra University.

The combination of the most amiable and progressive personalities of Prof. V.S. Krishna, Prof. C. Mahadevan, Prof. P. N. Ganapthi (Prof. R. Gopala Aiyer had by then left the University after doing a yeoman's job in building up the Zoology Department) and Prof. Savour, created the best possible environment for the growth of marine sciences in India. By their combined efforts, the Departments of Zoology, Botany, Geology, Chemistry and Meteorology took keen interest in the oceanographic programmes initiated by Prof. LaFond.

Soon after arrival, at a meeting of heads of departments with the Vice-Chancellor, the stage was set for launching the first oceanographic cruise into the Bay of Bengal. At the request of the Vice-Chancellor, the Indian Navy arranged for the loan of minesweepers for cruises, and Commanding Officer Soman and other naval officers extended full co-operation. Under the cruise leadership of LaFond, fifty-one research cruises were organized on board the Indian naval minesweepers, mainly, INS *Konkan*, INS *Madras*, INS *Bengal*, INS *Rajputana*, INS *Rohilkhand* and the harbour tug *Rana Pratap* (see Table 1.4).

During these cruises, teachers at the Andhra University and others deputed from sister institutions in the country had, for the first time, the opportunity to go to sea with oceanographic equipment and collect data such as vertical serial temperatures, water transparency, water samples for salinity, nutrients, pH and other chemical properties; plankton net and dredge hauls for biological studies, bottom snapper and core samples for sediment properties, etc.

The results of these cruises and shore work were published by the Andhra University as *Memoirs I & II*, containing a total of 35 research papers. This apart, several doctoral theses were accepted by the University for the award of PhD/D.Sc degrees. These theses are, perhaps, the first in Indian oceanographic science and are listed in Table 1.5.

By the time LaFond left India in 1957, not only had Andhra University established for itself a new name and fame as a centre for oceanographic research, but had also attracted young scientists from all over the world to study oceanography, with bright and fruitful career possibilities in various marine and fisheries institutes being set up and expanded by Panikkar in the early 1960s. Prof. and Mrs. LaFond's services to the cause of oceanography in Indian waters will always be remembered with gratitude and deep appreciation by the Indian scientific community working in marine sciences.

Annamalai University: Located at Chidambaram, a famous pilgrim centre, 200 miles south of Madras, the Annamalai University and Prof. R. V. Seshaiya (during 1945-1965), Head of the Centre for Advanced Studies in Marine Biology, are synonymous. Seshaiya was a person singularly devoted to marine biological research, particularly its biochemical aspects. Working around bureaucratic hurdles, he was able to establish a well equipped marine biological laboratory at Porto Novo, a picturesque coastal village, a few kilometres from Annamalai University campus at Chidambaram. He attracted a large number of students for post-graduate studies thus laying a firm foundation for the advancement of marine biological research, and acquired a 50-foot (15m) oceanographic vessel for work on estuaries. Recognizing this, the University Grants Commission upgraded this laboratory to a Centre for Advanced Studies in Marine Biology and provided adequate funds for its activities.

The Universities at Trivandrum-Cochin and Bombay were relatively active in marine biological research. They had well organized departments of zoology and botany for post-graduate education and research, but none of them, including Annamalai, had sea-going oceanographic vessels. The minesweepers of the Indian Navy used by the Andhra University were also not suitable and comfortable for marine research, but served the purpose for initial training of Indians in the collection of oceanographic data from the open seas.

Table 1.4 Andhra University's first oceanographic cruises, 1952-1957

Cruise No.	Vessel	Date	Station Nos.	Location
		1952		
1	Rana Pratap	21 Oct.	1-13	Off Waltair/Visakhapatnam
2	INS Konkan	29-31 Oct.	14-36	Off Waltair/Visakhapatnam and south
3	INS Madras	19-26 Nov.	37-88	Visakhapatnam to Madras
4	INS Madras	29 Nov.	89-90	Off Waltair/Visakhapatnam
5	INS Madras	1-11 Dec.	91-100	Off Waltair/Visakhapatnam/Calcutta
		1953		
7	INS Bengal	17-18 Feb.	126	Off Waltair/Visakhapatnam (anchored)
8	INS Bengal	19-20 Feb.	127-144	Off Waltair/Visakhapatnam
9	INS Bengal	24 Feb.	145-150	Off Waltair/Visakhapatnam
10	INS Bengal	24-27 Feb.	151-154	Visakhapatnam to Kakinada
11	INS Bengal	2 Mar.	183-191	Off Waltair/Visakhapatnam
12	INS Bengal	4-5 Mar.	192-219	Off Waltair/Visakhapatnam
13	INS Bengal	8-29 Mar.	220-256	Visakhapatnam to Madras to Cochin
14	INS Rajputana	10 Apr.	257-264	Off Waltair/Visakhapatnam
15	INS Rajputana	13 Apr.	265-268	North of Visakhapatnam
16	INS Rajputana	17 Apr.	269-284	Off Waltair/Visakhapatnam
17	INS Rajputana			
18	INS Rajputana	24 Apr.	309-317	Off Waltair/Visakhapatnam
19	INS Rajputana	27 Apr.	318-326	Off Waltair/Visakhapatnam
		1954		
22	INS Rohilkhand	2 Apr.	340-352	Off Waltair/Visakhapatnam
24	INS Rohilkhand	21 Apr.	356-385	Visakhapatnam to Vamsadhara
26	INS Rohilkhand	15 May	418-423	Off Waltair/Visakhapatnam
		1955		
30	INS Rohilkhand	1 Dec.	446-452	Off Waltair/Visakhapatnam
31	INS Rohilkhand	9 Feb.	453-465	Off Waltair/Visakhapatnam
32	INS Rohilkhand	2-4 Mar.	466-278	Off Waltair/Visakhapatnam
33	INS Rohilkhand	27-28 Oct.	479	Off Waltair/Visakhapatnam (anchored)
34	INS Rohilkhand	10-11 Nov.	480-499	North and south of Visakhapatnam
35	INS Rohilkhand	15-17 Nov.	500-518	Visakhapatnam to Madras
36	INS Rohilkhand	9 Dec.	519-523	Off Waltair/Visakhapatnam
37	INS Rohilkhand	13-14 Dec.	524-524G	Off Waltair/Visakhapatnam
		1956		
38	INS Rohilkhand	18-20 Jan.	525-569	Visakhapatnam to Kakinada
39	INS Rohilkhand	24 Jan.	570-573	Off Waltair/Visakhapatnam
40	INS Rohilkhand	9-10 Feb.	574-587	Off Visakhapatnam and Bhimilipatam
41	INS Rohilkhand	17 Feb.	588-592	Off Waltair/Visakhapatnam
42	INS Rohilkhand	21-22 Feb.	593-606	Off Waltair/Visakhapatnam
43	INS Rohilkhand	7 Mar.	612-620	Off Waltair/Visakhapatnam
44	INS Rohilkhand	13-14 Mar.	621-638	Off Waltair/Visakhapatnam
45	INS Rohilkhand	20-23 Mar.	639-662	Visakhapatnam to Puri area
46	INS Rohilkhand	13-14 Apr.	663-682	Waltair/Visakhapatnam and south
47	INS Rohilkhand	24-25 Apr.	689-709	Off Waltair/Visakhapatnam
48	INS Rohilkhand	4 May	710-715	Off Waltair/Visakhapatnam
49	INS Rohilkhand	15 May	716-718	Off Waltair/Visakhapatnam
		1957		
51	INS Konkan	21-27 Sep.	720-744	Visakhapatnam to Barua area

Table 1.5 List of Andhra University doctoral theses and their authors, 1954-1972

Author	Year	Thesis
M. Poornachandra Rao	1954	Some aspects of marine geology in certain parts of the Bay of Bengal. Sources of sediments, their characteristics and distribution along the east coast of India.
U. Aswathanarayana	1954	Radioactivity of some Indian rocks, sea floor sediments and minerals. Distribution patterns of radioactivity of east coast sediments, correlated with lithological zone patterns.
T. S. Satyanarayana Rao	1955	Studies on the Chaetognatha in relation to the biology and hydrography of the Indian seas. Distribution of these organisms relative to the water properties, light, temperature, salinity, depth, etc.
A. A. Ramasastry	1955	Some aspects of the physics of the atmosphere and the sea. Time and space calculations of heat, evaporation, and other air/sea interaction properties over the Bay of Bengal.
R. Prasada Rao	1956	Beach studies in some parts of Andhra coast. Beach sand movements and composition related to storms, tides, waves, rainfall and current.
R. Nagabhushanam	1957	Studies of marine woodboring mollusca of Visakhapatnam harbour. Environmental characteristics of various woodborers on a variety of wood submerged in the harbour.
C. Balarama Murty	1958	Studies of the physical oceanography of the western Bay of Bengal. Currents, temperature structure, and other properties of the water off the east coast of India.
J. Sivarama Sastry	1958	Some aspects of the shoreline processes and physical oceanography. Studies of wave refraction along the shoreline between Visakhapatnam and Kakinada, and the resulting changes in the shoreline.
V. V. R. Varadachari	1959	On some meteorological and oceanographic studies of the coastal waters off Waltair in relation to upwelling and sinking. Seasonal cycles in current and water structure in relation to the monsoon wind and radiation.
M. Subba Rao	1959	Some aspects of continental shelf sediments off the east coast of India. Physical and chemical properties of shelf sediments, deposition of sediment in Kakinada Bay.
Y. Radakrishna	1963	The systematics and ecology of bottom fauna. The study deals with the systematics and ecology of the bottom fauna off the Waltair coast and the estuarine region of the Guatami-Godavari River near Kakinada.
D. V. Subba Rao	1965	Studies on the hydrography, phytoplankton and primary production off the Waltair coast, Bay of Bengal.
P .V. Bhavanarayana	1969	Studies on the hydrography and the distribution of pelagic tunicates in the western part of the Bay of Bengal. The distribution of the different groups of pelagic tunicates.
G. R. Lakshmana Rao	1972	Studies of the heat balance parameters of the north Indian Ocean. The study deals with the dominant heat balance parameters: total incoming solar radiation, net radiation, sensible and latent heat fluxes and the energy storage in the ocean. The influence of the monsoons and the annual variation of seawater temperature.

By the year 1960, research and educational capabilities in oceanography and related sciences were fairly well established in India. The Central Marine Fisheries Research Institute had experts dealing with zooplankton, productivity, systematics of marine fauna and flora, fishes and fishery survey. The Institute was keen to understand the fishery resources discrepancy between the Bay of Bengal and the Arabian Sea, the latter contributing 60-70 percent of the marine fish landings in the country. The question to be answered was why the Bay of Bengal was so poor when compared with the Arabian Sea fishery resource. There were other interesting problems seeking solutions, such as red tides and mud banks off the Kerala coast, the location of spawning grounds of the Indian sardine and mackerel, the distribution of shrimp fishery resources, and so on. The cause and effect of the alternating monsoons on ocean currents and productivity were also not known at this time.

The other organizations were involved marginally in oceanographic and related research in India. The Zoological Survey of India and the Indian Museum at Calcutta had some excellent collections of fishes and other marine organisms, thanks to the efforts of surgeon-naturalists working on HMIS *Investigator*, which contributions have already been mentioned. From among the scientific staff of the Zoological Survey, the important contributions during the years 1941-57 came from B. N. Chopra on the prawn fisheries, and S. L. Hora on fishes and fish geography. Hora also read a paper entitled *Oceanographic Studies in Indian Waters* before the Zoological Society of India in 1950.

The Indian Navy established a Hydrographic Office at Dehra Doon, with a fleet of ships under its command and a Naval Physical and Oceanographic Laboratory at Cochin to undertake the investigations required by the Navy. The Forest Research Institute located at Dehra Doon started an all-India research programme on marine fouling and boring in collaboration with some of the maritime universities. In addition, the central government established a central Board of Geophysics with an oceanographic wing attached to it for conducting research in marine geophysics. While all these activities were going on in India, none of the countries along the Indian Ocean coastline had any major marine science programmes, except for South Africa, Kenya and Egypt on the western side and Australia on the eastern side. However, most of these countries had fishery departments and efforts were being made to launch long-range programmes to survey the fishery resources for development and exploitation. In this connection, mention should be made of the great contribution of Norway in fishery survey and research in the Arabian Sea. Under the Indo-Norwegian project, a detailed fishery survey programme was initiated on-board the Norwegian research vessel *Varuna* during the 1960s with its headquarters located at Cochin. As a result, the annual marine fish catch increased substantially thanks to the introduction of mechanized fishing and to the location of new fishing grounds, as also to the application of scientific post-harvest technology and marketing strategies.

This was the state of the art in marine sciences in India and the other coastal countries overlooking the vast expanse of the Indian Ocean, when a major international oceanographic programme was about to be launched.

The International Indian Ocean Expedition (1959-1965)

The genesis

The genesis of the International Indian Ocean Expedition (IIOE) may be traced back to 1937, when T. W. Vaughan, in his report on the *International Aspects of Oceanography*, brought to the attention of the oceanographic community, the fact that very little was known about the Indian Ocean and it was time to plan and bridge this big gap in our knowledge. World War II and the subsequent turmoil interfered with such plans for almost two decades. It was only in 1957 that Lloyd Barkner, the American geophysicist who played a leading role in organizing the *International Geophysical Year (IGY)* and who was then President of the International Council of Scientific Unions (ICSU), requested Roger Revelle, Director of the Scripps Institution of Oceanography, to appoint a Special [later, Scientific] Committee on Oceanic Research (SCOR) so that oceanographers could play a major role in the proceedings and plans of the ICSU. Accordingly, Revelle formed a committee consisting of about 15 members; among them were G. E. R. Deacon, Director of the Institute of Oceanographic Sciences (UK), Columbus O'D. Iselin, Director of the Woods Hole Oceanographic Institution (USA), Gunther Boehrecke, Head of the German Hydrographic Office (Federal Republic of

Germany), Lev Zenkevich, a Soviet marine biologist, Maurice Hill, the leader of the Marine Geophysics Group at Cambridge (UK) and N. K. Panikkar, Fisheries Development Advisor to the Government of India.

At the first meeting of the SCOR, held at the Woods Hole Oceanographic Institution (WHOI) 28-30 August 1957, it was decided to plan an international expedition to the Indian Ocean. Roger Revelle, who presided over the meeting, appointed a Working Group under the Chairmanship of Columbus Iselin to prepare a plan for such an expedition. The members of the working group included representatives from Australia, France, Federal Republic of Germany, India, Japan, South Africa, UK, the USSR and the USA. SCOR also appointed Robert Snider as Co-ordinator, whom Revelle called 'a born expeditor'.

SCOR, during its meeting at WHOI, considered three long-range problems seeking a solution by way of the proposed expedition, all of them important for the future of mankind. First, to know the Indian Ocean potential for fishery resources, since most of the countries bordering the Indian Ocean were deficient in proteins in their diet; second, to assess the role of the northern Indian Ocean in effecting the monsoonal changes, which are vital for agricultural operations in the Indian sub-continent, but which also influence the current patterns, upwelling systems, productivity and the carbon-dioxide cycle; and third, to determine the limits to the use of the oceans for dumping human wastes, including spent nuclear fuels etc. It was also suggested that, during the first two years of the expedition, participating countries should encourage standardization of equipment and methods of analysis and data logging so that the results obtained by different ships would be comparable. SCOR also recommended that, during the third and fourth years of the expedition, as many as 16 ships should simultaneously cruise in the Indian Ocean and make a combined assault on the largest unknown area of the Earth: the deep waters of the Indian Ocean and its sea bed. With SCOR's endorsement of the Expedition, scientists from different countries began to discuss and plan their participation. There were also discordant voices, doubting whether such a large programme was feasible. There were others who very much wanted to study the Arabian Sea in particular for its reversing monsoons, the Somali upwelling and high rates of productivity. Soon there came into being a publication called the *Indian Ocean Bubble* at irregular intervals, and the editor invited freelance discussion on the proposed IIOE. The first issue contained a lengthy letter from Henry Stommel, the renowned oceanographer from WHOI, recommending detailed studies of the Arabian Sea and, to quote, 'The question which we would like to resolve is how much does the internal density structure of one of these semi-enclosed basins respond to the variations of wind stress. A clear-cut observational answer would be an interesting test of theoretical ideas about the oceanic circulation ...'. Stommel also brought up the Somali current off Somalia and, again to quote, 'according to ship observations, the current flows toward the south during the northeast monsoon and toward the north during the southwest monsoon. It appears to be strong, intense and narrow – ideal for repeated hydrographic sections, season by season. Welander's computations indicated that this ought to be the world's most strongly oscillating current system – the difference in south and north flows amounting to about 61 million cubic metres per second'.

The same issue of the *Bubble* contained a letter from another scientist, asking 'Do you think it would be possible for some of those interested in surveying the Indian Ocean to meet in a bar or other relaxing place, during one of the less enthralling sessions of the Oceanographic Congress in New York next September, 1959?'.

In the second issue of the *Bubble* in February 1959, R.B. Montgomery wrote a letter expressing the hope that the Indian Ocean Programme would be so designed as to aid directly the development of one or more oceanographic research centres in the underdeveloped countries bordering the Indian Ocean. This wish was fully realized in India and Pakistan, where national centres for oceanography were established after the IIOE.

In the third issue of the *Bubble*, which appeared on 10 May 1959, a letter from Martin J. Polak was published in the same vein as above. While endorsing the proposed IIOE programme, he said, 'probably the primary need is to fill in some of the large gaps in the geographical distribution of hydrographic stations ... It seems that it should be possible to combine some of the required reconnaissance surveys with special studies of the circulation patterns. For instance, a seasonal study of the monsoon regimes in the Arabian Sea and the Bay of Bengal would serve a dual purpose: these two areas are virtually untouched by subsurface thermometers'.

How bad the Indian Ocean situation was can be grasped from a sarcastic postscript added by Montgomery to his letter in the *Bubble*, 'In case anyone should think of a use for them, I have a set of noon temperatures of water, air, and wet bulb made from a passenger vessel from Singapore to Suez in December 1958'.

Stommel made some further comments, stating that he and Fuglister had examined bathythermographic data obtained across the equator in the Atlantic, and saw similar features to those reported for the Cromwell Current in the Pacific Ocean and if such a current exists in the Atlantic Ocean it may also exist in the Indian Ocean, and this was worth investigating. The results of the International Indian Ocean Expedition later showed that the Cromwell Current did exist in the Indian Ocean, as predicted by Stommel.

Further support came from George Wüst, Head of the Institute of Marine Science at the University of Kiel, Federal Republic of Germany, who submitted a plan for the survey of the Indian Ocean to SCOR. In the history of oceanography, Wüst is well remembered for his classical work on board the German Research Vessel *Meteor* which criss-crossed the Atlantic Ocean 14 times, from 20°N to the ice edge of the Antarctic Ocean. He suggested that the Indian Ocean be investigated from about 30°N south to the Antarctic on a grid basis, stations to be occupied at 8° intervals. His plan was appended to the IIOE prospectus prepared and issued by SCOR for wide circulation and comment. The prospectus was prepared by about 40 scientists invited by SCOR, representing different disciplines in oceanography, and was finalized by a group of 3 eminent scientists, namely, Roger Revelle, of the United States, George Deacon, of the United Kingdom, and Anton Bruun, of Denmark, who had been the leader of the second Danish *Galathea* Deep-Sea Expedition of 1950-52 and then the first Chairman of the Intergovernmental Oceanographic Commission (IOC) of UNESCO. Recalling those days, Behrman (1981), in his excellent book on the IIOE, quotes Revelle as remembering 'that July and August of 1960 were the most difficult months in my scientific life. After the subcommittee met in July, George Deacon, Anton Bruun and I had to put together their reports, synthesizing them into a coherent document... Nobody had studied the Indian Ocean. This was to be an exploration in the old-fashioned sense. There were so many scientific problems and the Indian Ocean was so far away from all our institutions that no one felt that his territory was being usurped... The Indian Ocean expedition was a pioneering effort in international oceanographic planning. It was like the International Geophysical Year, but on a much bigger scale. We learned how difficult the task was. We had to accommodate conflicting interests, for this was a political operation in which people had to be persuaded'.

In a letter to the fourth issue of the *Indian Ocean Bubble* in July 1959, LaFond wrote:

'I have been interested in the various discussions of the proposed oceanographic studies for the Indian Ocean appearing in the Indian Ocean Bubble. To me, the problem is not what to do, but rather, who in the Indian Ocean Region can be rounded up to do it? Everyone should be reminded that this is the Indian Ocean, and not the Woods Hole or Scripps ocean.

To spread the gospel and attain any lasting results, the work has to be carried on partly by the scientists of the Indian Ocean area. This does not mean just coming along for a ride, but actually give a major share in planning, analysis, and reporting. Most Asian students will be enthusiastic about the work if given the opportunity to collect data for thesis material. This opportunity and encouragement should be the primary goal for the expedition.

All plans dealing with the expedition and their execution will progress very slowly in Asia due to lack of authority and the complex restrictions on travel, money exchange, imports, immigration, and numerous other necessities of the program. Thus, planning with Indian Ocean scientists for participation should start early to avoid some of the inevitable delays.

Unfortunately, it is not possible to deal directly with students. It is necessary to go down through the chain of command. The most promising approach would be to contact high-level people, such as heads of scientific organizations, naval laboratories, fisheries, or universities, explaining the proposed program. Eventually, through these contacts, it may be possible to assemble some good Indian Ocean scientists and get them started in oceanographic research in their ocean.'

It was important, therefore, to involve developing countries, so that the expedition would not appear to be what Revelle called 'A club of rich countries that wanted to do oceanography'. Here, help came from the late N. K. Panikkar, an Indian scientist on SCOR, whom Revelle remembers as 'very sensible and very enthusiastic'.

Shortly after the plan for the IIOE was unfolded in the prospectus issued by SCOR at its meeting in

Helsinki in August 1960, Robert Snider left on a mission round the world to meet and persuade all countries interested in the expedition to contribute, plan and execute the programme. In this task he got full support from the members of SCOR national committees in various countries. He particularly made it a point to meet 'political activators' and influential members of the governments and local parliaments. He also carried with him colourful charts showing the proposed cruise tracks of the IIOE and, more importantly, an IIOE emblem the use of which on letterheads and packages would facilitate customs clearance of scientific equipment, exempting them from payment of duties, etc., for all countries participating in the International Indian Ocean Expedition. In India, he approached Homi Bhaba, the eminent nuclear scientist, who in turn obtained the support of Jawaharlal Nehru, the then Prime Minister of India. The American participation in the IIOE had already been endorsed by President Eisenhower and later by President Kennedy.

In all these activities, SCOR, being a nongovernmental group, had a fairly free hand to co-ordinate the expedition programme till the end of 1962, at which time, the newly formed IOC of UNESCO assumed co-ordinating responsibilities from Snider and SCOR. At the Woods Hole meeting of SCOR in August 1957, an estimate of the cost of the International Indian Ocean Expedition was projected (Table 1.6).

Nobody seemed to have paid much attention to these cost estimates, for the simple reason that there was just no money for such a major venture, either through UNESCO or any other country. In a way, this was for the best, because individual countries had to fend for themselves to obtain the necessary funds from their own governmental agencies. Recognizing the importance of the Expedition, and the fact that economic conditions were improving worldwide, many countries found funds and support from their governments for their participation in the IIOE.

At the Helsinki meeting of SCOR in August 1960, plans for the IIOE and general guidelines were issued. The SCOR working groups on the Indian Ocean gave some directions to the scientific programmes. The panel for physical oceanography, meteorology and chemistry recommended some areas for intense study: the Arabian Sea in summer and winter and the waters northwest of Australia; the Red Sea and the Persian Gulf for their heat budget and, similarly, the southern part of the Bay of Bengal. Emphasis was laid on correlating weather charts with oceanographic conditions wherever possible.

In chemistry, it was urged that all participating ships undertake a minimum common programme, such as the collection and analysis of water samples from standard depths for estimating dissolved oxygen and nutrients at each station. The panel on geology requested that all ships carry precision echo-sounders and continuously run them to record the water depth and to share such information with all other participating ships. Geomagnetic and gravimetric studies were also planned. In the biology programme, it was recommended that plankton samples be collected at all stations using an Indian Ocean Standard Net, designed by Currie, and make a vertical haul from 200m depth to the surface. Also, phytoplankton samples and productivity measurements should be taken wherever possible, and more particularly along the meridians 62°, 78° and 95° in the north-south direction. Another recommendation was to establish a sorting centre for zooplankton in India, where all the zooplankton samples could be sorted group-wise and then distributed to specialists. All the participating ships were requested to send the plankton samples to the proposed sorting centre.

Another very important programme endorsed and executed by SCOR and UNESCO related to the intercalibration of equipment and standardization of analytical procedures. Accordingly, three such tests were conducted, the first of which was organized in September 1960, at Honolulu, on board the Australian ship *Gascoyne*, the Soviet vessel *Vityaz* and at the laboratories of the University of Hawaii. The second test was held off Freemantle, Australia, again on board the *Vityaz*. And finally, a third series of chemical tests was organized on board the British ship *Discovery* in 1964. Against this background, an exciting drama to unfold the scientific secrets was played on the vast stage of the Indian Ocean. In view of its strategic location, India played a significant role in the overall operations and co-ordination of the IIOE, as well as being an active participant.

The Indian response to the IIOE

A person in the right place at the right time was N. K. Panikkar in India. As already mentioned earlier in this chapter, Panikkar held a very influential position as Fisheries Development Advisor in the Ministry of Food and Agriculture, in Delhi. He also represented India on SCOR. Besides, Panikkar knew, perhaps

Table 1.6 IIOE participants, ship time, estimated cost (in US$) etc.

1. **Countries participating**:

   (a) Ship-operating countries: Australia, France, Germany (Fed. Rep.), India, Indonesia, Japan, Pakistan, Portugal, Republic of South Africa, Thailand, USSR, UK, USA.

   (b) Other participants: Burma, Ceylon, China, Ethiopia, Israel, Italy, Malagasy Republic, Federation of Malaya, Mauritius, Sudan.

   Number of ship-months: 320 (approximately)

   Distribution: Australia (37), France (20), Germany (Fed. Rep) (6), India (24), Indonesia (3), Japan (20), Pakistan (18), Portugal (3), Republic of South Africa (13), Thailand (2), USSR (20), UK (35) and the USA (119).

2. **Area of study**: Indian Ocean including adjacent seas.

3. **Period of study**: 1959 to 1965, peak at 1962, 1963 and 1964.

4. **Object of study**: Complete survey of the Indian Ocean, including descriptive physical, chemical, biological oceanography, marine geology, geophysics and meteorology.

5. **Co-ordinator**: The Secretary, IOC

6. **Principal sponsors**: UNESCO, SCOR and IOC

7. **Other interested organizations**: International Meteorological Centres, Indian Ocean Biological Centre, FAO, WMO.

8. **Estimated costs** (US$):

(a)	Cost of approximately 16 ships each operating in the Indian Ocean for 8 months	2,400,000
(b)	Training of 25 scientists from Indian Ocean area, each for one year @ $3,000 per head	75,000
(c)	Special equipment for each ship @ $24,000 per ship	384,000
(d)	Salaries of the scientists working on ships (100 people for 1 year)	550,000
(e)	Working up the scientific results (100 people for 1 year)	600,000
(f)	Publication of results	30,000
(g)	TOTAL COST	4,039,000
	Less estimated contribution from normal operating costs and salaries	2,000,000
	Estimated extraordinary cost	2,039,000

From UNESCO (1963)

better than anyone in India, the marine research capabilities that the country could muster for taking part in the IIOE. Having participated in the SCOR-UNESCO meetings in Paris and Helsinki in 1960, Panikkar decided to act and, on his advice, the Government in Delhi appointed an Indian National Committee on Oceanic Research (INCOR) with the following terms of reference:

1. to draw up a co-ordinated plan for India's participation in the IIOE;
2. to advise on the allocation of a programme between governmental departments, research organizations, universities and other institutions;
3. to consider and approve detailed plans for research in the several scientific disciplines related to India's participation and to recommend financial grants;
4. to further and co-ordinate research programmes;
5. to advise the Government generally on all matters connected with India's participation in the Expedition.

In the light of these terms of reference, INCOR, besides being responsible for India's participation in the IIOE, also became the focus for all developments and research projects connected with oceanographic research in the country. The Committee and its working groups included almost all the important scientists representing various Indian institutions concerned with different branches of oceanography. The results of the Committee's deliberations led directly to the establishment of an Indian Ocean Expedition Directorate as a department of the Council of Scientific and Industrial Research, and to the allocation of sufficient funds and staff to ensure the full participation of India in this Expedition. The subsequent establishment of the International Meteorological Centre at Colaba (Bombay) and the Indian Ocean Biological Centre at Cochin is well known to marine scientists the world over.

The Expedition sought to explore in detail the oceanography of the Indian Ocean and to make the area as well known as the Atlantic and the Pacific. It was also fashionable for all who ever spoke on the programme of IIOE to draw attention to the enormous population increase of India and neighbouring countries and the inadequacy of food and protein supply for the undernourished people and to plead fervently for the exploration and exploitation of the food resources of the sea. If the speaker was a marine meteorologist, he would say how interesting are the reversing monsoons and how unpredictable or unreliable they are for Indian agriculturists. A geophysicist would stress the possibility of locating oil resources in the shelf areas off India. In this way, all sought to justify the many cruises, ships and nations participating in the joint enterprise of the IIOE. India's response to the call of the IIOE, as that of other countries in the region, was positive.

India's exploitation of fishery resources has been growing every year and it was reasonable to expect that, given more vessels and men, it should be possible to double the gross tonnage of fish landed on India's coasts. The most interesting aspect of the problem, however, was the fact that 67-75% of marine fish landed annually in India came from the west coast. This was pointed out by Panikkar and Jayaraman (1956) at the 8th Pacific Science Congress, and the picture remains the same even today. The reason for the apparent scarcity of fishery resources off the east coast of India and the rest of the Bay of Bengal should be urgently investigated and considered quite separately from the problem of whether we are fully exploiting the available fishery resources off the west coast. Although, in recent years, mechanization of fishing vessels has increased, the fact remains that fishing is done mostly in nearshore waters while the vast shelf off the west coast remains totally unexploited. For example, off the Bombay and Gujarat coasts, the shelf extends out for nearly 200 miles and, but for a few vessels of the deep-sea section of the Food and Agriculture Ministry, no large-scale commercial trawlers are commissioned to fish these vast areas. Regular mapping of coastal areas rich in fishery resources was one of India's immediate requirements.

Next, we may consider the mineral resources of the sea. India lacks natural deposits of fertilizer salts and rich deposits of fossil fuel. One of the ingenious suggestions was that the possibility of extracting nutrient salts, such as phosphates and nitrates, from the vast quantities of marine sediments dumped by the great river systems into the Bay of Bengal and the Arabian Sea should be considered. The possible accumulation of oil under the shelf (offshore areas), particularly off Cambay, was also considered suitable for investigation, particularly since the nearby area of Ankaleswar had yielded exploitable deposits of gas and oil.

Again, the mapping of coastal currents and the bathymetry of nearshore areas, and their importance for coastal navigation, defence, harbour construction, and so on, cannot be minimized. With the increase in industries and the number of nuclear reactors, the question of pollution and waste disposal takes on added significance. Our knowledge of coastal bathymetry and bottom topography was inadequate and the physical oceanography of the coastal waters was almost unknown. The use of modern instruments was essential.

Finally, there was the study of the monsoons. Two problems were involved: first, how does the reversal of the monsoons affect the oceanic circulation in the northern Indian Ocean, and secondly, how may the onset and intensity of the southwest monsoon be predicted. This second problem was of great agricultural importance since most of the *ryots* (farmers) in India depend on monsoon rains for the cultivation of summer crops (*kharif*).

These were some of the problems in which India was interested at the time of launching of the IIOE (Rao, 1967). Quite rightly, therefore, the planners of the Indian Programme concentrated their efforts in these directions and constrained their cruises and observations so as to obtain substantial information on the coastal areas in the Arabian Sea and the Bay of Bengal. With the inauguration of the 1st Scientific Cruise of INS *Kistna* on 9 October 1962, by Professor Hymayun Kabir, Minister for Scientific Research and Cultural Affairs, the Indian Programme of Work during the IIOE was officially launched. Besides INS *Kistna*, the Indian Programme included scientific cruises by RV *Varuna*, of the Indo-Norwegian project, RV *Conch*, of the University of Kerala, and FV *Bangada*, an exploratory fishing vessel of the Ministry of Food and Agriculture, Government of India. All the cruise tracks and programme of work were co-ordinated so that a complete coverage of important coastal areas in the Bay of Bengal and the Arabian Sea was effected.

The participation and programme of the INS *Kistna* in the IIOE are unique in many respects. The Indian Navy should be congratulated for placing the frigate at the disposal of the Indian National Committee on Ocean Research solely for oceanographical work. This was a most welcome development for the future of ocean sciences in India since, by this gesture of co-operation, the Indian Navy's full support of IIOE was assured.

Commencing in October 1962, INS *Kistna* completed 28 scientific cruises and, had it not been for the unfortunate Indo-Pakistan conflict in the middle of 1965, she would have successfully accomplished the rest of the cruises planned for the autumn of 1965. The vast amount of data collected by INS *Kistna* were later analysed at the data and planning division of the National Institute of Oceanography.

Meanwhile, the International Meteorological Centre (IMC) at Colaba, Bombay, functioned from 1 January 1962, with Prof. C. S. Ramage of the University of Hawaii as Director. This Centre was financed by the Council of Scientific and Industrial Research and was manned by the Indian Meteorological Department. The United Nations Special Fund provided an IBM 1620 electronic computer for data processing. The US National Science Foundation gave liberal assistance in the form of equipment and other services. The extended Indian Ocean chart in use at IMC covers the whole of the Indian Ocean plus adjacent areas. Reception of about 75 daily radio teletype broadcasts from Canberra, Nairobi, Singapore and Pretoria provided the bulk of the southern hemisphere coverage. Data exchanged with Tokyo and Moscow and some 40 radio teletype/carrier-wave broadcasts received from Karachi, Aden, Colombo, Jakarta and Saigon, supplemented by collections at the meteorological communication centre, formed the coverage of the northern hemisphere. Ships' reports obtained over radio teletype circuits from Mauritius by the Indian Navy added to the coverage of the southern Indian Ocean. On a typical day the total coverage amounted to: Surface reports – 1155; Ships – 384; Upper air – 429. Aircraft reported from long-distance international flights on three or four air routes.

At the IMC, synoptic charts were prepared for two principal times – 00 and 12 hours Greenwich Mean Time – for surface and standard isobaric levels; namely, 50, 100, 200, 300, 500 and 700mb. Back plotting was also done after reception of additional information/data from other centres. During the IIOE, perhaps the most important observations on the monsoons were carried out by specially instrumented research aircraft of the US Weather Bureau Research Flight Facility and by the Woods Hole Oceanographic Institution. In addition, an automatic weather station was anchored in the Bay of Bengal half-way between Madras and the Andamans in April 1964 but this was soon lost. An Automatic Picture Transmission (APT) receiving equipment on loan from the US National Science Foundation was installed at IMC in December 1963, and this picked up pictures of cloud cover from *Tiros* VII and *Nimbus* meteorological satellites during their orbits over the Indian subcontinent. It is too early to say that the IMC has solved the problem of the development of the monsoon or is able to predict the arrival of the monsoon accurately, but it has accumulated a vast quantity of information and the preliminary analysis of the data has improved our knowledge of the circulation pattern of the monsoon winds. In fact, for the first time, meteorologists in India were able to get data from over the oceans for their studies and have come to realize that the weather pattern of the Indian sub-continent is greatly influenced by conditions in the sea.

The Indian Ocean Biological Centre (1962-66)

The establishment of the Indian Ocean Biological Centre (IOBC) at Ernakulam (Cochin) marks a very important milestone in the history of marine biology in India. The Centre was established by the Council of Scientific and Industrial Research in co-operation with UNESCO. The main considerations which led to the selection of India for the location of the Centre were:

- geographical location of India at whose ports many of the ships participating in the expedition were likely to call;
- the very considerable interest in biological and taxonomic studies in India at scientific and university institutions;
- the availability of a large number of trained biologists who could take on the work;
- the advantage of a centre of this type in South Asia which would stimulate marine biological studies in the Asian region.

The principal functions of the Centre were:

- maintenance of a named reference collection of Indian Ocean material and duplication of it for laboratories throughout the world;
- sorting zooplankton samples taken by standard methods
- examination of the sorted standard material or sending it to specialists throughout the world;
- sorting of zooplankton samples at the request and expense of participating laboratories;
- training.

The development of this Centre in India has provided a unique opportunity for the training of biologists from India and other countries in the region.

The US IIOE programme in biology

Woods Hole is a picturesque village located on Cape Cod in the State of Massachusetts in the United States. Geomorphologically, Cape Cod is like a hooked finger projecting from the northeast American coast and appears to beckon people from other countries, which is a fact if one looks at the number of foreign visitors during the summer months at the famous Woods Hole Oceanographic Institution (WHOI) and the Marine Biological Laboratory (MBL). As already indicated, it was at WHOI that a decision to plan the IIOE was made in August 1957 by a group of eminent oceanographers who met under the chairmanship of Roger Revelle.

The US National Committee acquired the Presidential Yacht *Williamsburg* for conversion into an oceanographic vessel (the ship was 243 feet (73m) long and displaced 1700 tons) and was renamed the *Anton Bruun*, after the famous Danish scientist who led the second *Galathea* Deep-Sea Expedition round the world in 1950-52. Also, as the first Chairman of the Intergovernmental Oceanographic Commission of UNESCO, at that time, he influenced the UN bodies concerned and the Member States to support the proposed International Indian Ocean Expedition. The National Science Foundation named John Ryther, a renowned scientist at WHOI, as Director, and Edward Chin, as Associate Director, of the US Programme in Biology in the IIOE and gave them a free hand to organize the *Anton Bruun* cruises in the Indian Ocean.

Ryther had a good understanding of the problems in biology the Expedition was seeking to solve. Behrman (1981) quotes Ryther's writing in 1963 that 'for the systematics, the Indian Ocean represents a world of which only tantalizing glimpses have been obtained. A few fortunate individuals have taken part in expeditions to some of the more remote, exotic island groups (the Seychelles, Maldives, Laccadives, Comores and Chagos) and have brought back a wealth of new material. Just enough is known of the flora and fauna of these areas to whet the appetite of the taxonomist with the desire to make a thorough and exhaustive study of the entire region ...

For the ecologist, there are reports of many fascinating phenomena of unknown nature and origin. Vast fish mortalities in the central Arabian Sea are perhaps produced by the overturn of water from mid-depths reportedly devoid of oxygen and laden with hydrogen sulphide. The central Bay of Bengal may at times have similar properties. Are these anoxic layers related to the biological productivity of the overlying surface waters? Do they reflect stagnation implying lack of vertical or horizontal circulation for long periods of time? Notorious outbreaks of discolored water, sometimes also producing mass mortalities of marine life, are frequently reported along the coasts of India and Africa. Are these "blooms" of dinoflagellates similar to the causative agent of the Florida red tide? Are they the result of fertilization of the coastal waters from upwelling processes...? Huge meadows of blue-green algae extending for many hundreds of square miles are known to occur in the Arabian Sea. What makes these plants grow in this particular region? Where do they get their nutrients? How does their presence affect other forms of marine life? These are just a few of the problems, probably unique in the Indian Ocean, which will require a combination of physical, chemical and biological information to answer.'

The *Anton Bruun* had a crew of 30 and accommodation for 28 scientists. Of these, 8 staff scientists were responsible for the basic programmes such as casting of water bottles, BT lowerings, plankton collections, primary-production determinations, meteorological observations, depth soundings, etc. In early 1962, Ryther held a planning conference in New York and, because of his previous work in the Bay of Bengal and good contacts with Indian scientists, asked Eugene LaFond to head the first cruise of the *Anton Bruun* and to develop a programme for the Bay of Bengal and Andaman Sea areas. Since the Andhra University oceanographers had established the time and space variables on the western side of the Bay of Bengal, LaFond wanted to establish the same variables on the opposite side of the Bay. For example, when upwelling occurs off Waltair, would upwelling exist on the other side or would we find sinking? Would the immense dilution due to river discharges extend as far down the east side as it does in the west? There were other questions dealing with the monsoon circulation gyres, such as: are the currents in the Bay one large eddy or are they broken up into smaller gyres? To answer these questions, a pattern of stations was laid out along a track crisscrossing the Andaman Sea, and the eastern, northern and western parts of the Bay of Bengal.

Prior to the beginning of the cruise, Panikkar asked the LaFonds to visit Indian marine science centres, give lectures, explain the IIOE programme and, especially, the objective of the US Programme in Biology and the *Anton Bruun* cruises (Table 1.7). A number of Indian scientists were invited to participate in the cruises. The LaFonds visited and lectured at:

- Bombay Academy of Sciences, in Bombay
- Taraporevala Marine Biological Research Station, in Bombay
- Osmania University, in Hyderabad
- Andhra University, in Waltair
- Navy Physical Laboratory, in Cochin
- Oceanographic Research Wing of National Geophysical Research Institute, in Cochin
- Maharaja's College, in Ernakulam
- Kerala University, in Trivandrum
- Central Marine Fisheries Research Institute, in Mandapam
- Annamalai University, in Chidambaram
- Maharaja Sayajirao University, in Baroda
- Gujarat University, in Ahmedabad
- Physical Research Laboratory, in Ahmedabad
- Delhi University, in New Delhi

These lectures and discussions informed the Indian scientific community of the oceanographic programmes about to take place in the Indian Ocean, and the members of this community were invited to participate in the scientific cruises.

The *Anton Bruun* arrived in Bombay in early March 1963 after occupying a series of stations in the Arabian Sea. She soon refueled and departed on 12 March on the first of 9 cruises (see Table 1.7) with all scientific billets filled.

Table 1.7 *Anton Bruun* cruises in the Indian Ocean during the IIOE

Cruise No.	Dates	Area covered
A	2/24/63-3/04/63	Gulf of Aden to Bombay
1	3/12/63-5/10/63	Bay of Bengal
2	5/22/63-7/23/63	Along 70°E and 80°E
3	8/08/63-9/20/63	Along 60°E
4	9/25/63-12/10/63	Western Arabian Sea
5	1/26/64-5/04/64	Along 55°E
6	5/16/64-7/16/64	Along 65°E
7	7/29/64-9/10/64	East African coast
8	9/25/64-11/09/64	East African coast
9	11/18/64-12/28/64	Somalia and Red Sea

The initial programme consisted of taking underway BT observations from Bombay across the southern Bay of Bengal to the Andaman Sea. In the Andaman Sea, the ship stopped for full-fledged oceanographic stations. The normal procedure in occupying a station was to make the following measurements and collections:

- hydrographic cast using Nansen bottles;
- 200m haul with the IIOE plankton net;
- phytoplankton trawl;
- deep-water fish trawl;
- vertical series of water samples for primary production;
- meteorological observations;
- bathythermograph lowering to 900 feet (275m);
- coring and dredging of the sea floor;

The Indian scientists for the first leg were:

C. Poornachandra Rao (meteorologist), International Meteorological Centre, Bombay
S.P. Anand (chemist), Indian Research and Development Centre, New Delhi
K. Balasubrahmanyan (marine biologist), Annamalai University, Marine Biological Station, Porto Novo
V. Chalapati Rao (marine biologist), Andhra University, Waltair
R. Varadarajulu (meteorologist), Andhra University, Waltair

Others who joined the first *Anton Bruun* cruise on one or more later legs were:

P. W. Backar (photographer), Ministry of Information and Broadcasting, Bombay
P. Chandramohan (marine biologist), Andhra University, Waltair
C. M. Gupta (marine geologist), University of Baroda, Baroda
R. M. Kidwai (marine geologist), Andhra University, Waltair
G. R. Lakshmana Rao (physical oceanographer), Andhra University
G. H. Madhusudana Rao (marine geologist), Andhra University
K. V. Nair (marine planktonologist), Indian Ocean Biological Centre, Ernakulam
N. K. Panikkar (marine fisheries biologist), Council of Scientific and Industrial Research, New Delhi
B. Ramareddi (meteorologist), Andhra University, Waltair
M. Sakthivel (marine planktonologist), Indian Ocean Biological Centre, Ernakulam
V. N. Sankaranarayan (marine chemist), Indian Ocean Biological Centre, Ernakulam
T. S. S. Rao (marine biologist), Scientific Liaison Office, Bombay
R. V. Unnithan (marine biologist), Indian Ocean Biological Centre, Ernakulam
V. V. R. Varadachari (physical oceanographer and marine meteorologist), Andhra University, Waltair
S. Varma (photographer), Ministry of Information and Broadcasting, Madras
A. B. Wagh (marine planktonologist), Institute of Science, Bombay
P. K. Das (physical oceanographer), National Geophysical Research Institute, Cochin
K. Krishnamurty (marine zoologist), Annamalai University, Marine Biological Station, Porto Novo

In addition to the Indian oceanographers and the eight staff scientists, there were two Thai scientists:

Thumnoon Supanich (marine biologist), Chulalongkorn University, Bangkok
Mahn Bhovichitra (fisheries biologist), Chulalongkorn University, Bangkok

Needless to say, enormous quantities of oceanographic data were obtained from the network of stations throughout the Andaman Sea and Bay of Bengal.

In this connection, it may be stated that very little information was available on the responses of other countries in the Indian Ocean area, once the IIOE was over. One could infer from later publications, that Pakistan and some Gulf States had established marine research institutions. For its part, India went ahead to advance marine research in a big way.

The benefits of the IIOE to India

The IIOE achieved for itself the distinction of being one of the best examples of co-operation between many nations from East and West. Indian scientists visited many neighbouring countries and the Indian ship INS *Kistna* visited Singapore. India played host to scientists from many nations, resulting in deep and abiding friendships. Many of the participating foreign research ships, as the RV *Anton Bruun*, RV *Argo* and RV *Horizon* (USA), the RV *Vityaz* (USSR) and the RRS *Discovery* (UK) provided facilities for ship-board training and research for many Indian scientists.

The expedition also provided opportunities for organizing seminars in which many young scientists from different parts of the country and senior scientists from abroad participated. An All-India Seminar on Marine Science was held in Waltair, 26-27 April 1963, sponsored by the Andhra University, Waltair, the Indian National Committee on Ocean Research, the US Programme in Biology and the US Information Service, to present results from the first *Anton Bruun* cruise. It was well attended and was a great success. At the time of the visit of RV *Horizon* and RV *Argo* to Cochin and then again at Calcutta during the visit of the US Coast and Geodetic Survey Ship *Pioneer*, seminars were arranged at which visiting scientists and their Indian counterparts participated in discussions. In July 1965, an International Symposium on the Meteorological Results of IIOE was held in Bombay and this was attended by a large number of foreign and Indian participants.

As a finale to all this activity, a training programme was organized at the postgraduate level to train junior scientists in the practice of oceanography as a multi-disciplinary science, during January-March 1966 at Bombay. This was jointly sponsored by UNESCO and CSIR (Council of Scientific and Industrial Research). Twenty-five trainees were recruited from among the applications submitted by India and adjacent Asian countries to UNESCO. The break-down figures for the trainees were: India 20, Thailand 2, Singapore 1, Ceylon 1, and Malaysia 1.

While the training programme marked the end of IIOE activities, it also saw the birth of the National Institute of Oceanography (NIO) of India. The Indian Government approved the establishment of this institute as one of the national laboratories under the CSIR and appointed Dr. N.K. Panikkar, Director of the Indian Programme under the IIOE, as Director of the new institute. All the activities started under the Indian Programme of the IIOE, including the Indian Ocean Biological Centre, have now been merged into the National Institute of Oceanography.

Table 1.8 Principal atlases resulting from the International Indian Ocean Expedition

- *International Indian Ocean Expedition Plankton Atlas*. Panikkar, N.K., editor (1968-1973).
  1(1): Maps on total zooplankton biomass in the Arabian Sea and the Bay of Bengal (1968).
  1(2): Maps on total zooplankton biomass in the Indian Ocean (1968).
  2(1): Distribution of copepod and decapod larvae in the Indian Ocean (1970).
  2(2): Distribution of fish eggs and larvae in the Indian Ocean (1970).
  3(1): Distribution of Crustacea (Cladocera, Ostracoda, Cirripedia, Mysidacea, Cumacea, Isopoda, Amphipoda, Euphausiacea, Stomatopoda) and Insecta (Halobatida) in the Indian Ocean (1972).
  3(2): Distribution of planktonic mollusca of the Indian Ocean (1972).
  4(1&2): Distribution of Platyhelminthes, Tomopteridae and other pelagic Polychaeta, Trochophores and Sipunculida of the Indian Ocean (1973). Distribution of Actinotrocha, brachiopod larvae, Chaetognatha, Copelata, Pyrosoma, salps and doliolids and Amphioxus of the Indian Ocean (1973).
  5(1&2): Indian Ocean Biological Centre, National Institute of Oceanography, Council of Scientific and Industrial Research, New Delhi.

- *Oceanographic Atlas of the International Indian Ocean Expedition*. Wyrtki, K. (1971). National Science Foundation, Washington DC, USA.

- *Meteorological Atlas of the International Indian Ocean Expedition*. National Science Foundation, Washington DC, USA.
  Ramage, C.S., F.R. Miller and C. Jefferies (1972). *The surface climate of 1963 and 1964.*
  Ramage, C.S. and C.V.R. Raman (1972). *Upper Air.*

- *Geological-geophysical Atlas of the Indian Ocean*. Udintsev, G. B., editor (1975). Academy of Sciences, Moscow.

- *Phytoplankton Production Atlas of the International Indian Ocean Expedition*. Krey, J. and B. Babenard (1976). Institut für Meereskunde, Kiel.

The main results of the IIOE

The IIOE ended officially in 1965. More than 40 oceanographic research vessels belonging to 13 countries surveyed the Indian Ocean and collected useful data in almost all disciplines in the marine sciences, except perhaps in fishery research and marine microbiology. The scientists' work begins when the expedition ends and sometimes it takes several years for the results to be published. Since the completion of the expedition proper, hundreds of papers have been published and some of them reprinted and included in the 8 volumes of collected reprints of the International Indian Ocean Expedition published by UNESCO. This apart, a set of atlases was published which is unique for the Indian Ocean (Table 1.8).

The introduction of computers for data logging and analysis made a world of difference in the handling of large amounts of data collected during the various cruises. Many countries established national oceanographic data centres for storage and dissemination of information. Two World Data Centres were also created, one in Washington, DC, and another in Moscow. Moreover, sorting centres for handling fauna and flora, including plankton, came into existence in Cochin (Indian Ocean Biological Centre), in Tunis (The Mediterranean Sorting Centre), and the National Oceanographic Data Centre at the Smithsonian Institution in Washington, DC. Some time later, a sorting centre was also started in Mexico. For biology specialists, the sorting centres were of immense help, since the drudgery of mechanical sorting of animals and plants was done elsewhere, so that the biologists could concentrate on their specialized work.

Perhaps the most important thing was that oceanography became an eligible science for support and funding by the governments, particularly those of developing countries, and the interest shown in the IIOE by developed countries by their extended participation in an area so remote from their home, triggered a kind of paradigm for the developing countries. The benefits to coastal countries in the Indian Ocean region include the training of their scientists aboard research vessels of the developed countries, as the *Discovery* (UK), the *Meteor* (Germany), the *Atlantis* and the *Anton Bruun* (USA). This apart, oceanographic and marine biological research institutes were either newly started or the existing ones were strengthened. The oceanographic institutes established at Karachi, Pakistan, and Goa, India, the marine stations at Pukhet in Thailand and Nosy-bé in Madagascar, and the East African Marine Fisheries Research Institute in Zanzibar are some of the examples. What was even more important was the mingling of scientists with different backgrounds, both in scientific status and culture, for the common cause of Indian Ocean exploration; they got to know each other and continued their contacts for many more years after the Expedition.

From the technical point of view, the six years of the IIOE (1959-65) marked a watershed in the state of the art in oceanographic instrumentation. Some of the new research vessels built and commissioned for survey during the expedition had better winches and echo-sounders and far more accurate navigational instruments; a satellite navigation system was available to the *Atlantis* II of the Woods Hole Oceanographic Institution. Narrow-beam precision echo-sounders, magnetometers and gravimeters were also made available on most of the oceanographic vessels.

Going through the literature and the excellent atlases to assess the results is a long but exciting task. Many interesting findings are briefly mentioned in the following account.

Geology and geophysics: The major discoveries relate to the complexity of the mid-Indian Ocean ridges and the famous triple junction, like an inverted Y, found south of the Seychelles, where the southern end of the Carlsberg Ridge (sometimes called the Mid-Indian Ocean Ridge) meets the Southwest Indian Ocean Ridge and the Southeast Indian Ocean Ridge which extend eventually into the mid-Atlantic Ridge and the South Australian Ridge, respectively. Until comparatively recently, geologists had no clear proof of the theory of continental drift first proposed by Wegener in 1910. However, in 1959, coinciding with the IIOE, Ewing, with Heezen and Tharp, published a paper describing a mid-ocean ridge running continuously in all the three major oceans for a total distance of 60,000km, 30-400 km wide and rising to 3000-5000m above the ocean floor. The significance of these long ridges was realized when several ships all made successful and successive cruises to study the geological and geophysical aspects of the mid-Indian Ocean ridges. They were: the USSR's *Vityaz* in 1960-62 and 1964-65, the UK's HMS *Owen* and *Dalrymple* in 1961 and 1963, and the RRS *Discovery* in 1963, the USA's RV *Argo* and RV *Horizon* in 1960-64, as well as the RV *Vema*, RV *Chain*, the

Conrad and the *Pioneer*, and Germany's *Meteor*. The results indicated that these ridges are the site of basaltic upwelling and sea-floor spreading to form the new ocean floors. This phenomenon is now called plate tectonics. Also discovered was the 90°-East Ridge in the eastern part of the Indian Ocean. This straight ridge rises out of the Bay of Bengal sediments, about 1000km north of the equator and extends in a straight line for 4000km to the south; its crests lie 1800-3000m below the ocean's surface. This ridge is unconnected with other ridges and is aseismic; its origin is an enigma.

The hot holes in the Red Sea were another surprising discovery of the IIOE. Although the *Dana* and the *Atlantis* had reported finding some unusually warm waters in the bottom of the Red Sea during their cruises in 1947 and 1958, little did they expect that what they had passed over was really a rich mineral source at the bottom of the sea. To quote Swallow in one of the issues of *Oceanus*: 'We were expecting something unusual on Discovery Station 5580 on September 11, 1964. It was near 21°N in the middle of the Red Sea, very close to the place where both the *Atlantis* in 1958 and the *Atlantis* II in 1963 had found abnormally hot salty water near the bottom. We had anchored a radar buoy in water about 2200 metres deep, and were putting down a closely-spaced cast of water bottles. Approaching the bottom, the one-second pinger on the wire below the bottles had gone out of step and then re-synchronized itself with the echo-sounder – a sure sign (with that particular pinger) that it had gone through a sudden change of temperature. But even then we found it hard to believe the thermometers when the bottles came up. All quite normal, around 22° C. to within 200 metres of the bottom, then 26° C. then both thermometers went off scale (over 35°C), then again both protected thermometers went off scale but the unprotected [thermometer] showing 58°C. And so on. We did a second dip using only 60° unprotected thermometers on the deeper bottles – the only means we had of measuring the high temperature of the bottom water, which we found after correction to be about 44.3°C. This was far in excess of the 25.8°C found previously and which in itself had seemed abnormally high'.

Then the salinity turned out to be equally surprising. When water was being drawn from bottles that had been near the bottom, it seemed to run out more slowly than usual, and salinity at these holes was estimated to be extremely high, measuring about 270 parts per thousand, while the normal salinity of the Red Sea water varies between 38-40 parts per thousand. The origin of these warm-water and high-salinity holes was traced to the great rift valley which originates in East Africa and extends through the Red Sea and Gulf of Aden, since all this is an area of intense sea-floor spreading and the birth of an embryonic ocean. The overlain sediments in these hot holes were estimated to contain a high percentage of iron, manganese, zinc and copper. A Sudanese-Saudi joint commission, in collaboration with Germany, was looking for the possibility of exploiting these sediments for these metals.

Physical oceanography and meteorology: The scientists were astounded to see the see-saw game played by the monsoon winds with the surface currents. During the strong southwest monsoon season, the Somali current was raging towards the north and then northeast near 8°-10°N. The current was strong, up to 7 knots (≈3.6m/s), and its inner edge was close to the Somali shore, where the temperature of the surface waters was sometimes 16°C or less when the rest of the Arabian Sea was at nearly 30°C. It was also noticed that a strong set of the Somali current preceded the onset of the southwest monsoon along the west coast of India by almost a month, thereby indicating a possible relationship between the two events. Meteorologists and physical oceanographers now descended on the East African coast with many ships to study and unravel the Somali events vis-à-vis the Indian monsoons.

Biology and chemistry: The biologists were quite anxious to learn what sustainable yield of fish the Indian Ocean could provide and what were the productivity estimates and their translation into biological resources. Here again, the ship surveys covered upwelling regions and reported highest productivity rates in the Arabian Sea and off the northwest coast of Australia. However, no new fishing grounds were discovered, but the biological surveys indicated areas for further detailed study. As a result, the Food and Agricultural Organization of the United Nations (FAO) constituted an Indian Ocean Fisheries Commission to help assess the fishery potential of the Indian Ocean.

At first sight, the massive Somali upwelling area appeared to promise rich fishing grounds, as is the case with other boundary currents elsewhere, as the Benguela, Peru and California currents, but it would appear that the high productivity recorded off the

Somali and Saudi Arabian coasts failed to reach massive production at the tertiary level: in practice, fish. Why? Perhaps we need to plan and execute another IIOE soon, to answer this question.

The chemical oceanographers did not have much to say about the nutrient distribution in the Indian Ocean, as compared to other oceans. A chemical front was recognized around 10°-20°S through which an increase in the nutrient content of the waters took place as one proceeded towards the northern Indian Ocean. Another important feature noticed was the existence of an oxygen-deficient layer (between 200 and 800m depth), in the Bay of Bengal and Arabian Sea, for which a convincing explanation is still to be found. In some places in the Arabian Sea, even the presence of hydrogen sulphide was recorded. It would appear that the development and persistence of an oxygen-minimum layer in large parts of the Bay of Bengal and the Arabian Sea is a curse on the northern Indian Ocean's productivity.

Conclusion

The IIOE – International Indian Ocean Expedition (1959-1965) – marks a watershed in the pursuit of knowledge of the Indian Ocean. The main results, briefly described above, further attracted the attention of scientists the world over to begin more detailed studies of selected areas, such as the Somali current, the mid-Indian Ocean ridges, the effects of monsoonal winds on surface currents, the productivity of the upwelling areas, geochemistry and geophysics etc. Some of these phenomena and processes are described in the following chapters.

The following three chapters deal with the environmental setting of the Indian Ocean: the origin and structural geology; the hydrography; and the nutrients.

CHAPTER 2

GEOLOGICAL ORIGIN AND EVOLUTION OF THE INDIAN OCEAN

Geological setting and limits of the Indian Ocean

Comprising only 20 percent of the world's ocean surface, the Indian Ocean has some unique features as a result of its geological and geographical setting. It is roughly triangular in shape and its greatest breadth, of nearly 10,000km (about 6200 miles), separates the cape of southern Africa from Australia around the 35°S latitude. It is the third largest ocean in the world and, including the Red Sea and the Persian Gulf, it is estimated to have a total area of about 73,000,000km^2 (28,000,000mi^2).

The average depth of the Indian Ocean is 3890m (12,760ft) and contains an average of 292,131,000km^3 (70,086,000mi^3) of water. All types of shores and beaches are found, such as wave-eroded, sedimentary and coral.

The Indian Ocean is bounded to the north by Iran, Pakistan, India and Bangladesh, to the west by Arabia and Africa and to the east by Australia, the Sunda islands of Indonesia and the Malayan peninsula, Myanmar (Burma) and Thailand, and to the south by the Sub-tropical Convergence (see next paragraph). In the southwest, the Indian Ocean opens into the Atlantic Ocean near the southern tip of South Africa. Eastwards, it opens into the Pacific Ocean through straits and marginal seas, such as the Makassar and Sunda Straits, the Timor and Java Seas, and the Great Australian Bight.

The oceanographic limits of the Indian Ocean are somewhat hazy, particularly to the south and east. Although there is the unhindered stretch of water from the Persian Gulf to the Antarctic coastline, it was assumed, for the purposes of the International Indian Ocean Expedition, that the southern limit of the Indian Ocean would extend to 40°S which coincides approximately with the Subtropical Convergence, south of which lies the Southern Ocean surrounding the Antarctic continent. The southern Ocean has a distinct hydrographical regime that is quite different from that of the Indian Ocean. Therefore, the southern boundary of the Indian Ocean stretches from Cape Agulhas at the southern tip of Africa to the Cape of Tasmania.

The Indian Ocean's eastern boundary is still not very clear. Bass Strait, between Tasmania and Australia, is sometimes considered part of the Pacific, and at other times as part of the Indian Ocean. Similarly, the boundary line between the Pacific and Indian Oceans is suggested to run across the Torres Strait between New Guinea and Australia; however, there is another view that considers the Arafura and Timor Seas as part of the Pacific Ocean. Through these island channels, the western Pacific is in close contact with the eastern Indian Ocean thus constituting the largest tropical oceanic area in the world.

The Indian Ocean has fewer seas than the Atlantic and the Pacific. To the north, extending up to 30°N, there are the semi-enclosed seas: the Red Sea and the Persian Gulf. They open into the Indian Ocean through the Gulf of Aden and the Gulf of Oman, respectively. The Arabian Sea, the Bay of Bengal and the Andaman Sea are marginal seas. So

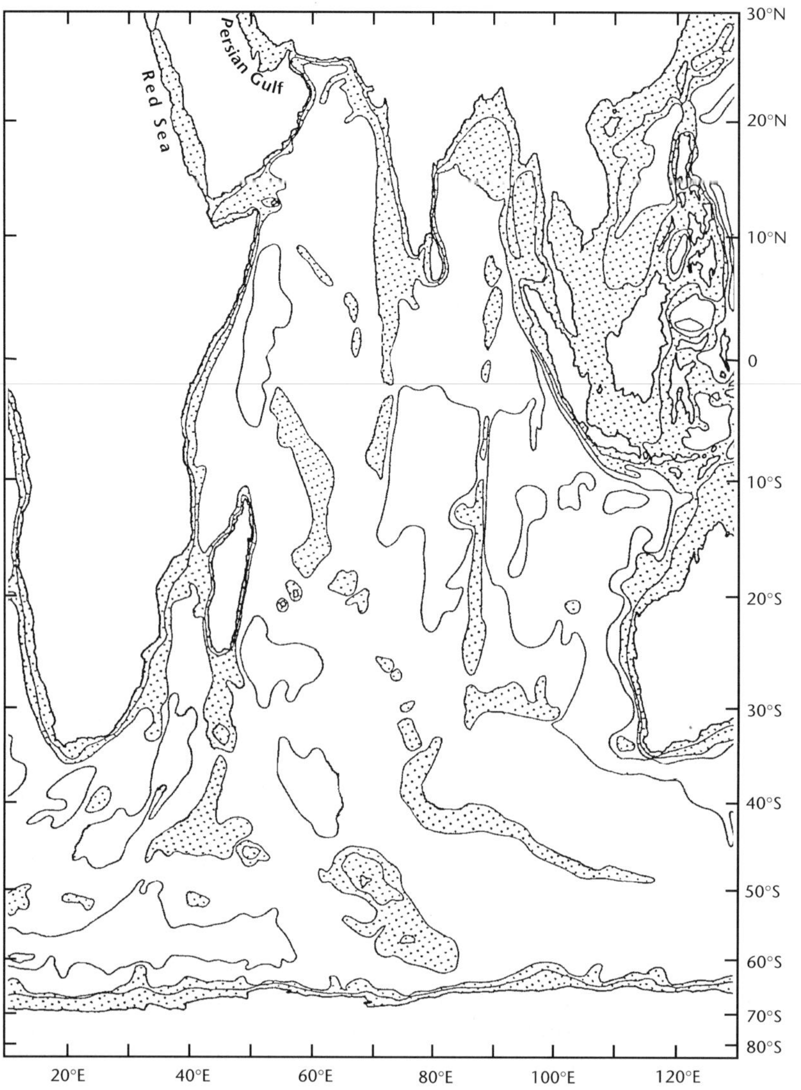

Figure 2.1
Chart showing the bathymetry (1000m, 3000m, 5000m isobaths) of the Indian Ocean; depths less than 3000m are stippled.

also are the Gulf of Carpentaria and the Arafura Seas to the southeast, and the Indian Ocean is widely open to the Great Australian Bight.

Among the islands in the Indian Ocean, Madagascar, Sri Lanka, Seychelles, Mauritius and Socotra are of the continental type. The islands of volcanic origin are Kerguelen, Crozet, Prince Edward, Amsterdam and St. Paul which lie in the Southern Ocean. In the tropical northern Indian Ocean there are many coral atolls and islands such as the Laccadives, Maldives, Amirante, Farquhar, Cocos group and the Chagos Archipelago. Some of the volcanic islands, as the Mascarenes and Comoros, are ringed by coral reefs.

The geological exploration of the Indian Ocean was virtually nil during the 19th and early 20th centuries. The famous *Challenger* Expedition of 1872-76 did not survey the Indian Ocean. The *Valdivia* (1888-89) mainly concentrated its surveys in the East Indies. Since the Danish *Dana* Expedition of 1928-30, investigations in the Indian Ocean gained momentum and the cruises of the *Snellius* (1929-30), the *Mabahiss* (1933-34), the *Albatross* and *Galathea* (1950-52) and the *Ob* (1956-58), and many others, have added considerable information. Then followed the International Indian Ocean Expedition, in which some 40 survey ships from 13 countries took part. Extensive geological and geophysical data were collected by these ships and the results included the publication of a remarkable bathymetric map of the Indian Ocean by Heezen and Tharp (1965) and of the Geological and Geophysical Atlas of the Indian Ocean, by the Academy of Sciences of the USSR in 1975 under the direction of Udintsev (Figure 2.1).

Subsequently, the French survey vessels, the *Marion Dufresne* (1972), *Le Suroit* (1977) and the *M.S. Galliene* (1978) made 18 cruises dealing mainly with geology and geophysics, particularly of the western Indian Ocean between the Gulf of Aden, 60°S and 80°E. A magnificent programme of deep-sea drilling was undertaken by the Americans and their colleagues from other countries on board the DSDS *Glomar Challenger*, from 1968 onwards and in all the oceans (legs 22 to 27 were in the Indian Ocean). The initial results are spectacular and most useful for our understanding of the structure and the age of the sea floor. Extensive papers/reports have been published on the geology and geophysics of the Indian Ocean based on the surveys made by the above-mentioned ships.

Origin and evolution of the Indian Ocean

Primeval history

No one really knows how or when exactly the earth and oceans were formed, though there are many theories about their origin (York, 1993). Among the planets of the Sun, Earth is the only one that has enough water to cover 70 percent of its own surface, with an average depth of about 3.6km (2.25mi). Where did all this water come from and why is it the other planets do not have any water, except for some traces locked up in their rocks? These questions are yet to be fully answered.

According to radiometric dating, Earth is about 4.5 billion years old, an age inferred from meteorites rather than rocks on the Earth itself, where the oldest are about 3.7 billion years old. Since some of these are also sedimentary rocks, large water bodies must have already been present on the Earth at that time. It would therefore appear that during the first 1000 million years of the early Earth's history, land and water were formed, besides micro-organisms whose fossil records can be traced in the sedimentary rocks.

The continents carry the record of the Earth's history much more completely than the sea floor which is not more than 160-190 million years old (Anderson, 1986).

The oceans, therefore, are very old, possibly 4 billion years, and the waters seem to have come mainly from volcanic activity, during which process, the hydrated rocks would have released the water in the form of water vapour, which ultimately condensed to form the oceans. So, even as early as 4 billion years ago, the Earth's surface was apportioned between the oceans and the land, although their proportions are not clear. Since there is no obvious additional source of water, however, the present 70 percent coverage of the oceans of the Earth's surface must be equally old.

It is assumed that, in Palaeozoic-Mesozoic times, all the present-day continents were assembled into two supercontinents: Gondwanaland (Figure 2.2) and Laurasia, situated in the southern and northern hemispheres, respectively. Surrounding them were

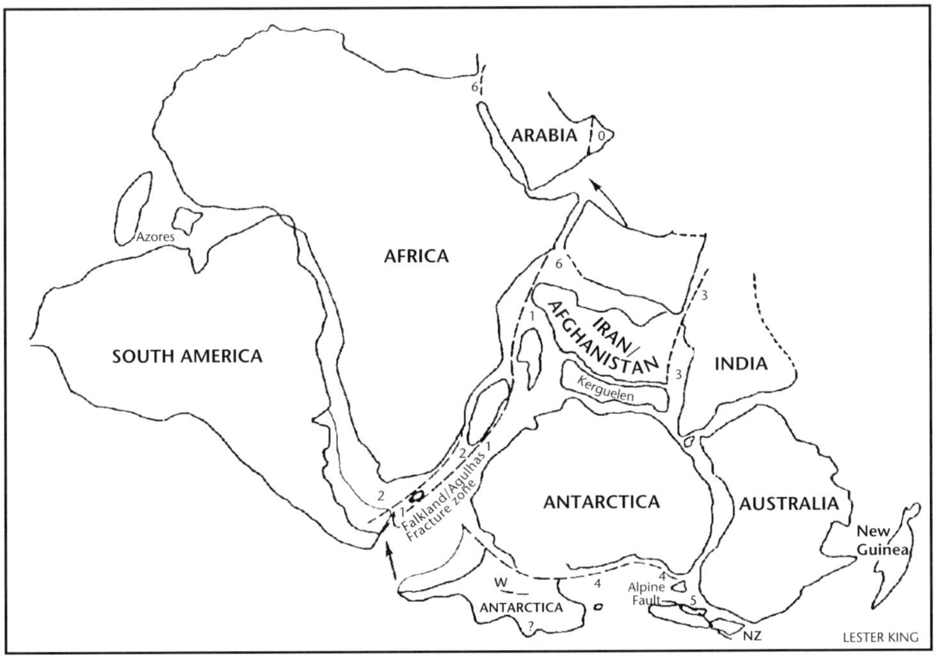

Figure 2.2
Reassembly of Gondwanaland from stratigraphic structural data; Mesozoic period about 150 million years ago. (After King, 1983)

the two oceans, the proto-Pacific and Tethys. As time passed, both supercontinents moved through different climatic zones attracting or developing different fauna and flora, as is evident in their fossil records. During the late Mesozoic, a break-up of the supercontinents took place; as a result, what are today called the southern continents, namely Australia, Antarctica, Africa and South America, drifted apart into their present positions. India, however, rafted all the way across the equator to collide with the Eurasian Plate, a part of Laurasia.

Sea-floor magnetic anomalies and the sea-floor spreading rates provide fairly accurate information about the ages of the various regions of the Indian Ocean. There are many outstanding contributions in this field. Mention may be made of Vine and Hess (1968), Heirtzler *et al.* (1968), Le Pichon and Heirtzler (1968), Bullard (1969), Laughton *et al.* (1969), Fisher *et al.* (1971), McKenzie and Sclater (1971), Norton and Sclater (1979), Nairn and Stehli (1982), Girdler (1984), Haq and Milliman (1984), Hocult (1987), Haq (1988) and Munsch and Schlich (1989). Besides these, there are many books, old and new, dealing with the marine geology of the world oceans: *Submarine Geology* by Shepard (Third Edition, 1973), *Wandering Continents and Spreading Seafloors* by King (1983), *Ocean Basins, Their Structure and Evolution* by the Open University Team (1988), *Marine Geology* by Anderson (1986), *Oceanography* by Gross (6th edition, 1990), and many others. The following account is based on the foregoing literature and cross references contained in them.

It would appear that the geological events that shaped the origin and development of the Indian Ocean commenced during the Mesozoic (about 150 million years ago) and was completed when the Indian Plate encountered the Eurasian Plate during the mid-Eocene period, about 50 million years ago. The oldest identifiable magnetic anomaly in the Indian Ocean, recorded in the Mozambique Channel, is of late Jurassic age (M 22), about 140 million years ago. This is based on the concept of plate tectonics and sea-floor spreading which was first propounded by Dietz (1961) and later by many others.

Plate tectonics

To understand the origin and evolution of ocean basins and their sea floors, a brief description of the theory of plate tectonics would be useful. The starting point for the modern concept of plate tectonics is Wegener's theory of continental drift (Wegener, 1924). The most attractive geomorphological fit of the east coast of South America and the west coast of Africa was the focal point for Wegener's theory. His main argument was that South America fits snugly into the west African coast because they once formed a single continent which somehow broke and drifted apart as two continents, carrying with them identical rocks, fauna and flora. In later years, similar explanations were given for the drifting apart of the southern continents of Australia, Antarctica and India which share, with South America, common stratigraphic systems. Although this theory attracted the attention of geologists, it was hard for them to provide a scientific explanation for the continental drift. However, Wegener believed 'that final solution of the continental drift problem would come from geophysics, since only that branch of science provides sufficiently precise methods' (from the foreword to the last revision of Wegener's book on the origin of continents and oceans). And, in fact, a solution did come from geophysical investigations of the sea floor.

The theory of plate tectonics or sea-floor spreading proposes that the Earth's crust is broken into 7 large plates and many smaller plates, called lithospheric plates, which move and float over the molten surface of the mantle called the asthenosphere. New lithospheric plate is formed at the mid-ocean ridges, which occupy the central axis of all the oceans, forming nearly 23 percent of the sea floor. When the plates meet, any of the following three things can happen: (i) one plate may go under another plate by subduction and get destroyed; (ii) crustal shortening and folding may take place elevating mountains, as the Himalayas; or (iii) the plates may slide past each other, creating transform faults.

As the newly formed plate moves away from the ridge, new lithosphere is formed either continuously or at frequent intervals. At the time of their formation, magnetic minerals in the rocks align themselves in the direction of the Earth's magnetic field and freeze in the same position, as the lithospheric rocks (lava) cool and solidify. It is not known why the Earth's magnetic field reverses at irregular intervals, but these reversals thus recorded make of the ocean floor a gigantic magnetic tape, preserving the record of changes in polarity in the Earth's magnetic field, over millions of years. It is indeed a wonder of nature that sea-floor spreading occurs symmetrically on both sides of the central ridge, at the average rate of 2-3cm per year (Figures 2.3 and 2.4).

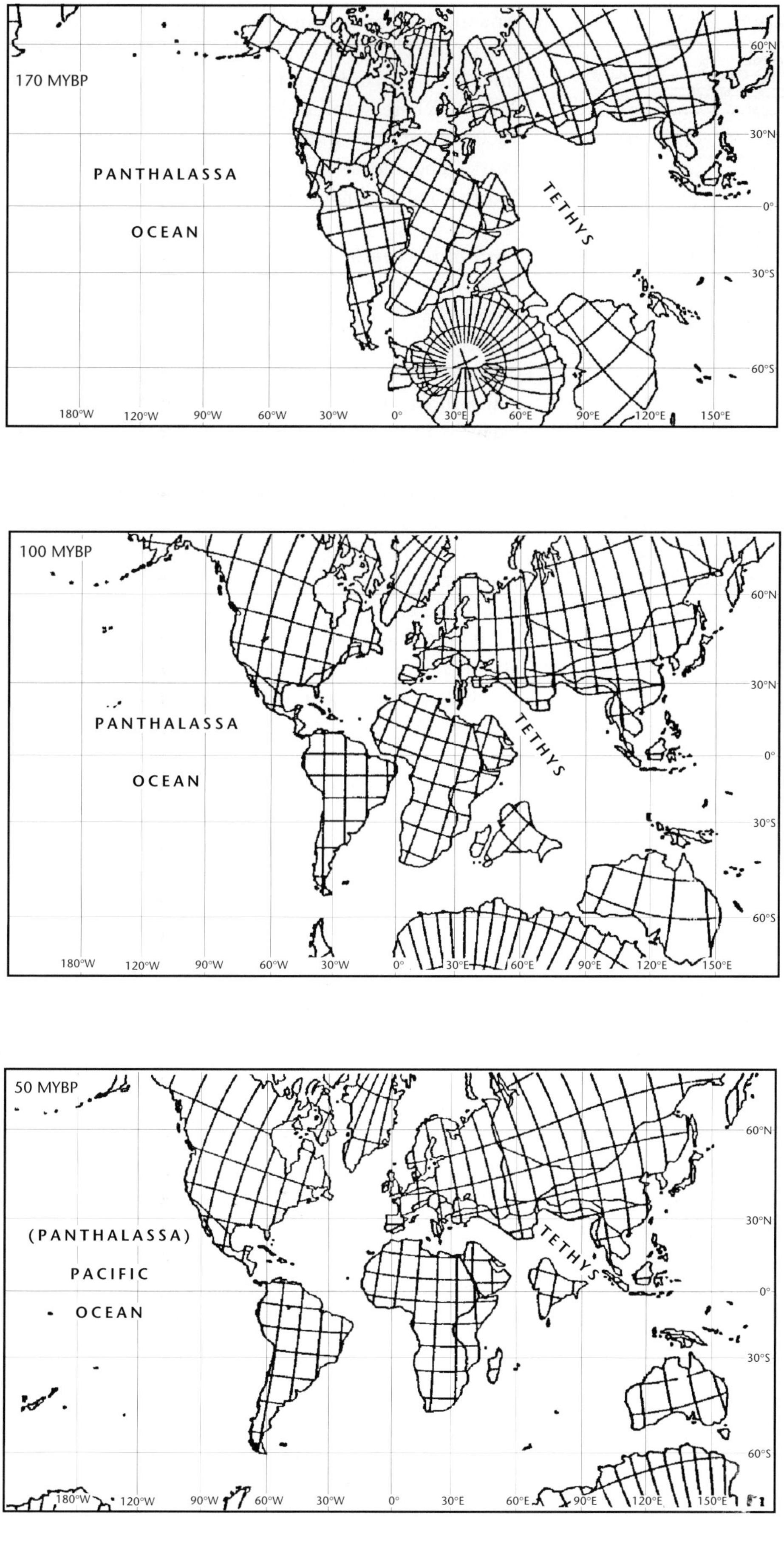

Figure 2.3
Palaeographic reconstruction showing the disposition of the continents 170 (top), 100 (middle) and 50 (bottom) million years ago; the present coastlines have been retained to facilitate recognition of the various continents. (Open University, 1988)

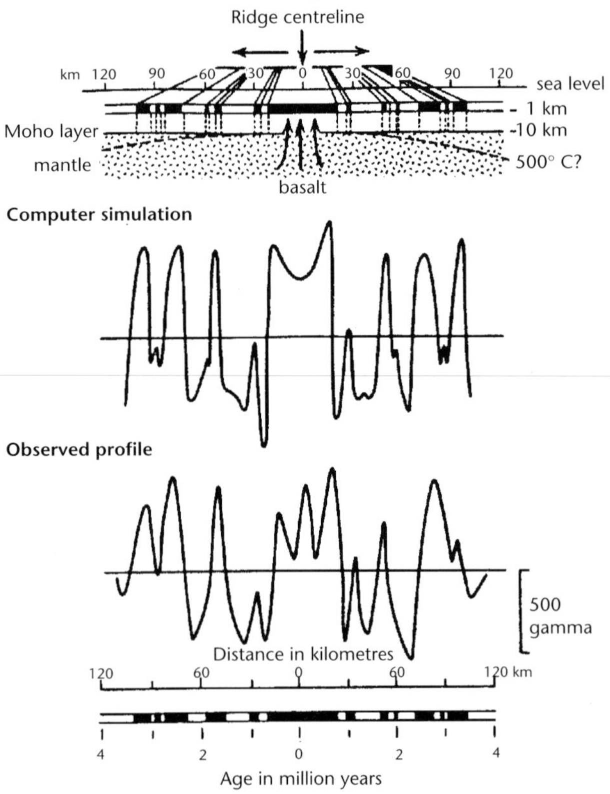

Figure 2.4
Diagrammatic representation of sea-floor spreading about a mid-ocean ridge (the Juan de Fuca Ridge model). (From Anderson, 1986)

Time scales based on radiometric dating have been established and each magnetic anomaly (i.e., reversal of the polarity) has been dated and numbered (Haq, 1988; see Table 2.1) and is used in the oceans for estimating the age of the sea floor. Besides, the new ocean floor occupies a higher level, since it is hot and less dense, but, as it moves away from the ridge, it cools off and become denser and therefore subsides to greater depths. Therefore, the older the crust, the deeper the ocean floor.

Because of the existence of magnetic records in the sea-floor rocks, it is possible to estimate not only the age of the rocks, but also to locate to some extent the future events that may take place, such as volcanic eruptions and earthquakes which normally occur at the crustal plate boundaries. Accordingly, the magnetic records show that most of the Atlantic and Indian Oceans were formed during the past 80 million years or so. However, the rifting of the southern continent (Gondwanaland) may have started much earlier, since the oldest rocks in the north Atlantic are 160 million years old, whereas the oldest in the South Atlantic and the Indian Ocean are about 140 million years old. The crest of the mid-ocean ridges is considered to be young and occurs at a depth of about 2500m, whereas the oldest sea floor, at a depth of 5000/6000m, may indicate an age of 150-160 million years. The oldest oceanic crust is in the western North Pacific which is 190 million years old.

Drift of the continents

As already stated, the breakup of the supercontinent of Gondwana in the Mesozoic period and the subsequent northward movement of the crustal plate carrying India, Australia and New Zealand was a great event in the geological history and evolution of the present-day continents and seas. Of all the constituent continents of Gondwanaland, India moved farthest north.

Haq (1988) has given an excellent, but brief, summary of the geological evolution of the Indian Ocean (see Table 2.1).

It would appear that India rifted from Gondwana about 90 million years ago. The initial rifting and northward movement was slow, but became faster around 80 million years ago. Half-spreading rates between magnetic anomalies 33-28 was about 6cm per year and around anomaly 28 (64 million years ago) the rates increased to between 7 and 8cm year. At the time of anomaly 24 (about 53 million years ago), further acceleration, of 18-19cm per year, took place, and thereafter the rate slowed down dramatically. It would appear that, after anomaly 21 (40 million years ago), India's northward migration came to a standstill. This event perhaps marks the first encounter of the Indian plate with the Eurasian plate. After a short time, the spreading at the mid-Indian Ocean ridge commenced during anomalies 17 and 19 (about 35 million years ago).

It may also be significant inasmuch as subduction along the Eurasian plate facing the Indian plate came to a halt at the same time. The consequence of this collision of India with the Eurasian plate resulted in: (i) crustal shortening and elevation of the Himalayas and adjacent mountain ranges; and (ii) the closing and destruction of the Tethys Ocean, which had hitherto marked the northern boundary of the Indian plate. The suture of the former Tethys can be traced in Indonesia, the Himalayas and Iran. On the eastern side, it starts at the Indonesian trench, passes into Myanmar (Burma) and continues north of the Himalayas, then into Baluchistan and then east along the Makaran subduc-

Table 2.1 Major events in the evolution of the Indian Ocean

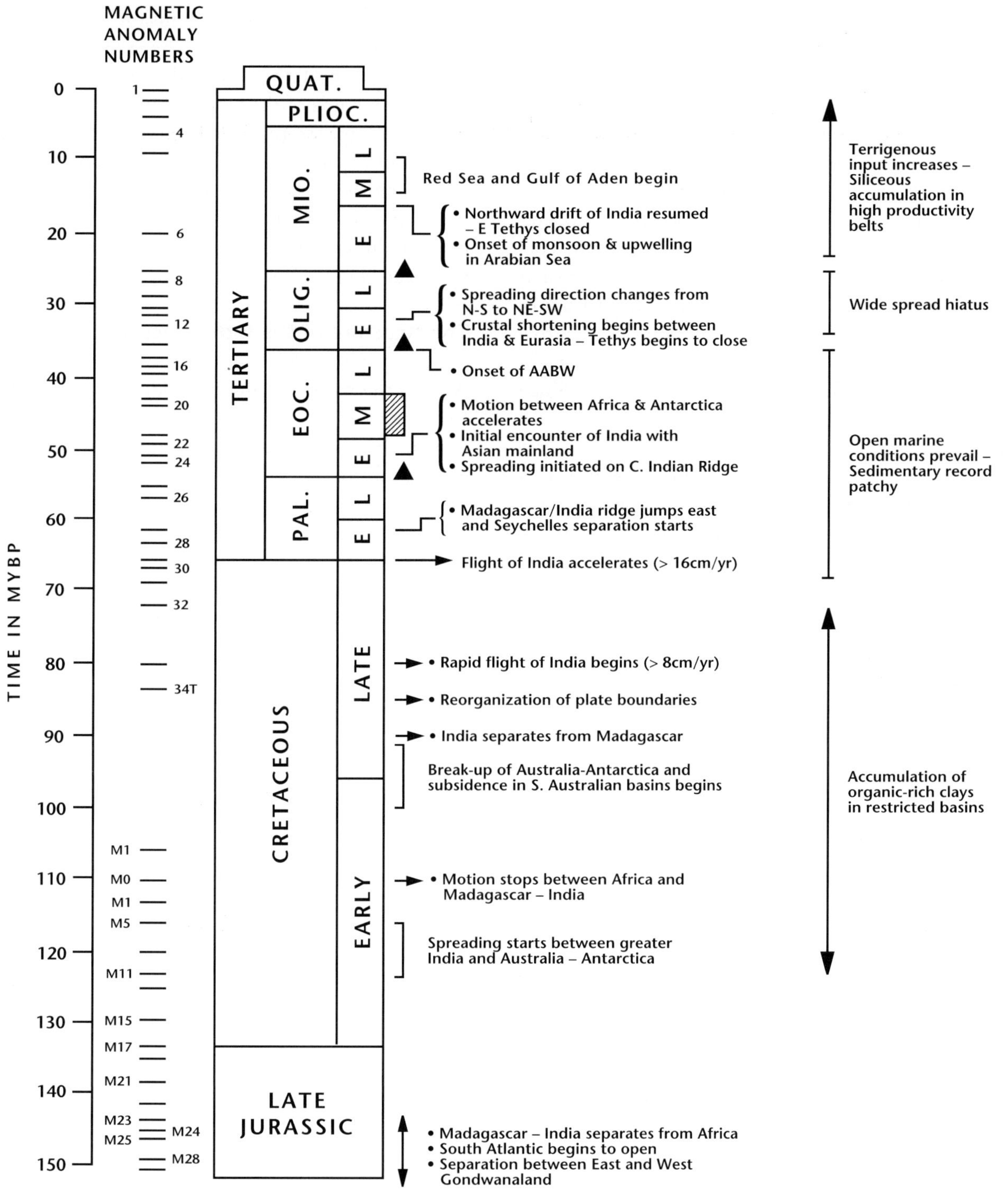

▲ Times of widespread hiatuses

▨ Time of widespread cherts

tion coast and beyond. Major thrusting and vertical elevation of the Himalayas began at the close of the Eocene epoch (36-37 million years ago).

Destruction of the Tethys Ocean

Surrounding the eastern part of the Gondwanaland and bordering the Indian and African coasts to the west and south and the Eurasian coastline to the north was an extensive ocean, Tethys. In some respects, it was a circum-global ocean connecting with the proto-Pacific (Pan-Thalassa) through epi-continental and rift channels between the Eurasian and Gondwana continents. This was the situation during the Mesozoic. During the Jurassic, Tethys extended westward towards Europe through what is now Afghanistan, Iran and the Mediterranean. Since the passage between South and North America was still blocked, a circum-global Tethys current was yet to be established. However, this passage was breached at some time in the mid-Jurassic, connecting the eastern Tethys with the Pacific through the proto-North Atlantic.

With the breakup of Gondwanaland and the movement of the constituent continents in different directions, much of the Tethys sea floor has been lost to subduction, which occurred to the north and northeast, and to the crustal shortening and over-thrusting that followed the collision of the Indian-Arabian plates with Eurasia. The story of the birth and death of the Tethys Ocean is well documented in the fossil records of the Siwalik Himalayas.

To sum up, the Indian Ocean was probably born as a series of rift valleys of the Gondwanaland, very much like the present day Red Sea, 150-170 million years ago, in the southern hemisphere and reached its present dimensions and location about 25 million years ago (Figure 2.5).

The structure of the Indian Ocean

Among the world oceans, the Indian Ocean presents the most complex structural features. As mentioned earlier, there have been numerous geophysical and geological studies. The present account is mainly based on the review papers by Laughton *et al.* (1969) and Haq (1988). The outline bathymetric chart of the Indian Ocean published by Heezen and Tharp (1965) presents the principal structural and morphological features of the Indian Ocean (Figures 2.1 and 2.6).

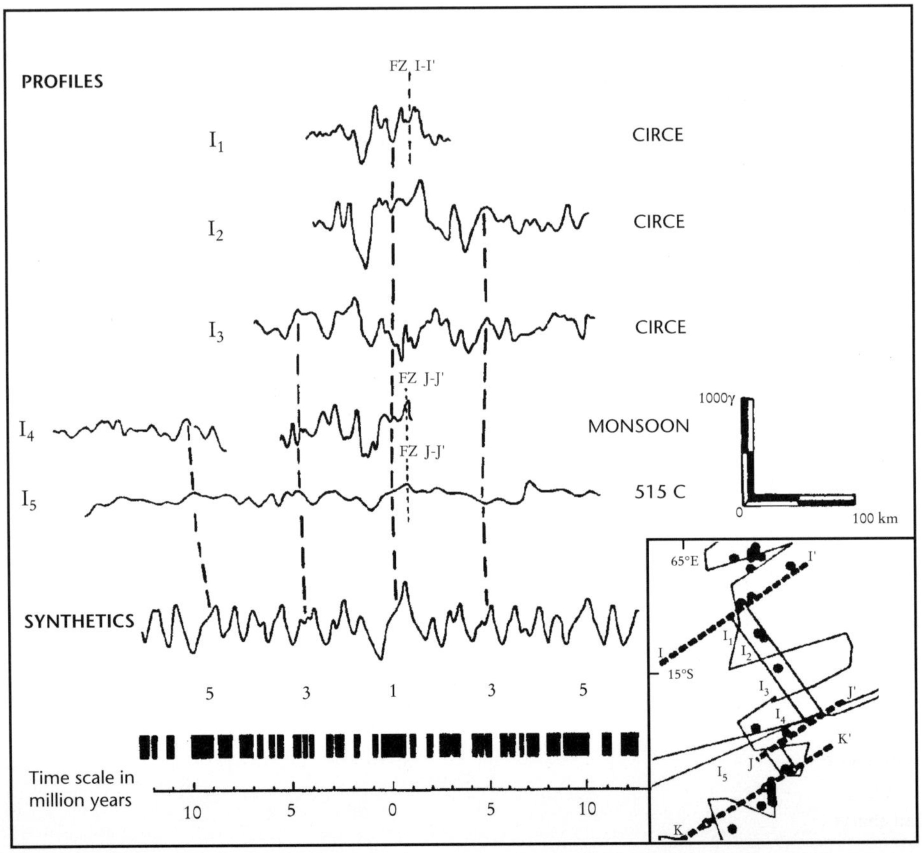

Figure 2.5
Results of magnetic survey of the Central Indian Ocean Ridge, showing magnetic anomalies between 14°S and 18°S; four sections across the ridge are shown, and the spreading rate (half) was assumed to be 1.8cm per year. (After Fisher *et al.*, 1971)

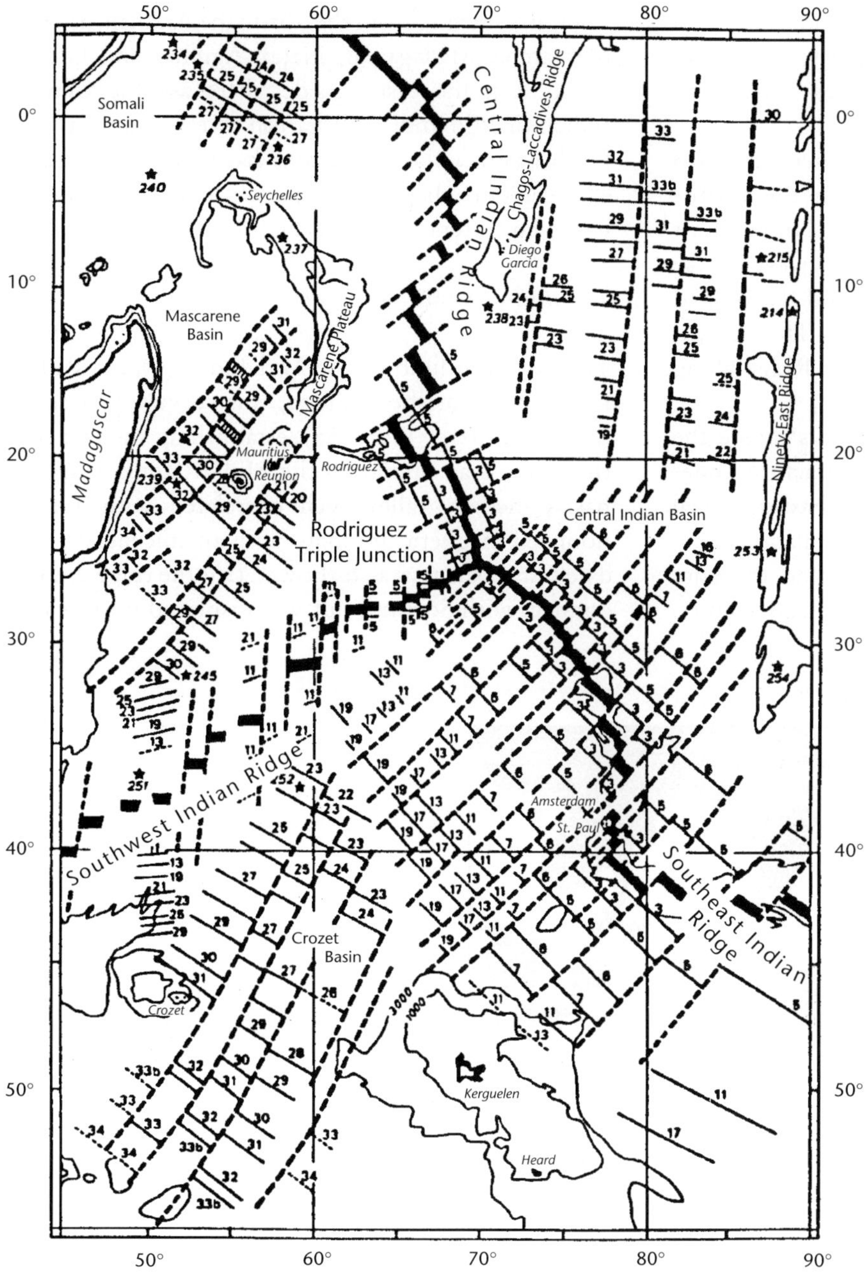

Figure 2.6
The Rodriguez triple junction: fracture zones (broken lines) and magnetic lineations (thin solid lines, if confirmed; dotted lines, if questionable) to either side of each ridge (solid black lines) are also shown. The numbers follow magnetic time scale proposed by Lowrie and Alvarez (1981); Deep-Sea Drilling Programme sites are marked by stars (★). (From Munsch and Schlich, 1989)

One of the most prominent among them is the three-limbed mid-Indian Ocean ridge in which the three limbs meet at a what is known as the Rodriguez Triple Junction south of the Rodriguez Ridge. One of the limbs of the mid-Indian Ocean Ridge extends north towards the central Arabian Sea and is called the Central Indian Ocean Ridge which continues in a northwesterly direction as the Carlsberg Ridge which finally joins the Sheba Ridge in the Gulf of Aden and the rift valley of the Red Sea. The second limb is the Southeast Indian Ocean Ridge which circumvents the Australian land mass to join the ridge system of the west Pacific. The third limb, the Southwest Indian Ocean Ridge, proceeds due south and southwest to reach the Mid-Atlantic Ocean Ridge near the Bouvet Fracture Zone.

All these three ridges are seismically active and exhibit spreading centres at many locations. The half spreading rates of these ridges (displacement of a point on one side of the ridge – or rift – relative to the axis of the ridge/rift itself) range from 1cm per year, on the Southwest Ridge, to 2-3cm per year on the Central and Southeast Ridges. These ridges are also cut and offset by many prominent fracture zones, such as: the Owen Fracture Zone across the Carlsberg Ridge; the Vema, Argo and Marie Celeste

Fracture Zones across the Central Indian Ocean Ridge; and Prince Edward, Discovery, Indomed, Atlantis and Melville Fracture Zones across the Southwest Indian Ocean Ridge.

Besides these three main ridges, there are two prominent aseismic ridges trending in the north-south direction. One of them, called the Ninety-East Ridge, is a straight feature originating west of the Andaman and Nicobar Islands around 10°N and proceeding due south to about 35°S. The second aseismic ridge is the Chagos-Laccadive Ridge along the 65°E meridian.

These main ridges and their offshoots divide the Indian Ocean into several basins, the fracture zones providing a connecting link between the basins, as well as serving as channels for deep-water circulation.

On the western side of the Indian Ocean, there are many basins named after their nearest land feature. They are, starting from the north: the Arabian, Somali, Mascarene, Madagascar, Natal, Agulhas and Crozet basins. On the eastern side, there are the Central Indian or Mid-Indian, Wharton (West Australia) and Perth basins. All these basins have a depth range of 4000-5000m. The deepest area of the Indian Ocean is in the Java Trench (7450m; Figure 2.7).

The mid-ocean ridge system

The northern limb of the Mid-Indian Ocean Ridge, named the Carlsberg Ridge, was surveyed in detail by HMS *Owen* (1957-58), establishing the fact that this ridge was similar to the Mid-Atlantic Ridge. It has a median valley well correlated with a pronounced magnetic anomaly with an epicentral zone. The *Owen* surveys also showed that the ridge did not connect with Socotra, but was displaced northwards

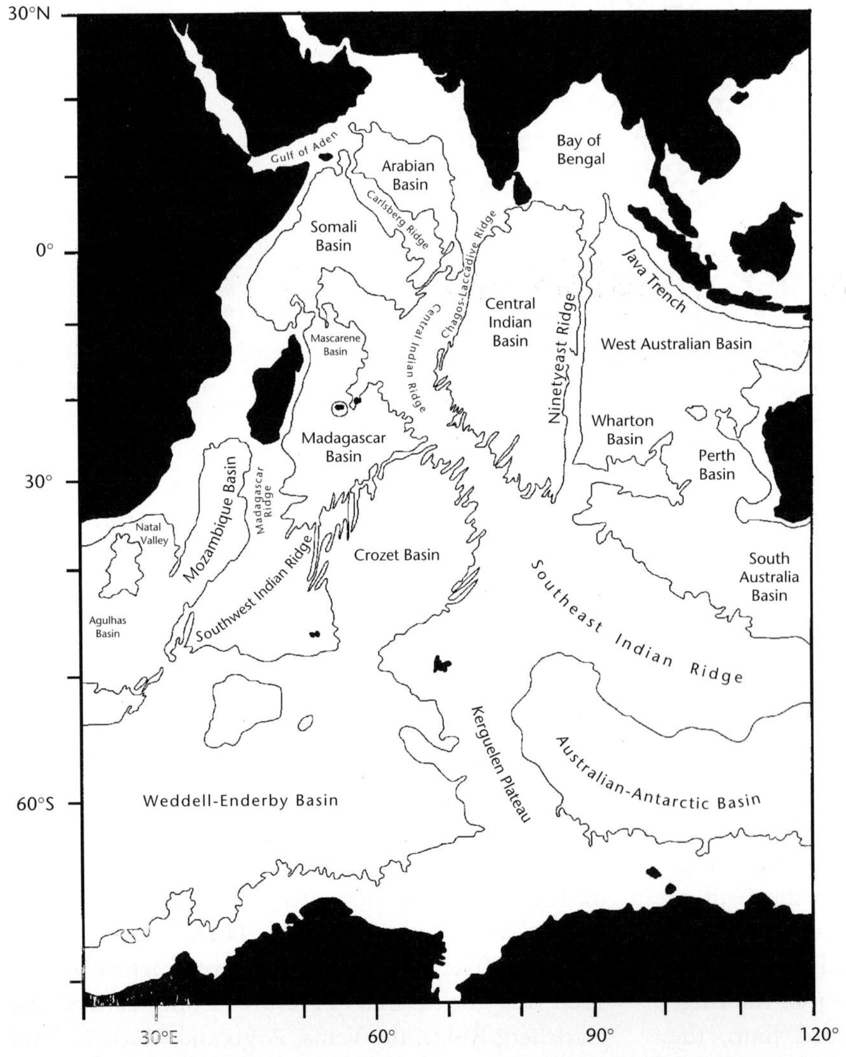

Figure 2.7
Map showing the principal basins and ridges of the Indian Ocean and the associated part of the Southern Ocean. (Adapted from Tomczak and Godfrey, 1994)

by about 300km by the Owen Fracture Zone which runs west from Pakistan to the Somali Basin. The continuation of this ridge into the Gulf of Aden was demonstrated by RRS *Discovery* in 1967 when evidence of a well developed median valley and magnetic anomaly were found.

The Central Indian Ocean Ridge runs from north of the equator more or less due south to 21°S where it is intersected by the E-W Rodriguez Ridge. The Central Ridge is cut across by many fracture zones lying mainly in the NE-SW direction. The main cleft of each fracture zone is flanked by single shoal lips or ridges that run parallel to it for several hundred kilometres. Seismic reflection profiles have indicated that some of the troughs do not contain any sediment fill, strongly suggesting their recent movement or formation. Studies made at 5.5°S showed that the ridge was active, as evidenced by the occurrence of ultra-basic rocks which probably belong to the upper mantle. Similar findings were made by the RV *Argo* in 1968 between 8°S and 28°S, on the same ridge.

The Vema Trench (9°S, 67.5°E) is located in a fracture zone of the same name and has a depth of about 6400m, the greatest known depth in the western part of the Indian Ocean. The relief from the tip to the trough floor reaches about 4300-4500m in this fracture zone which seems to have offset the median valley by about 180km as indicated by the location of earthquake epicentres in this zone. Similar offsets are found elsewhere along the Central Ridge; as a result, the active median ridge becomes much disturbed and discontinuous.

As already stated, the Southwest Indian Ocean Ridge runs from the triple junction at 20°S around South Africa and joins the southern part of the Mid-Atlantic Ocean Ridge. The physiographic map of Heezen and Tharp (1965) shows that this ridge is cut by a series of fracture zones that are continuous with or parallel to the Mozambique and Madagascar Ridges. As a result, the Ridge has become discontinuous and does not show the central magnetic anomaly. This is attributed to the preponderance of fractures and related shearing and igneous activity without active spreading. Besides, it is reported that, at low spreading rates (about 3cm per year), typical patterns of magnetic anomalies may not show up (Vine, 1966).

The Southeast Indian Ocean Ridge, which proceeds south and then east to meet the West Pacific Ridge around Australia, has a very rugged and irregular topography near the triple junction. There are also several fracture zones cutting across the Ridge as well as running parallel to it and, besides, no median valley has been identified. Extremely fresh basalts were dredged in the central zone and high heat flow has been recorded. The central region is also devoid of any sediments. All this points to the Ridge being an active seafloor spreading zone with a central magnetic anomaly. Beyond the Amsterdam Fracture Zone, the ridge continues in an east-southeast direction and its topography becomes smoother and broader. No median rift valley has been identified in this zone of the Ridge; however half spreading rates of about 3cm per year have been found and, in this respect, it resembles the East Pacific Rise.

As already stated, the Rodriguez Triple Junction (Figure 2.6), in the shape of an inverted Y, is perhaps the most prominent geological feature in the Indian Ocean. Its structure and evolution are reviewed by Munsch and Schlich (1989) based on a detailed survey made in 1989 on board the French research vessel the *Jean Charcot*. The junction is located near 25°30'S, 70°E and 900km southwest of Rodriguez Island. From this point, the Southeast Indian Ocean Ridge extends southeast through the islands of Amsterdam and St. Paul and joins the Macquarie Ridge south of Tasmania and then the Pacific Antarctic Ridge. The half-spreading rate of the Southeast Indian Ocean Ridge close to the Rodriguez Triple Junction is estimated to be 3cm per year and progressively increases to 3.7cm per year near the Macquarie Ridge. The Central Indian Ocean Ridge extends north of the junction to reach the Carlsberg Ridge and exhibits a spreading rate of 1.8-2.5cm per year. The Southwest Indian Ocean Ridge extends up to the Bouret Triple Junction in the South Atlantic after rounding South Africa, and its spreading rates are somewhat less than 0.8cm per year. As a result of this spreading, the Indian plate is moving away from Antarctica and Africa.

The Gulf of Aden and the Red Sea

The rates of spreading from the mid-ocean ridge system in the world's oceans appear to be similar everywhere. The rift and fracture zones expose lower portions of the oceanic crust which is close to the upper parts of the mantle. Analysis of the rocks dredged from various locations of the ridge system has indicated that they are basalts of the low-K_2O oceanic Tholierite-magma type. Russian geologists have reported from the Triple Junction area the occurrence of serpentinized harzburgites with minor serpentinized dunites and chromites.

The Afar Triangle is a meeting place of an extraordinary nature. Here, the East African Rift Valley, the median valley of the Red Sea and the Mid-Indian Ocean Ridge system run into each other. The Carlsberg Ridge continues into the Gulf of Aden as Sheba Ridge. Magnetic surveys have established the presence of linear magnetic anomalies parallel to the axis of the Sheba Ridge right across the Gulf of Aden, and seismic refraction studies have indicated the oceanic nature of the Gulf of Aden. It is suggested that the Gulf has formed as a result of the separation of Arabia and Africa, driven by the same crustal processes as those that underlie the creation of mid-ocean ridges. The origin of the Red Sea is a similar case; whether or how far the continental separation is involved in the formation of the Red Sea is debated.

Aseismic ridges and plateaus

The Indian Ocean floor is characterized by the occurrence of many ridges and plateaus that are free from earthquake activity. These are called micro-continents or oceanic ridges. Morphologically, they can be grouped into those that show considerable elongation (mostly in a S-N direction), such as the Chagos-Laccadive Ridge, Ninety-East Ridge, Mozambique Ridge, Madagascar Ridge, Kerguelen Ridge and the Mascarene Plateau, and those that are roughly equidimensional, as the Agulhas Plateau, Crozet Plateau, Broken Plateau, Wallaby Plateau and Naturalist Plateau.

Chagos-Laccadive Ridge: This is composed of three major groups of coral atolls that cap shallow, steep-sided plateaus on the Ridge, which is 2700km long and gently slopes down from about 2000m at the base of the plateaus to 4000m in abyssal depths. To the south and east of Chagos Islands, there is a well defined trench running parallel to the Ridge. The northern end of the Ridge disappears under the continental shelf off Bombay.

The atolls show evidence of growth on a steadily subsiding platform. Geophysical surveys made in the area reveal that the atolls sit on a ridge which is continuous and has a history of volcanism associated with the Deccan Traps of peninsular India. The Ridge has volcanic material that is 4-5km thick, overlying an oceanic crustal layer at the base, which is 17-20km deep. The volcanism in this Ridge appears to have commenced in the south in the early Cretaceous and reached a peak of activity in the north during the early Eocene. Subsequently, this Ridge has been seismically quiet.

The Ninety-East Ridge: This Ridge is the longest and straightest aseismic feature in the world's oceans. It consists of a single ridge, asymmetric at its southern end, about 200km wide, in some places flat-topped and at others fractured and mountainous, and stretches 4800km as a strikingly linear N-S feature. Crestal depth ranges from 1800 to 3000m. To the north, the Ninety-East Ridge disappears beneath the Ganges Cone and, in the south, it terminates in the foothills of the Southeast Indian Ocean Ridge. It is reported that this Ridge is a relic of a transform fault with volcanic leakage, forming a trace of the northward movement of the Indian Plate. Its age progressively decreases in the north-south direction, from late Cretaceous (about 85-65 million years ago) in the north to Oligocene (about 30-35 million years ago) in the south.

Indian Ocean Plateaus: The Mascarene Plateau is a faulted composite arc extending southwards from the Seychelles Islands for nearly 2300km right up to the volcanic island of Mauritius. The granitic rocks of the Seychelles are unique in the ocean environment and dated to an age of 650 million years, belonging to the Precambrian period. The granite and dolomite massif is cut by an alkaline granite ring complex estimated to be 50 million years old. The volcanic structures of Saya de Malha, Nazareth, Mauritius and Reunion Islands are underlain by volcanic foundations.

The Seychelles Bank is considered to be a Precambrian granitic microcontinent. The northern environs of the Mascarene Plateau represent a continental fragment that was separated from Madagascar in the late Mesozoic. In the southern portion of the Mascarene Plateau, the underlying volcanic piles have provided a base for the thick coral-like limestones and broad atolls of Saya de Malha, Nazareth and Cargados Carajos.

Mozambique and Madagascar Ridges trend southwards continuously from continental blocks with depths between 1000 and 2000m. Agulhas Plateau lies close to the South African Shelf and is capped by relatively unstratified sediments 400-500m thick, with the cored sediments ranging in age from Cretaceous to Miocene. The Crozet Plateau lies EW and comprises two separate blocks. The eastern block is topped by extinct andesite volcanoes and the Crozet Island. The basement of the submarine plateau is smooth and covered by sediment more than 500m thick, in contrast to the adjacent Southwest Indian Ocean Ridge which has little sediment cover near the centre.

The Kerguelen Plateau is at the northern end of the Kerguelen Ridge and extends southeast for over 2000km toward the Antarctic Shelf. It is reported that the structure of Kerguelen Plateau consists of Tertiary limestones hidden by a sheet of basaltic lavas and is considered as being a fragment of the continental platform.

On the eastern side of the Indian Ocean are located three possibly continental-type aseismic plateaus. The Broken and Naturalist Plateaus form the northern boundary of the Diamantina Fracture Zone and are considered as having once been part of the West Australian Shield. Wallaby and Cuvier Plateaus around 23°S, 107°E are also considered as being continental structures. All these plateaus appear to be capped by sediments varying in age from Cretaceous to Recent.

Continental margins

A knowledge of the extent of continental margins is essential for our understanding of the evolution of the Indian Ocean. A review of this subject is found in some detail in Laughton *et al.* (1969) and Shepard (1973). Proceeding from Africa, it would appear that the Agulhas Plateau is entirely separated from the African continent whose edge is at the base of the slope. Between the Republic of South Africa and Mozambique, the shelf is very narrow (<10km) except for the Agulhas Bank. Coming to the Mozambique Channel, between Africa and Madagascar, not much is known except for the accumulation of sediments 1-1.5km thick. Madagascar is bounded on the east by a linear fault zone which has been active since the Cretaceous and, perhaps, that marks the edge of the Precambrian block.

Off the Kenya coast, the bathymetric profiles do not show any shelf, slope or rise and support the idea of thick accumulation of sediments hiding the edge of the continent which may be somewhere half-way between Africa and the Seychelles. Seismic refraction studies made between Kenya and the Seychelles have shown that, whereas the western side of the section shows abnormally thick crust of 9-19km, comprising 4-12km of sediment, the eastern section has abnormally thin crust (3km) with typical oceanic magnetic anomalies. This situation supports the concept that African continental structures extend far into oceanic areas after the downwarping of the continental margin.

A normal configuration of shelf, slope and rise is found in the North Somali Basin; however, the shelf here is narrow (10-20km) and the sediment accumulation is very thick (>2km) and is confined by the Chain Ridge. The source of these sediments may be the Arabian coast before the formation of the Gulf of Aden.

Seismic refraction studies across the continental margins of northwestern India and in the northern Arabian Sea show that the basement under the ocean is covered by 5-8km of sediments. A great sedimentary cone which thins from 2.5 to 0.5km has been found in the Arabian Basin north of the Carlsberg Ridge. These sediments may have reached the present location by turbidity currents via the Indus Canyon from the erosion of the Himalayas. Ewing *et al.* (1969) estimated an average sedimentation rate of 17cm per thousand years, if the Himalayan uplift originated in the mid-Miocene.

A profile across the Chagos-Laccadive Ridge to Cochin has indicated that the ridge is separated from the Indian continental slope by deep water. There is also the fact that, at its northern end, the Ridge cuts into the shelf near Bombay and lies close to the seaward extension of the Deccan Traps. The existence of a volcanic layer beneath the Ridge is linked to the volcanism of the Traps. Rao (1976) has made similar observations. The sediments over the continental shelf off Bombay are relatively thin, but over the slope they attain a thickness of over 4km. No detailed geophysical studies have been undertaken along the southern and eastern Indian coasts.

The Indonesian Island Arc is a dominant structure straddling the continental margin on the eastern side of the Indian Ocean. The active belt starts in Myanmar (Burma), passes through the Andaman-Nicobar range and then to the islands west of Java. Parallel to this belt runs the volcanic mountain chain through Sumatra and Java. The area has been studied in great detail and indicates that structural trends of northern Sumatra and Myanmar (Burma) could be traced southwards across the Andaman Sea. Gravity anomalies over the Andaman-Nicobar area reveal a continental thickness of the crust. At the eastern end of the Indonesian Arc, the Timor Trough separates the arc complex from the continental shelf off northern Australia. The Sahul Shelf is abnormally wide (400km) and consists of a bottom topography of erosional type, perhaps originating in the middle to late Tertiary period.

Bathymetric charts off western Australia reveal that the broad continental shelf extends southwards to 21°S and the slope is rather gentle, at least in its upper part. The Wallaby Plateau and other highs lie westwards of the continental margin and the Australian continent and may continue in their direction at 200-2000m depth. The Naturalist Plateau, which lies off southwestern Australia, appears to be a continuous extension of the continent. The Diamantina Fracture Zone runs parallel to the south coast of Australia and appears to represent a normal transition from continental to oceanic structure.

To complete the geological picture, studies along the Antarctic coast in the Indian Ocean sector have revealed three zones of the continental shelf: (i) a coastal zone with a hillocky shelf formed by sporadic disturbances; (ii) a cone of deep faults oriented *en echelon* relative to the coast and caused by changes of ice loading during the Quaternary; and (iii) a marginal zone of old shelf plains that were uplifted during the Holocene and became tilted toward the continent. Here, a normal transition is found, from a continental crust 35km thick to an oceanic crust of 10-14km thickness, beneath the continental slope.

Sedimentation

One of the earliest papers on the nature of the seabed and the overlying deposits in the Indian Ocean area comes from Sewell (1925) working in the Andaman Sea and the Bay of Bengal on board HMIS *Investigator*. He reported that, in the northern Bay of Bengal, the bottom consists of mainly terrigenous deposits, at that time identified as brown or blue mud, whereas to the south, the deposits were of a pelagic type named Globigerina ooze in the ship's records. In addition, Sewell also reports on the occurrence of volcanic sand frequently occurring at depths greater than 2000m in the Great Channel between Sumatra and Great Nicobar Island and it is possible that their source was the volcanic regions of Sumatra and Java. Another observation of importance is the changes in the percentage composition of calcium carbonate in the mud samples; its concentration apparently decreases with depth and is inversely related to the silicon content.

Other sources of our information on the bottom deposits in the area are the reports of the *Valdivia* Expedition (1888-89) and the *Siboga* Expedition (1889-90). Subsequently, there have been sediment surveys by many ships from different countries, as mentioned earlier in this chapter. Besides, there are some excellent reviews and papers on sedimentation in the Indian Ocean (eg. Ewing *et al.*, 1969; Kidd and Davies, 1978; Kolla and Kidd, 1982; and others). However, research of an epochal nature was carried out from 1968 onwards by the Deep-Sea Drilling Ship (DSDS) *Glomar Challenger* in all three major oceans, including the Red Sea and the Mediterranean, to study the nature and thickness of sediments as well as the structure of the sea bed down to great depths, in some places exceeding a few kilometres. In the Indian Ocean, the ship drilled more than 50 holes during legs 22-25 in 1972 and repeated them in 1987. The initial reports of the project record a wealth of details of the sediment layers and their age based on fossil remains of organisms distributed in the core layers. Perhaps the most important discovery relates to the age of the oldest sea-floor rocks, which does not exceed 160-190 million years, in contrast to some of the land rocks which are dated to 3.7 billion years.

The following account is mainly derived from the above-mentioned papers and the initial DSDP reports.

Surface sediments

The sediment types and their distribution are shown in Figure 2.8. The thickness of the sediment is shown in Figure 2.9. Pure terrigenous sediments are naturally confined to areas peripheral to land masses, more particularly estuaries and river mouths, except in the Bay of Bengal and to some extent in the northern Arabian Sea where the terrigenous influence extends into deep and distal areas. The dispersal of the terrigenous sediments is mainly dependent on the prevailing ocean currents and local physiographic features specific to the Indian Ocean. Carbonate sediments are generally laid down in shallow areas along the Australian and African coasts, on the shallow ridges (rises) of the western Indian Ocean, on the Ninety-East and Broken Ridges and in the shallower regions of the Somali Basin. Normally, in the deep ocean basins, carbonate sediments are absent. This is because the contribution of the carbonates is more than offset by their dissolution. This happens at a depth of about 4000m in the Indian Ocean as indicated by the analysis of sediment cores. This depth is known as the Carbonate Compensation Depth (CCD). It is deepest (>5000m) in the equatorial region between 10°N and 10°S; southwards, say between 20° and 30°S, it becomes shallow and reaches

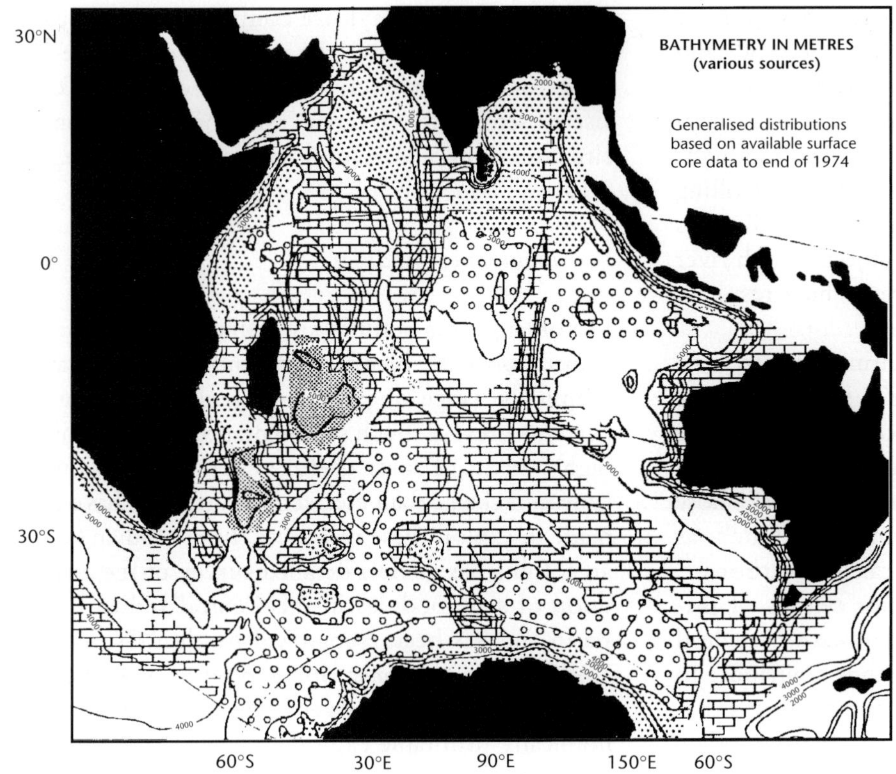

Figure 2.8
Sediment distribution in the Indian Ocean (bricks = calcareous ooze; o = siliceous ooze; • = terrigenous sediments; light stippling = pelagic clay; dark stippling = other clays; === = volcanic sediments). (After Kidd and Davies, 1978)

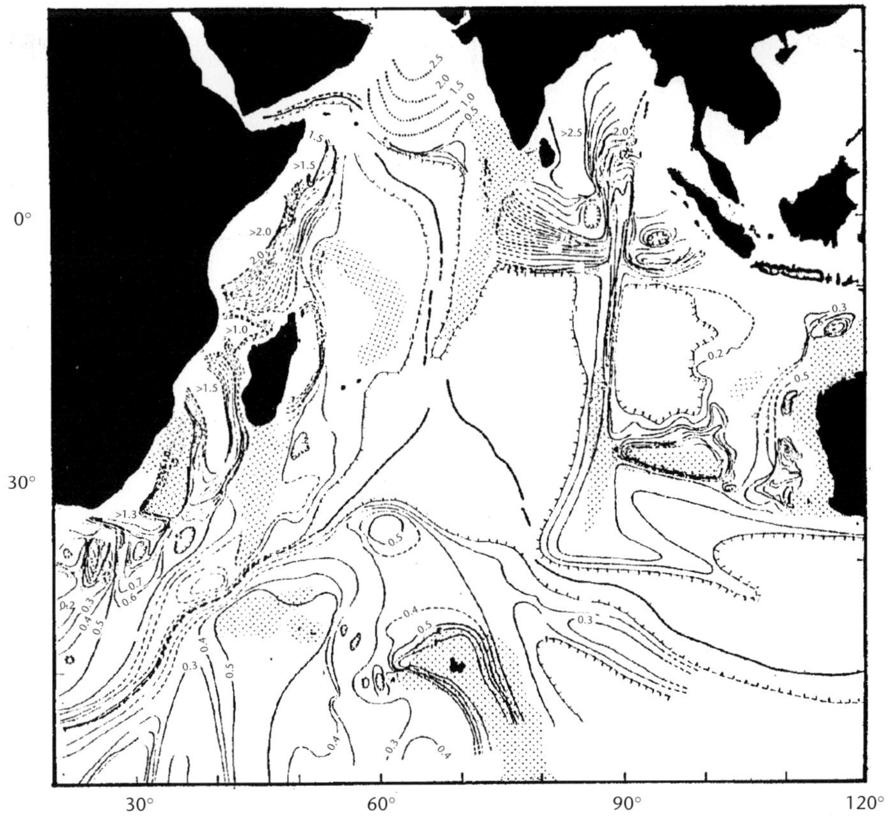

Figure 2.9
Chart showing isopachs of unconsolidated sediments; contours are in tenths of a second, two-way reflexion time (0.1s is approximately equivalent to 100m). Shaded areas represent aseismic ridges and plateaus; the axes of the main ridges are shown by thick black lines (where well defined) or thick black dashed lines (where not well defined). (After Ewing *et al.*, 1969)

a depth of 4600-4800m; it is shallowest (about 3900m) in the southernmost region between 50° and 60°S. It is also observed that high carbonate production, mostly shells of Foraminifera in the equatorial regions, and low production in subtropical and high-latitude areas are mainly responsible for controlling the CCD's depth. Also, high-carbonate sediments are not found close to land areas subject to high river discharges and warm temperatures, which result in increased dissolution rates.

Siliceous sediments are confined mostly to upwelling regions in the equatorial belt and in the high latitudes of the Southern Ocean. Although the ultimate source for silica is river runoff, a major silicon pathway in the sea is through diatoms whose outer shells are made of silica, very much as foraminiferan tests are made of calcium carbonate. Since the upwelling areas are major centres of primary production, they also become areas where dead diatoms sink to the sea floor and thus contribute to the siliceous nature of the sediment. Similar situations exist in the Southern Ocean. In the Indian Ocean, high silica abundance on the Somali shelf and slope is attributed to the intense upwelling taking place there, induced by the SW monsoon.

Brown or red clays are confined to deep basins far from land, below the CCD and outside the zones of high productivity. These clays may consist of sediments of terrigenous origin or from in situ alteration of volcanics within the ocean (Figure 2.10).

Ferromanganese nodules

Ferromanganese nodules are found in deep basins in association with brown clays and in shallow areas within the ridge or plateau systems where bottom currents prevent much sedimentation. Because of the high content of some useful minerals, such as manganese, nickel, cobalt etc., serious attempts have been made to exploit these resources. In the Indian Ocean area, India has made a detailed survey of the distribution of such nodules in the Central Indian Ocean Basin and has been recognized by the UN Sea-Bed Authority as a *pioneer* for exploiting the nodules. But the technology has yet to be developed for economically profitable exploitation of these resources. These nodules generally grow around a nucleus of some material such as a shell or tooth fragment as a site of mineral precipitation from sea water. The rate

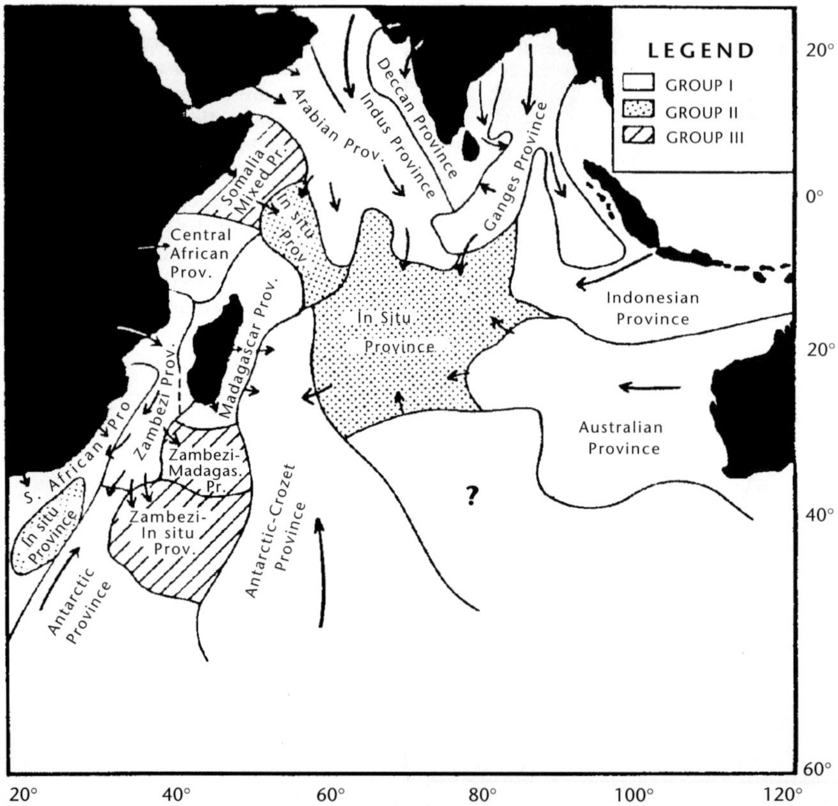

Figure 2.10
Clay mineral provinces in the Indian Ocean: Group I – rich in minerals derived from continents of Southern Ocean volcanic areas; Group II – rich in smectite; Group III – mixed type with low mineral content. ? = not known. (From Kolla and Kidd, 1982)

of growth is very slow – 111mm per million years, according to one estimate.

In the Indian Ocean, the nodules cover an area of 10-15 × 10^6km^2 and the resources are estimated to be about 1.5 × 10^{11} tons and, in the central Indian Ocean, their concentrations average >5kg/m^2 and grade Ni + Cu + Co >2.47%, which does not meet the requirements of first-generation mining (Siddiquie *et al.*, 1984).

Phosphorites

Phosphorites are considered to be sedimentary deposits composed mainly of phosphate minerals, particularly fluorapatite and some rare elements such as uranium. They occur in the form of nodules, sands and encrustations, mostly in shallow waters with less than 1000m depth. The continental margin phosphorites are commonly associated with areas of upwelling and high biological productivity, where a rapid flux of organic material to deep waters takes place. The Agulhas Bank phosphorites are one of the largest known deposits in the world, spread in a belt about 600km long and 100km wide and at a depth of 100-500m. The deposits consist of phosphoritic nodules and phosphorized limestones.

Red Sea deposits

It was mentioned earlier that the Red Sea is part of a rift system with sea-floor spreading characteristics. As a result, the axial trough throughout its median line exhibits high heat flow and in certain locations called *deeps* the entrained sea water shows high temperatures (44°-56°C) and salinities (nearly 7.5 times greater than those of normal sea water). Moreover, the underlying sediments are rich in heavy metals. The first such *deep* was located and described by RRS *Discovery* and, subsequently, quite a number of research vessels have made a detailed study of the Red Sea brines and sediments (Degens and Ross, 1969). Among them are the German research vessel *Meteor*, the US research vessels *Atlantis*, *Atlantis II*, *Chain* and *Oceanographer*.

Emery *et al.*, (1969) have summarized the results of these investigations, particularly those of several workers using the data collected by the *Chain*, which was the first cruise directly intended to investigate the hot brine and its underlying sediments. The present account is based on their observations.

Among the several *deeps* located in the Red Sea, the *Atlantis II Deep*, centred on 21°23'N, 38°04'E, has a depth of 2009m and is the largest in area (60km^2). The sediment temperatures here are around 61.3°C and that of the hot brine above, 56.5°C, with a salinity of 257.76 (for comparison, the ocean water will have, at a depth of 2000m, a temperature of 2-3°C and a salinity of about 34.7).

The metalliferous deposits of the Red Sea are similar in content to those found in shallow, moderate and deep vein deposits in ore lodes, with their sulphides of iron, copper, lead and accompanying trace metals; however, the minerals in the *Atlantis II Deep* are distributed in a widespread blanket of sediments in the Red Sea. According to the above-mentioned authors, the economic metals in the *Atlantis II Deep* are far more concentrated than in any other known large marine sedimentary deposit. The top 10m of sediment is estimated to total 83 million tons on a brine-free basis and valued at 2.5 billion US dollars at 1967 prices (Table 2.2), but the cost of recovery still exceeds the value of what may be recovered.

Table 2.2 Value of metals in the sediment

Metal	Average assay (%)	10^6 tons in the top 10m	Value ($\$10^6$)
Copper	1.3	1.06	1,270
Zinc	3.4	2.9	860
Silver	0.0054	0.0045	280
Gold	0.00005	0.000045	50
Lead	0.1	0.08	20
Iron	29.0	24.3	-
			2,480

Initial mining tests were carried out in the Red Sea in 1979 by the Preussag Company of Germany; the mud concentrates produced from the *Atlantis II Deep* sediments contained 30-35% Zn, 14-18% Fe, 3.5-4.5% Cu. There are no recent reports of commercial mining of the Red Sea deposits. It is felt that mining operations have become bogged down in political/legal problems, particularly in light of the tenets of the new UN Convention on the Law of the Sea. Emery *et al.* (1969) conclude their summary by stating that 'there is a distinct possibility that lawyers will profit more from Red Sea deposits than will scientists or the metal industry.' This is rather an unhappy ending for an exciting discovery and for the hard work of so many research vessels and numerous scientists in the Red Sea.

Dispersal of deep-sea sediments

Most deep-sea sediments are fine-grained clay minerals; their dispersal and deposition provide the best means of studying their source and transport during the present and geological past. There are many factors that govern the distribution of clay minerals in the ocean, such as local climatology, geology, physiography, submarine volcanism and water circulation.

According to Kolla and Kidd (1982), three broad groups of clay mineral zones are distinguishable in the Indian Ocean (Figure 2.10). In group I provinces, the clay minerals are derived from land sources and transported along current/wind tracks. The Australian province is characterized by high amounts of kaolinite transported to the ocean by trade winds. The Indonesian area has clays rich in smectite resulting from the alteration of fine-grained volcanic dust blown by northeast trade winds from volcanic eruptions on the Indonesian islands. The Ganges and Indus provinces are rich in illite and chlorite originating in the Himalayas and transported through the north Indian/Pakistan river systems. The Deccan province is characterized by a high abundance of smectite supplied, again through rivers, from the Deccan basalt soils on both sides of the Indian peninsula. The areas close to Arabia show the transport of pahygorskite by winds. In group II provinces, the sediments are rich in smectite derived from the alteration of volcanoes within the ocean, there being no supply from land, and these sediments are generally confined to central parts of the Indian Ocean and are not generally transported long distances from their source. Group III areas combine features of groups I and II.

Deep-sea fans

These were primarily built by turbidity-current deposits during the Pleistocene and earlier times, fed by the major river systems. As a result, huge fan deposits accumulated across the northern Arabian Sea and the Bay of Bengal and are named the Indus and Bengal Fans, respectively. Both these fans were initiated in the Miocene and older times coincidentally with the Himalayan orogeny (Curry and Moore, 1971). The Mozambique Fan, though smaller than the Indus and Bengal Fans, also came into existence in the mid-Miocene.

The major factors for the development of these Fans are summarized by Kolla and Kidd (1982). The sediment supply to the Indus and Bengal Fans is enormous compared to that of the Mozambique Fan. The upper regions of the northern fans were laid down primarily as depositional submarine valleys, whereas in the case of the Mozambique Fan they were erosional in origin. Besides, its location is in a deep basin within the reach of the Antarctic Bottom Water currents which may have a significant effect on sedimentation.

Indus Fan: This Fan represents the most prominent physiographic feature in the Arabian Sea. It is about 1500km long, 960km maximum width and occupies an area of about 1.1×10^6km^2. It is bounded to the north by the continental margins of India and Pakistan, to the east by the Chagos-Laccadive Ridge, to the west by the Owen-Murray Ridge and to the south by the Carlsberg Ridge. The great Indus River system draining the Himalayas is the main source of sediments for this Fan. The water depth of the Fan ranges from 1400m at the foot of the continental slope to 4500m at its distal end towards the Carlsberg Ridge. The shelf width of the Indus delta is about 100-150km and the most pronounced bathymetric feature of the shelf is the Indus Canyon, with an average width of 8km, a depth of about 800m and a length of about 170km. At the edge of the continental slope and at a depth of 1400m, the Indus Canyon breaks out into several channels through which turbidity currents transport the sediments across the Fan.

The Indus Fan is composed of thick, terrigenous grey-green muds and turbidities deposited during the glacial intervals, interbedded with calcareous sediments deposited during the interglacial intervals of the Pleistocene underlying the Holocene sediments. The Quaternary sediments of the upper Fan are primarily muds, with fine-grained turbidities.

The thickness of sediments in the Indus Fan, based on two-way travel time, in seconds, reveals that, in the offshore basin, it exceeds 11km (5.5 seconds), whereas in the eastern and western basins, the sediment thickness is equivalent to a travel time of only 2 to 3 seconds (Coumes and Kolla, 1984).

Bengal Fan: Occupying the whole of the Bay of Bengal, and extending up to about 20°S (Figure 2.11), the Bengal Fan is the world's largest pile of sediments whose source are the Himalayas. The Fan is about

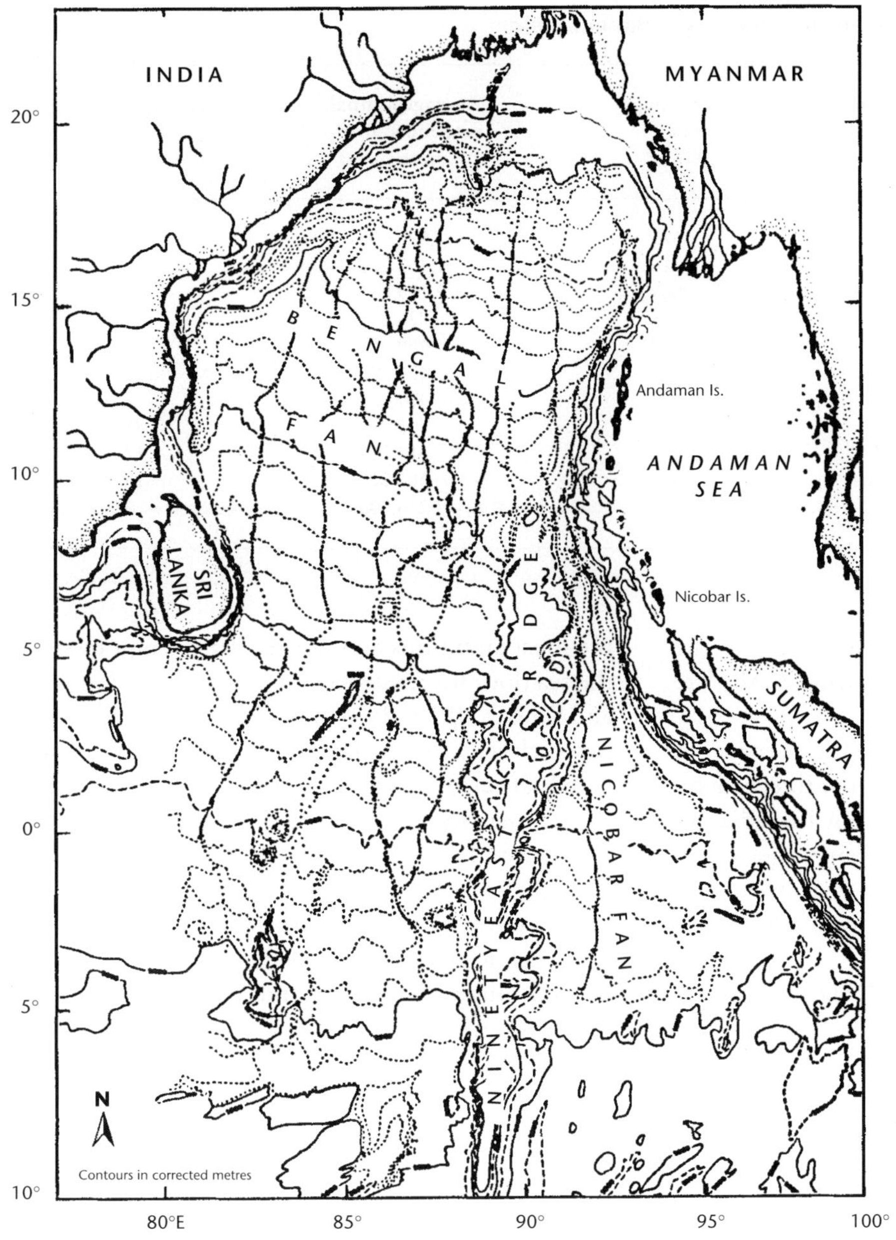

Figure 2.11
Bathymetric chart of the Bay of Bengal; fan valleys are indicated by dotted lines. (From Curry and Moore, 1971)

3000km long and 1000km wide and the sediment thickness at some places may exceed 12km. The sediments originate in the denudation of the Himalayan slopes by the Ganges (southern slopes) and the Brahmaputra (the northern slopes) and are transported through their delta and the Gangetic Canyon (also known as 'Swatch of No Ground') out into the various channels feeding the Fan across the Bay. The rate of erosion is estimated to be over 70cm per thousand years and the annual sediment load is the highest in the world (see Table 2.3). The composition of the sediments of the Bengal Fan is very similar to that of the Indus Fan.

Commenting on the structure and stratigraphy of the Fan, Curry and Moore (1971) suggest that the sediments fall into three categories: The uppermost sediments, arbitrarily termed 'W' sediments, are the younger sediments of the Fan; underlying 'W' sediments are the 'Y' sediments, which are exposed in outcrops over the Ninety-East Ridge and are deformed, faulted and folded in many areas; below the 'Y' sediments are the 'O' sediments, which are the oldest sedimentary rocks and basement rocks and probably volcanic. The general distribution and thickness of these different sequences are shown in Figure 2.12. It is estimated that the bulk of the sedi-

Table 2.3 Selected rivers of the world ranked by sediment yield

Name	Location	Drainage area $10^3 km^2$	Average annual suspended load 10^3 metric tons	Average discharge at mouth $10^3 m^3/s$
Yellow	China	673	1,890,000	1.5
Ganges	India	1116	1,452,000	11.7
Ganges/Brahmaputra	Bangladesh	2048	2,179,000	31.5
Brahmaputra	Bangladesh	935	726,000	19.8
Yangtse	China	1942	499,000	21.8
Indus	Pakistan	968	436,000	5.5
Ching	China	57	409,000	0.06
Amazon	Brazil	5773	363,000	181.1
Mississippi	USA	3220	312,000	17.8
Irrawaddy	Myanmar	430	300,000	13.6
Missouri	USA	1370	218,000	2.0

ments are modern 'W' sequences and belong to the Quaternary period, and the 'Y' sediments, to the late Miocene to Pliocene epochs.

There appears to be no place in the deep ocean where the topography has been so influenced by deposition of sediments coming from the adjacent land as has the Bengal Fan. The Fan slope has an almost even gradient of 1.5m/km for a distance of nearly 3000km and is criss-crossed by anastomosing channels with low levees.

The Deep-Sea Drilling Project (DSDP) is perhaps the most ambitious and spectacular attempt to study the sea floor for its structure and sediments (Simpson and Schlich, 1979). It is truly an international project in which scientists from many countries have participated since its inception in 1968 and which is still continuing. Studies made on the sediment cores have revealed and supported the sea-floor spreading theory and the fact that the floor of the ocean is very much younger than the ocean waters. The following is a brief account of the results of the drilling of a well through the distal end of the Bengal Fan

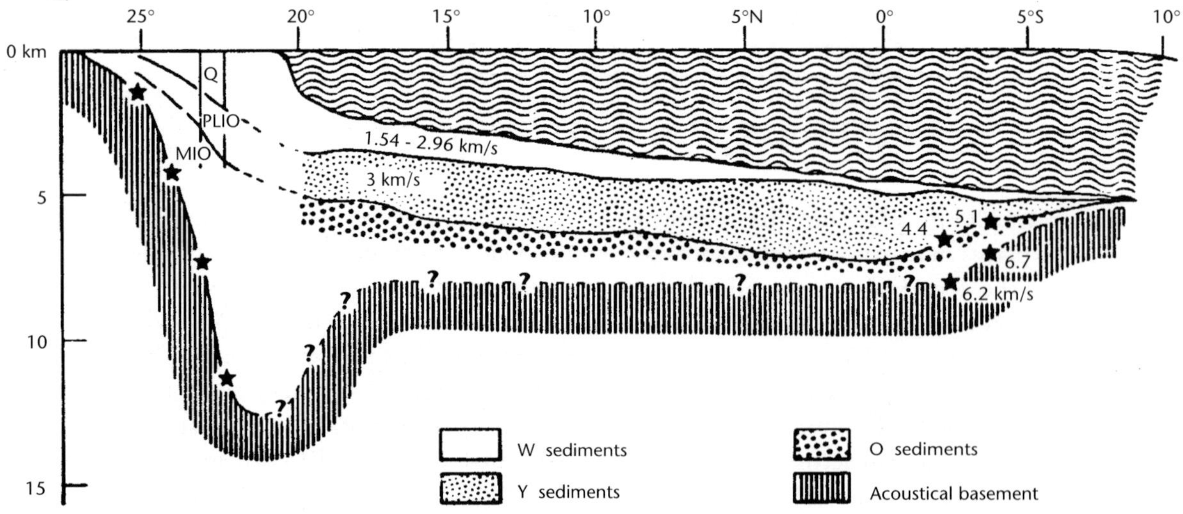

Figure 2.12
Hypothetical longitudinal section of the Bay of Bengal, from the Ganges-Brahmaputra delta to 10°S. W = Quaternary sediments; Y = Pliocene and late-Miocene sediments; O = pre-late-Miocene sediments. (After Curry and Moore, 1971)

(Proceedings of the Ocean Drilling Programme, Volume 116, issued in 1989). The drilling work was carried out at Site No. 717 during July-August 1987, at 1°S, 81°24'E, where the estimated thickness of sediments was between 1.5 and 2km. A total length of 814.8m was cored at a depth of 4734.7m; the deepest sedimentary unit cored was as follows:

Depth of sub-bottom: 533.2m
Nature: silt and silt mud turbidities
Age: Late Miocene
Measured vertical sound velocity: 1.7 to 2.0km/s

Sedimentation at Site No. 717 was dominated by fan sedimentation and consisted mainly of a sequence of turbidities. A thin layer of mud (5.5m) overlies a sequence of micaceous silt turbidities which accumulated rapidly during the late Pleistocene at a rate probably in excess of 350m per million years.

A very good record was indicated in the lithostratigraphic sequence at Site 717. At least three different sources of turbidities could be identified: silts and muds from the Ganges-Brahmaputra delta; dark grey organic-rich muds from the upper slopes of the Bay of Bengal; and greenish turbidities of biotic origin, probably from the Afnasy-Nitkin Sea Mount Group. Nanofossil records show that, although the site has been located close to or below the carbonate compensation depth for the last 10 million years, siliceous microfossils are almost completely absent, in spite of the location of the site within the supposed equatorial high-productivity zone.

There is further detailed analysis of sediment cores for lithostratigraphy, biostratigraphy, geochemistry etc. in the report. More than 50 holes were drilled in the Indian Ocean under the Deep-Sea Drilling Programme and it is desirable that a synthesis of the data be undertaken, to assist in increasing understanding.

Against this geological setting, the main hydrographic features, such as currents and water masses, are described in the following chapter.

CHAPTER 3

GENERAL HYDROGRAPHY, CIRCULATION AND WATER MASSES

Introduction

The roots of the tree of oceanographic knowledge of the Indian Ocean, as we have seen in chapter 1, go deep. The trunk was essentially in the nineteenth and twentieth centuries up to the time of the International Indian Ocean Expedition. Since then, the branches have multiplied and are now far too numerous to describe individually.

The phenomenal growth in the means of observation (remote-sensing satellites and aircraft, drifting and moored buoys with current meters, deep-sea pressure gauges, remotely operated vehicles carrying samplers, sensors and television, and so on), in the means of handling, evaluating and analysing the vast amounts of data, and the matching ingenuity of oceanographic and meteorological modellers of this Ocean and its volatile air have greatly increased our knowledge and understanding.

Some data are needed to design a first-generation model which, in turn, generates a need for more, better data and even for new types of data, with which to validate and improve the model. This is a spiralling process, until the model so well explains observable reality that it can be relied upon, in some cases, to predict the evolution of the system modelled and to fine-tune future data requirements; that is, reduce the amount and variety of data needed to keep the model 'up to date'.

Nevertheless, the objective of modelling is now rarely to describe the main features of the Indian Ocean hydrography, as we attempt to do in this chapter, but to evaluate its variability, and finally to predict that variability.

The intimate relation between the sea and the air at the sea surface determines the ocean's and the atmosphere's 'weather' and 'climate', although the ocean's natural 'signals' last for weeks, months, years, decades and more, whereas those of the atmosphere last only for a few hours to a few days (as the daily weather forecast proves). This means, therefore, that the Indian Ocean cannot now be dealt with in isolation, as we have been obliged to do here.

To understand the basic physical oceanography of the Indian Ocean (Sverdrup *et al.*, 1942) it is essential to note its geographical setting as well as its geomorphology, which were described in the previous chapter. The setting of the Indian Ocean is unique in the sense that it has no connection with northern polar seas, unlike the Atlantic and the Pacific. Next in importance to note is the complexity of the bottom topography which is dominated by the three branches of the Mid-Indian Ocean Ridge and by the Ninety-East Ridge which divide the deeper areas into several isolated basins, particularly at the depth of the 4000m isobath. This feature affects the deep-water movement in the Indian Ocean.

Superimposed on these particular geomorphological features are the reversing monsoons which, to a large extent, control the direction and strength of the surface currents in the northern Indian Ocean, north of 10°-20°S. A brief account of the meteorology of the Indian Ocean and its monsoons follows.

Meteorology of the Indian Ocean – the monsoons

The Eurasian continental land mass, which effectively blocks the Indian Ocean from the northern polar areas, presents a continuous mountain chain running mainly in the east-west direction and close to the northern boundary of the Indian Ocean. Some of these mountains, notably the Himalayas, are the world's highest. North of these mountains are the arid regions of China, Turkestan and Siberia. The Tibetan plateau occupies a central position, immediately north of the Hindukush and Himalayan ranges. These orographic and transcontinental features produce unique climatic conditions which influence the weather in the northern Indian Ocean.

During the northern-hemisphere winter, atmospheric high-pressure areas are centred on the Asian land mass, contributing to the creation of extremely cold and dry air masses which find a way through southeast Asia to become the NE monsoon over the northern Indian Ocean. These winds are weak and carry little moisture, except for some picked up from the Bay of Bengal and the South China Sea during their southwestward transit.

During the northern-hemisphere summer, the entire arc of desert areas of the continent surrounding the northern Indian Ocean, namely, Africa, Arabia, Pakistan, China and India, become very warm and create zones of low pressure which attract the moist winds from the southern parts of the Indian Ocean. These winds bring copious rainfall all over India and other south and southeast Asian countries; they are, essentially, the Southeast Trade Winds, which, after crossing the equator, become the SW monsoon.

We have, therefore, in the northern-hemisphere winter, the NE monsoon which blows from northeast to southwest and drives the surface ocean circulation anti-clockwise in the Bay of Bengal and Arabian Sea during the months of December-February. And in the northern-hemisphere summer, we have exactly the opposite, or the reversal of the circulation, when the SW monsoon winds prevail in the northern Indian Ocean.

The influence of the monsoons on the Indian Ocean is seen in the reversal of the surface circulation and in the hydrographical conditions of the surface waters down to 10-20°S (Figures 3.1 and 3.2). Beyond that limit, the southern Indian Ocean climatology is very similar to that of the Atlantic and Pacific Oceans at the same latitudes down to the Subtropical Convergence (around 40°S).

General hydrography

The boundaries of the Indian Ocean

The Indian Ocean comprises the Arabian Sea and the Bay of Bengal in the north and extends southwards to a quasi-permanent oceanographic feature, the Subtropical Convergence, in the southern hemisphere, and includes two marginal semi-enclosed seas, the Red Sea and the Persian Gulf. The western boundary is the east coast of Africa, but it is also arbitrarily bound seawards by the meridian 20°E which extends south from Cape Agulhas; its eastern boundary is also somewhat arbitrary but, for practical purposes, it is the coasts of Indonesia and Australia (from Cape Londonderry) and the 127°E meridian between them and to 147°E on the south side of Australia. Thus defined, the Indian Ocean covers an area of nearly $50 \times 10^6 km^2$.

The Subtropical Convergence, although usually fairly well developed, is not always as easily identified as it is in the Atlantic and Pacific Oceans. South of the Subtropical Convergence, the circum-Antarctic regime prevails.

The value of the Subtropical Convergence as a marker of the southern boundary of the Indian Ocean is supported by the fact that, as one proceeds south from the Convergence, the temperature and salinity between 40°and 50°S decrease rapidly from more than 15°C and 35.0 to less than 5°C and 34.0, indicating the

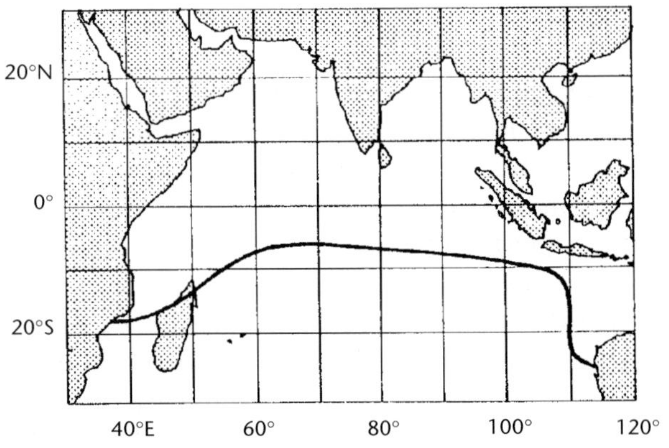

Figure 3.1
Southern limit of the monsoon regime in the Indian Ocean. (After Tchernia, 1980)

surfacing of the main oceanic thermocline and the development of a strong eastward circumpolar current aided by westerly winds. The strong inclination of this boundary, which covers a depth of more than 1000m to the surface, causes a powerful geostrophic current, the Antarctic Circumpolar Current (ACC). This is a permanent feature, centred roughly on 50°S, that maintains a strong front to the north, preventing warm sub-tropical water from flowing over the cold waters of the Antarctic.

Murray and Hjort (1912) reported that, out of the many ocean deeps known at that time, there were 57 with depths exceeding 3000fa (roughly 6000m), of which, 5 are found in the Indian Ocean. These deeps are named: Valdivia (partly in the Atlantic), with a maximum depth of 3134fa (nearly 6300m) covering an area of 1,136,000 square English miles (2,942,000km^2); Wharton, in the eastern Indian Ocean, with an area of 883,000 square English miles (about 2,287,000km^2), with a maximum depth of 3703fa (about 7400m); and Jeffreys, off the western Australian coast, with an area of 228,000 square English miles (about 590,000km^2); the two other deeps are called Melcar and Gardiner.

The following general description of the hydrology is derived from a number of classical sources and the recent summary of current knowledge by Tomczak and Godfrey (1994).

Sea-surface salinity (SSS) and temperature (SST)

The sea-surface salinity follows the P-E distribution (the difference between average precipitation and average evaporation): the P-E minimum near 30°S is reflected by a salinity maximum (after discounting the limited though obvious influence of the Red Sea and the Persian Gulf). The average SSS decreases southwards to the Southern Ocean where the effect of melting ice is present. The lowest annual mean SSSs are lowest in the northern subtropics: ≤ 33; in the inner Andaman Sea during the SW monsoon, the SSS is as low as 25. The highest annual mean SSSs are in the Arabian Sea: ≥ 36, although Red Sea and Persian Gulf SSSs are even higher. In the eastern tropical Indian Ocean, the values are close to those of the western Pacific: ≈ 34.5.

North of the equator, peninsular India divides the northern Indian Ocean into two large bays: the Arabian Sea and the Bay of Bengal. Oceanographically, they are very different but both interesting, reflecting their physical and chemical features, the influence of

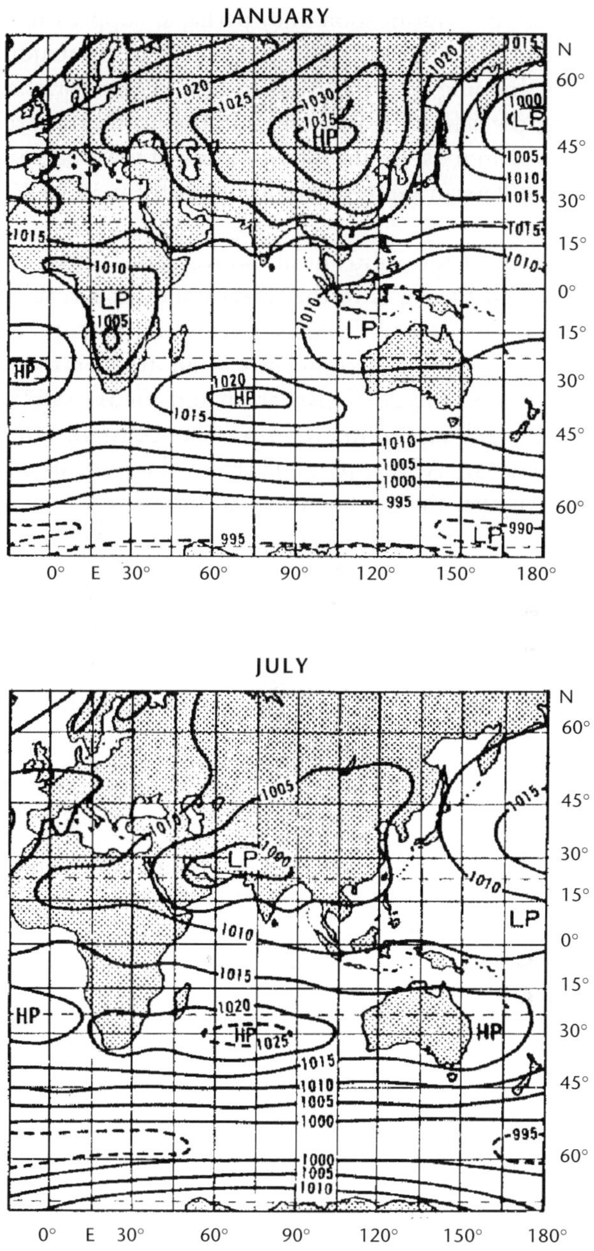

Figure 3.2
Distribution of high- and low-pressure systems in the Indian Ocean in January (above), between the NE and the SW monsoons, and in July (below), during the SW monsoon. (After Tchernia, 1980)

the reversing monsoons and the post-monsoon flooding, particularly along the Indian coastal areas. Many large rivers, namely, the Irrawaddy, Brahmaputra, Ganges, Mahanadi, Godavari, Krishna and Cauvery, debouch into the Bay of Bengal; as a result, the salinity of the surface waters is relatively low (30-34) over wide areas. This dilution also contributes to the marked stratification and, perhaps, suppression of vertical mixing in the Bay. Since there are no major rivers draining into the Arabian Sea, except for the Indus, the salinity is fairly high in the northern part and even more so in the interior areas of the Red Sea and the Persian Gulf, where it ranges from 36 to 40 or more.

North of the Subtropical Convergence, most of the Indian Ocean is in the tropics and, as a result, the northern Indian Ocean surface water is tropical in nature, having a uniformly high temperature, between 25° and 29°C (Reverdin and Fieux, 1987). However, along the southern Arabian coast and the east coast of Africa, particularly in the months of July-August, the temperatures are low (22°C), as a result of the Somali upwelling due to the prevailing SW monsoon (Fieux and Stommel, 1976). In the Bay of Bengal, the temperatures are low during February, under the influence of the NE monsoon. Moreover, the surface salinities of the Bay of Bengal are somewhat lower, owing to the heavy influx of river water during the months of August-October due to the wet SW monsoon.

The sea-surface temperatures over the Indian Ocean north of 15°S are similar to those of the western equatorial Pacific region (the highest mean SSTs in the Pacific Ocean): ≥ 28°C; only the Somali Current region has annual mean SSTs below 28°C, because the SW monsoon (which 'creates' the Somali Current) promotes upwelling which brings the SST down to ≤ 20°C.

There is no upwelling off the western Australian coast; on the contrary, there is a small poleward inflexion of the SST isotherms due to the Leeuwin Current. The poleward-flowing Agulhas Current also produces poleward inflexions of the SST isotherms, though much more strongly than does the Leeuwin Current; these boundary currents are briefly described below.

There is no equatorial SST minimum in the Indian Ocean as there is in the Pacific (where it is marked) and the Atlantic (less so).

Table 3.1 Temperature and salinity characteristics of the Indian Ocean water masses

Water mass	Temperature (°C)	Salinity
In the Upper Layers (0 - 500m)		
Bay of Bengal Water (BBW)	25 - 29	28.0 - 35.0
Arabian Sea Water (ASW)	24 - 30	35.5 - 36.8
Indian Equatorial Water (IEW)	8 - 23	34.6 - 35.0
Indonesian Upper Waters (IUW)	8 - 23	34.0 - 34.6
South Indian Central Water (SICW)	8 - 25	34.6 - 35.8
Persian Gulf Water (PGW) - surface	30 - 35 (summer)	36.4 - 42.0
	14 - 15 (winter)	36.4 - 42.0
Persian Gulf Water (PGW) - winter	23	40.0
Red Sea Water (RSW) on entry into Indian Ocean	≈22	38.0 - 40.0
In the Intermediate Layers (500 - 1500 m)		
Antarctic Intermediate Water (AAIW)	2 - 10	33.8 - 34.6
Indonesian Intermediate Water (IIW)	3 - 5	34.6 - 34.7
Red Sea - Persian Gulf Intermediate Water (RSPGIW)	5 - 14	34.8 - 35.4
Deep and Abyssal Waters (1500 - bottom)		
Circumpolar Deep Water (CDW)	0.1 - 2	34.62 - 34.73

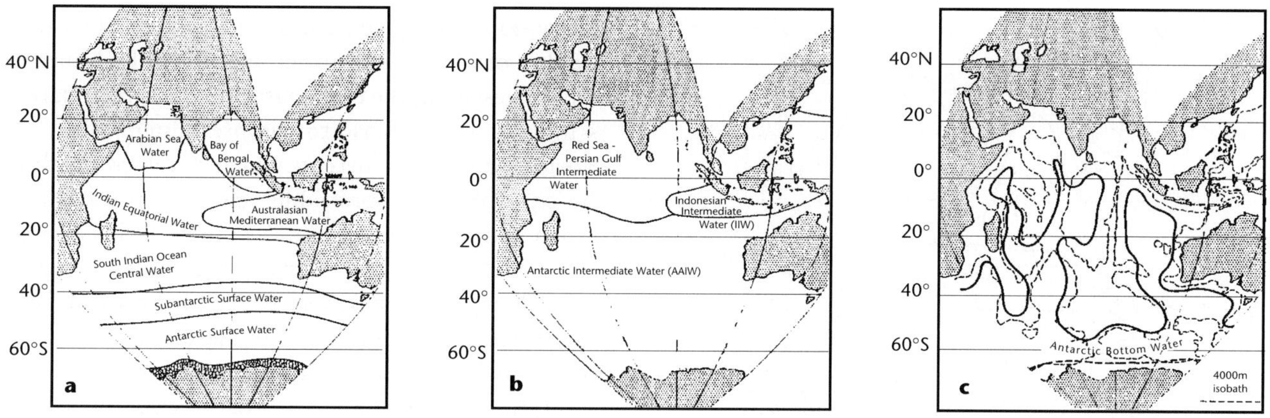

Figure 3.3
(a) Delimitation of Indian Ocean (and associated Southern Ocean) surface water masses (0-500m depth). (After Emery and Meincke, 1986). (b) Delimitation of Indian Ocean (and associated Southern Ocean) intermediate water masses (500-1500m depth). (After Emery and Meincke, 1986). (c) Delimitation of Indian Ocean (and associated Southern Ocean) deep water masses (1500m depth to the bottom). (After Emery and Meincke, 1986)

Water masses and circulation

The surface, intermediate and deep/abyssal water masses are briefly described in Table 3.1.

It is well established that most water masses acquire their characteristic properties at the ocean surface in particular areas and are transported by circulation either horizontally or to deeper levels all over the world oceans. Each water mass is marked by a specific range of temperature and of salinity, and therefore of a particular density, and is recognizable by these parameters wherever it exists. When the temperature (at specific depths) of a particular water mass is plotted against the corresponding salinity, the relationship is expressed as a T-S diagram which is characteristic of that water mass. Based on such diagrams, several authors have recognized a profusion of water masses in the Indian Ocean. However, Wyrtki (1971), in his Atlas of the Indian Ocean, and Emery and Meincke (1986) have summarized the water masses in the Indian Ocean and the present account is based mainly on their presentations. (See Table 3.1 and Figure 3.3). The various water masses may overlie each other along the appropriate density gradients and no rigid line of separation exists between adjacent (subjacent) water masses. The water masses may also mix, but precise data are not available to estimate the spatial coverage and distribution of the mixing zones (in effect, zones of overlap of the specific temperature and salinity ranges of contiguous water masses).

Surface layer (0-500m)

Wyrtki (1971, 1973) has summarized the salient features of the surface circulation in the Indian Ocean, as well as in the Indian Ocean sector of the Southern Ocean. On a large scale, he recognized three circulation systems, as follows:

1. The seasonally reversing monsoon gyres prevalent in the northern Indian Ocean, north of 10°S.
2. The sub-tropical anti-cyclonic gyre, similar to those at similar latitudes in the Atlantic and the Pacific.
3. The circumpolar current of the Southern Ocean, south of the Antarctic Convergence.

The main quasi-permanent and seasonal surface currents of the Indian Ocean are shown in Figure 3.4a, b.

The division into the three circulation systems suggested above is also reflected by the distribution of chemical properties, especially the phosphate content at 100m depth. The monsoon gyre in the northern Indian Ocean is marked by high phosphate content, above 1.0μmol/l, and it drops to less than 0.4μmol/l across the hydrographic front at about 10°S, the subtropical gyre having much lower values.

The monsoon gyres: In view of the seasonal reversal of the monsoon winds in the northern Indian Ocean, the direction of surface currents is determined mainly by the prevailing winds. The NE monsoon starts some time during November and reaches its maximum strength in December-February. The

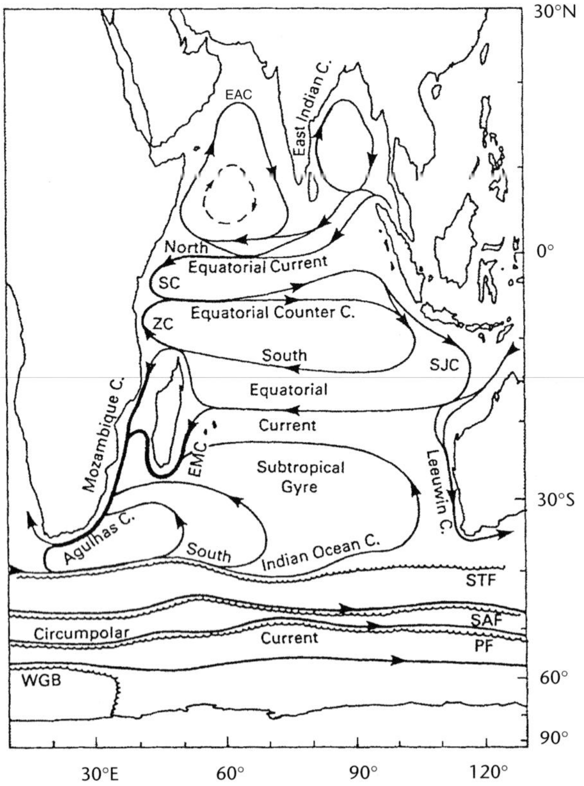

Figure 3.4a
Surface currents in the Indian Ocean in March-April (late NE monsoons). EAC = East Arabian Current; SJC = South Java Current; ZC = Zanzibar Current (also known as East African Coastal Current); EMC = East Madagascar Current; SC = Somali Current; STF = Subtropical Front (Convergence); SAF = Sub-Antarctic Front; PF = (Antarctic) Polar Front; WGB = Weddell Gyre Boundary. (After Tomczak and Godfrey, 1994)

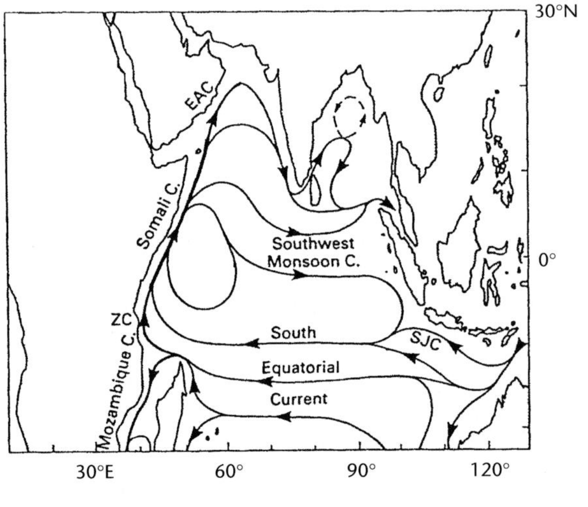

Figure 3.4b
Surface currents in the Indian Ocean in September-October (late SW monsoon); currents south of 20°S are as shown in Figure 3.4a. (After Tomczak and Godfrey, 1994)

NE winds pick up some moisture as they blow over the Bay of Bengal but, in the absence of high mountain ranges along the east coast of India, precipitation is limited. In response to the prevailing winds, the North Equatorial Current, which develops north of the equator, has appreciable velocities south of Sri Lanka and the Arabian Sea. A branch of this current flows north along the west coast of India, bringing low-salinity Bay of Bengal water to the west. With the collapse of the NE winds in March-April, the NE monsoonal circulation also collapses.

During the December-February period, being the height of the NE monsoon, with the winds blowing from NE towards SW, the surface waters everywhere show current direction to the west or southwest in the equatorial portion of the Indian Ocean. At this time, the North and South Equatorial Currents, as well as the intervening Equatorial Counter-Current, are well developed, much as the current systems in the two other major oceans. Anticyclonic circulation is developed in the Arabian Sea and the Bay of Bengal. In the Red Sea and Persian Gulf, the surface waters are found to move towards the north.

During the NE monsoon, the surface circulation consists of a west-flowing North Equatorial Current (NEC), north of the equator, then the south-flowing coastal current off Somalia which, together with the Zanzibar Current, feeds the Equatorial Counter-Current (ECC) moving across the Indian Ocean between the equator and about 8°S. In the east, it contributes to the South Java Current which feeds into the South Equatorial Current (SEC) flowing westwards across the Indian Ocean south of the Counter-Current at about 10°-15°S. These equatorial currents have been usefully described by Gonella (1983, 1984).

Fieux (1975, 1988) has summarized the sequence of oceanographic events involved in the triggering of the SW monsoon and its subsequent evolution (Figure 3.5), using data obtained in 1979 during FGGE, the First GARP Global Experiment (GARP = Global Atmospheric Research Programme, of WMO). The atmospheric element of FGGE was the Monsoon Experiment (MONEX); the associated oceanographic programme was the Indian Ocean Experiment (INDEX), of which the French component (in which Fieux played a leading role) was the Surface Indian Ocean Dynamic Experiment (SINODE; Reverdin *et al.*, 1982).

The SW monsoon picks up in May and, everywhere in the northern Indian Ocean and north of the equator, the surface waters flow in an eastward

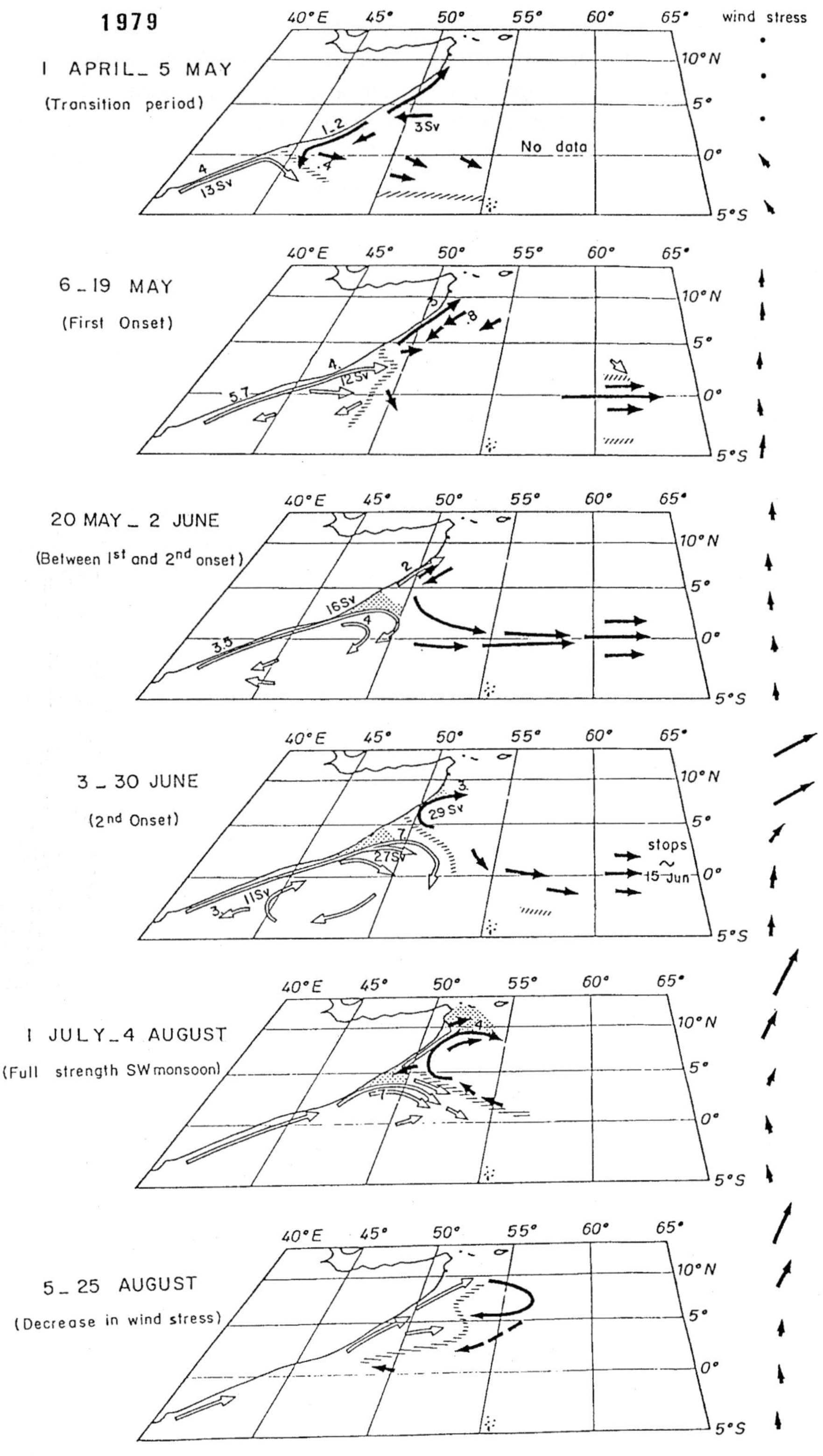

Figure 3.5
Evolution of the surface circulation during the triggering of the SW monsoon and subsequently, based on data obtained in 1979 during the international experiment INDEX. Current speed is in knots (1kn = 1.852km/hr); flux is in $10^6 m^3/s$ (Sv); open arrows represent low-salinity water; black arrows represent high-salinity water; stippling represents upwelling; //// represents hydrographic fronts. (From Fieux, 1988)

direction. By July, the monsoon current is well established north of the equator. At the same time, a north-flowing Somali Current is strongly developed as a continuation of the South Equatorial Current (which includes the northward-flowing Zanzibar Current); the SEC, after approaching Madagascar, directs the main part of its flow northwards along the east coast of Madagascar, turning more or less northeastwards between the northern tip of Madagascar (Cape Amber) and Farquhar Island (of the Seychelles group), feeding the Zanzibar (or East African Coastal) Current and the Somali Current for the most part, but also contributing to the southward-flowing Mozambique Current. The remainder of the South Equatorial Current turns southwards along the east coast of Madagascar (Fieux and Schott, 1988; Schott *et al.*, 1988). (The fate of this East Madagascar Current – EMC – is briefly described below in a subsection on anticyclonic circulation in the subequatorial zone).

On the eastern side of the Indian Ocean, near Sumatra, the SW Monsoon Current crosses the equator and joins the west-flowing South Equatorial Current. Thus the SW monsoon gyre is completed. This circulation is at its strongest during July-August, and thereafter it starts to weaken and is finally replaced by the NE monsoon circulation during October-November.

The Somali Current (SC), which is considered to be the western boundary current, is closely associated with strong upwelling along the northern coast of Somalia and the coast of Oman. Anticyclonic circulations are developed in the Arabian Sea and the Bay of Bengal and, as a result, a strong southerly current hugging the west coast of India develops. In the Bay of Bengal, a coastal current moves up the east coast of India as part of the anticyclonic gyre (Figure 3.4) and is called the East Indian Current. In the central portions of these two large seas, the surface currents are variable.

Table 3.2 French drifting-buoy programme in the Indian Ocean

Argos*	WMO	Deployment			End of transmission			Duration
Identity nos.		Date	Latitude	Longitude	Date	Latitude	Longitude	Months
01021	14621	12.01.79	23°48'S	53°95'E	08.02.80	30°44S	37°11E	13
01022	14622	07.01.79	23°69'S	59°21'E	04.01.80	35°87S	15°68E	11+
01027	14627	14.01.79	29°00'S	53°17'E	04.02.80	31°18S	33°65E	13
01028	14628	28.12.78	42°89'S	73°05'E	13.03.90	35°63S	94°75E	15
01029	14629	05.01.79	31°00'S	65°96'E	16.01.80	31°71S	59°53E	12
01030	14630	21.12.78	46°97'S	52°01'E	17.07.79	41°72S	93°02E	19
01031	14631	14.01.79	35°01'S	51°95'E	30.11.79	30°17S	45°59E	11
01032	14632	21.01.79	48°99'S	66°03'E	18.05.80	53°05S	164°60E	16+
01033	14633	16.01.79	42°00'S	52°01'E	09.09.80	17°13S	86°14E	20
01034	14634	30.01.79	36°21'S	74°74'E	08.05.80	29°60S	71°71E	15+
01036	14636	03.04.79	27°00'S	52°90'E	02.04.80	28°92S	48°64E	12
01039	14639	05.04.79	42°02'S	51°01'E	21.09.80	27°56S	83°45E	17+
01041	14641	09.04.79	46°69'S	51°92'E	01.09.80	22°46S	113°74E	17
01044	14644	20.04.79	44°98'S	72°83'E	29.07.80	29°31S	104°59E	15
01045	14645	21.04.79	39°95'S	75°87'E	10.10.80	27°43S	81°80E	18
01046	14646	26.04.79	35°00'S	70°97'E	20.10.80	34°04S	73°73E	18
01047	14647	28.04.79	27°99'S	64°03'E	21.02.80	25°49S	37°54E	10
01048	14648	04.04.79	34°82'S	52°33'E	23.07.80	22°44S	42°65E	14+
01049	14649	10.04.79	49°00'S	61°05'E	16.01.80	41°65S	119°07E	9
01051	14651	19.04.79	48°98'S	71°03'E	12.12.79	43°71S	125°64E	9
01090	14690	24.05.79	00°00'	62°05'E	16.02.80	00°40S	82°34E	9
01091	14691	11.05.79	00°00'	61°96'E	16.11.79	00°97S	100°35E	6
01092	14692	16.05.79	00°00'	62°00'E	04.01.80	05°06S	67°34E	8
01093	14693	20.05.79	00°00'	62°07'E	07.12.79	04°12N	96°13E	7
10110	14535	04.09.95	30°00'S	54°00'E				
10108	14536	06.09.95	40°00'S	52°00'E				

* *Argos is the French meteorological satellite-data receiving centre, now known formally as CLS (collecte-localisation-satellites). (Based on information kindly provided by Météo France, Centre de Météorologie Marine, Brest)*

The surface flow of water does not penetrate as deeply during the NE monsoon as during the SW monsoon, because of the relatively weaker winds during the NE monsoon.

Between the NE and SW monsoons, there exists a transitional period during which the surface circulations are variable north of the equator. The Somali eddy (gyre) generated during the SW monsoon may persist for some time after the cessation of this monsoon (Bruce *et al.*, 1981).

Subsequently, through an important drifting-buoy effort (Table 3.2), French oceanographers were able to track some of the major currents involved in the monsoon gyres (Reverdin *et al.*, 1982, 1983) and to establish the existence of equatorial Rossby waves (Gonella *et al.*, 1981, 1982). The trajectories of two buoys entrained in the Equatorial Countercurrent (also known as the Wyrtki Jet) remained surprisingly close together (Figure 3.6) between about 50°E and 85°E. Perhaps even more surprising was the trajectory of a drifting buoy, released at the equator at about 50°E, that was entrained in the southern part of the Equatorial Countercurrent, drifted just to the south of the equator to about 95°E, in five months, then returned with the South Equatorial Current, mainly along 10°S, to about 52°E, in nearly eight months, before turning back eastwards, mainly along 5°S, to about 88°E, in nearly six months, then returning westwards to about 50°S, mainly along 10°S, in about six months, before finally turning eastwards again to about 75°E, mainly along 5°S, in about two months, apparently to end its useful life (Figure 3.7). The average 'life' of the buoys given in Table 3.2 was, however, about 13 months.

Upwelling: The coastal areas of the Arabian Sea are major zones of upwelling during the SW monsoon (Currie *et al.*, 1973). Particularly off the Arabian coast, upwelling is observed to extend 400km offshore and runs parallel to the coast for nearly 1000km. This area shows the maximum abundance of phytoplankton and zooplankton in the Indian Ocean.

Off the southwest coast of India, upwelling starts even before the onset of the SW monsoon and continues till it ends in September. Major features of the upwelling are the upward displacement of the 20°C isotherm by nearly 100m and the invasion of the shelf off the west coast by nutrient-rich waters. These changes favour the occurrence of very rich zooplankton and enormous shoals of clupeid fishes (particularly sardines and mackerel).

Upwelling also occurs off the east coast of India, particularly off the Visakhapatnam coast. The main feature of this upwelling is the reduction of sea-surface temperatures by 4-5°C and the development of rich plankton blooms (LaFond, 1957; Varadachari, 1961).

Uda (1966) described upwelling of great intensity at about 15°S in the eastern Indian Ocean near the Wharton Basin. Wyrtki (1962) and Rochford (1969) reported upwelling off the islands of Java and Sum-

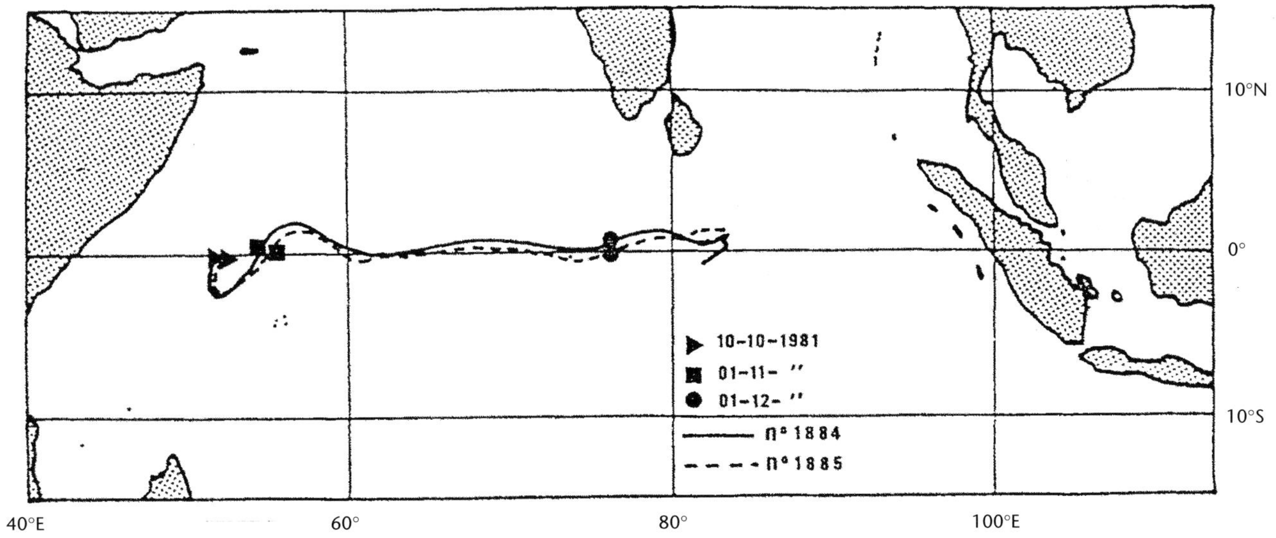

Figure 3.6
Trajectories of two drifting buoys entrained in the eastward-flowing equatorial or Wyrtki jet in October-November 1981. (From Fieux, 1988)

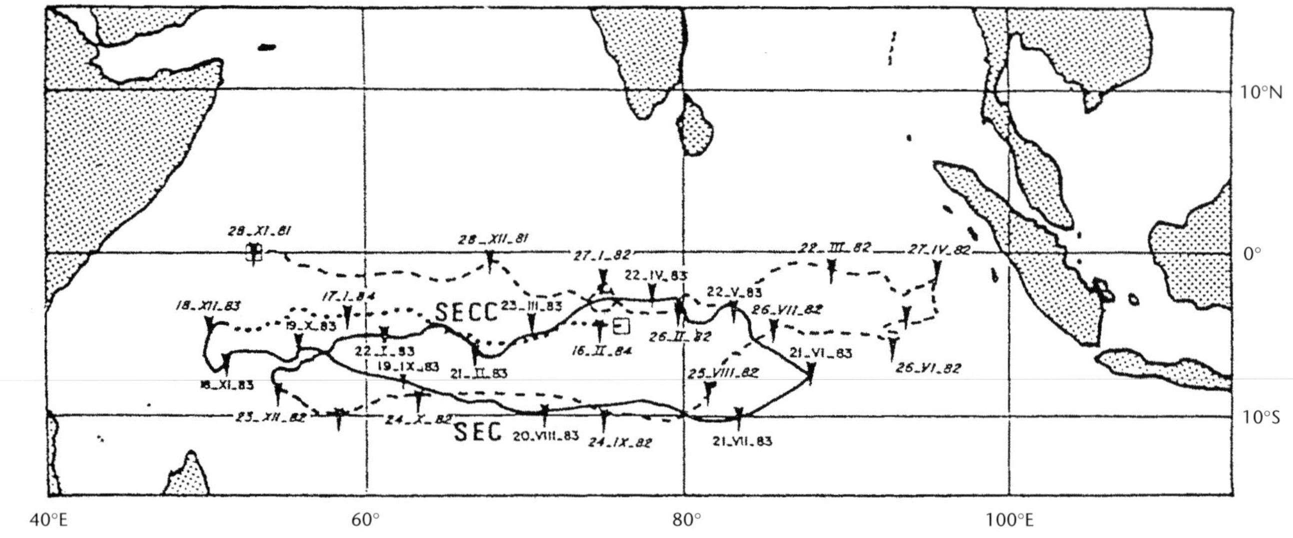

Figure 3.7
Trajectory of a drifting buoy initially released on 28 November 1981 into the eastward-flowing South Equatorial Countercurrent (SECC) and subsequently entrained in the westward-flowing South Equatorial Current (SEC), until it ceased to function on 16 February 1984, twenty-seven months later, after two and a half circuits across the equatorial Indian Ocean. (From Fieux, 1988)

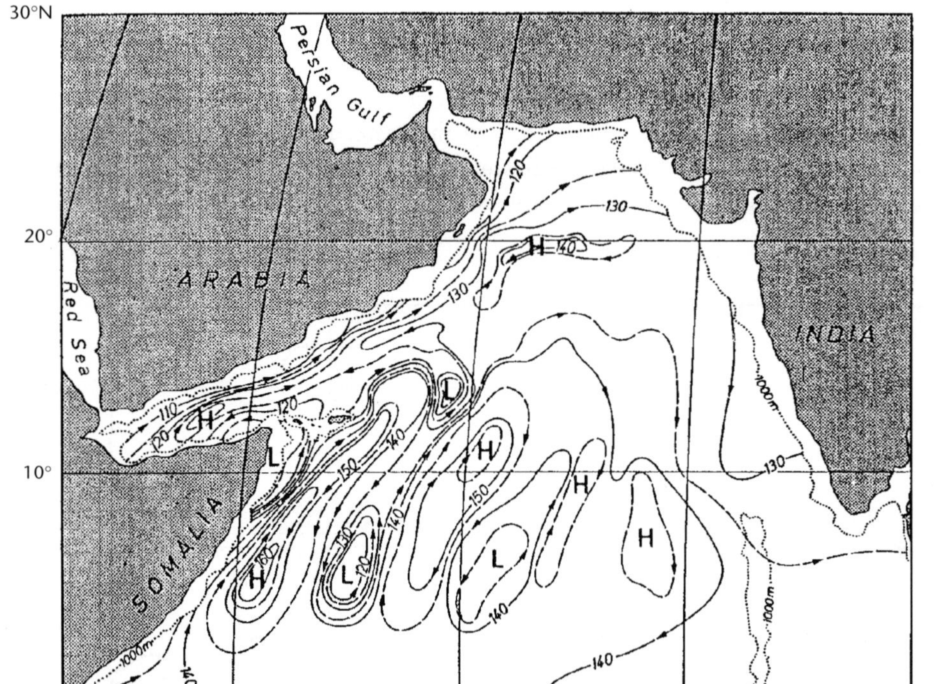

Figure 3.8a
Distribution of the dynamic height (in dynamic centimetres) of the sea surface relative to the 800 decibar surface. H = local maximum (centre of anticyclonic gyre); L = local minimum (centre of cyclonic gyre); dotted line = 1000m isobath. Geostrophic currents above the 800 decibar surface may be computed from such data. (After Duing, 1970)

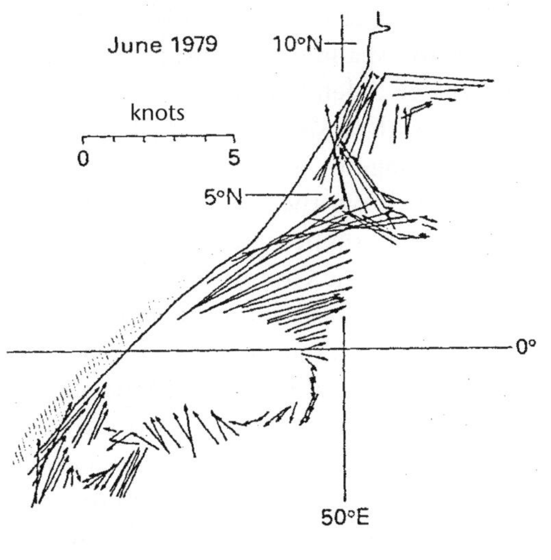

Figure 3.8b
Evidence of a two-gyre system in the Somali Current, from surface currents measured (1 knot ≈ 1.852km/hr) in June 1979 along the track of RRS *Discovery.* (From Swallow and Fieux, 1982)

bawa, and in the Arafura, Flores and Banda Seas, during the SE monsoon (May-September). This region is characterized by high inorganic phosphate concentrations at the bottom of the euphotic layer and by high plankton biomass.

Very strong upwelling systems are developed, particularly in the region of Somalia and Arabia, associated with the SW monsoon currents. According to Swallow and Bruce (1966), Duing (1970), Currie *et al.* (1973), Wyrtki (1973) and Schott *et al.* (1990), the Somali Current, as a strong western boundary current, causes structural adjustments in the baroclinicity along the Somali coast down to 1000m depth and the strong SW winds blowing parallel to the coast intensify the baroclinicity of the upper layer, and cause strong upwelling along the coast. This is most intense between 5° and 11°N and, as a result, the warm surface waters are entirely replaced by cold waters at less than 20°C (Warren *et al.*, 1966; Swallow *et al.*, 1983). Here, PO_4-P concentrations reach 10μmol/l and NO_3-N concentrations are also more than 10μmol/l, in contrast to values of less than 0.2μmol/l and 0.5μmol/l, respectively, in the offshore waters. Such high values are normally found at depths of 150m or more.

Much of the water of the Somali Current is recirculated in an intense eddy, the centre of which is about 300km offshore and stretches in a north-south direction for about 1000km parallel to the coast. To the east of this eddy, smaller eddies are also developed as part of the SW monsoon circulation (Figure 3.8a). There is also good evidence that the Somali Current is a two-eddy system at the beginning of the SW monsoon (Figure 3.8b) before generating subsidiary eddies in the Arabian Sea (Swallow and Fieux, 1982).

The northward flow of the Somali Current is terminated by the warm outflow of surface waters from the Gulf of Aden, which thus brings about the separation of the Somali and Arabian upwelling systems. During the SW monsoon, the strong wind that blows parallel to the Omani coast causes upwelling in the area between Ras Fartak and Ras al Hadd. The most important feature of this upwelling is the breadth of the upwelling zone and, because of this, water appears to be supplied from a much greater depth than is usual in other coastal upwelling systems (Currie *et al.* 1973).

On the west coast of peninsular India, the SW monsoon winds create a strong southward coastal current and the resulting structural adjustment of the baroclinicity in the area causes upward movement of the 20°C isotherm all along the west coast of India to a depth of 50m or above. This is considered to be weak upwelling; however, nutrient enrichment of the surface waters takes place, resulting in higher rates of productivity. Since these upwelled waters are poor in oxygen, there are reports of adverse effects and mortality of marine organisms along the west coast of India (Banse, 1968).

Surface water masses in the northern Indian Ocean: The processes described above are essentially wind-driven. They contribute to the creation of two different surface water masses in the northern Indian Ocean. One is the high-salinity Arabian Sea Water (ASW) and the other, the low-salinity Bay of Bengal Water (BBW). In the former, high salinity is due to excess evaporation over precipitation and, in the latter, the low salinity is due to large runoff from the various rivers of the Indian sub-continent. Besides, the high-salinity waters of the Red Sea and the Persian Gulf also contribute to the much higher salinity of the Arabian Sea water, compared to the Bay of Bengal water. As a result, ASW has a surface salinity value up to 36.8; the Bay water shows a decline in salinity from about 34 at 5°N to 31 in the northern part of the Bay. Some of this low-salinity water flows westwards to the north of the North Equatorial Current.

The main impact of the Red Sea Water (RSW) and the Persian Gulf Water (PGW) is, however, mainly

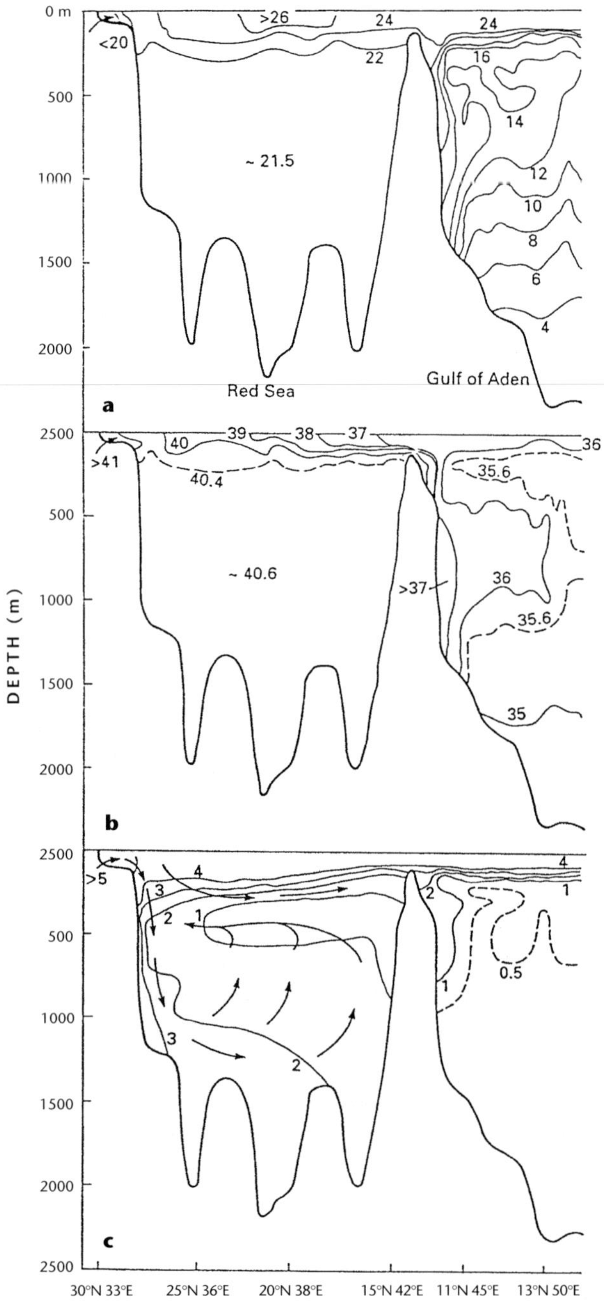

Figure 3.9
Hydrographic section along the axis of the Red Sea and the Gulf of Aden in winter. (a) = potential temperature in °C; (b) = salinity; (c) = dissolved oxygen in ml/l; arrows indicate flow of deep water. (From Wyrtki, 1971)

felt at intermediate depths (described below). It is, nevertheless, convenient to describe the properties of these two water masses here.

The Red Sea occupies a long narrow basin between latitudes 12° and 28°N. The depth averages 560m but, in the centre of the rift valley, the depth may reach 2900m. It is connected to the Mediterranean through the shallow, man-made Suez Canal and to the Gulf of Aden through a narrow strait, Bab-el-Mandeb, with a sill depth of about 100m, across which exchange of water takes place. As the surface water flows from the Gulf of Aden through the strait, there is a compensating subsurface flow from the Red Sea into the Gulf of Aden.

In the Red Sea region, surrounded by arid deserts and with very little precipitation, evaporation is very high; consequently, the salinity and temperature of the water are fairly high compared to those of the Arabian Sea water. Salinity reaches a value between 40 and 41 in the northern part during most of the year. However, with the decrease of temperature to about 18°C in winter, the whole of the Red Sea basin becomes filled up with a deep water with a high salinity and lower temperature. When a favorable situation exists, this Red Sea Water flows over the sill and spreads into the Arabian Sea at intermediate depths (Figure 3.9) and can be traced right down to Sri Lanka and beyond. This nearly isothermal outflowing water has a dissolved-oxygen level of 2-3ml/l. It flows out into the Arabian Sea, mainly during the winter, and can be traced throughout the Arabian Sea and beyond.

According to Premachand *et al.* (1986a), the core layer of the Red Sea Water is found at depths of 600-800m all over the Arabian Sea, particularly north of 16°N. As it spreads along the African coast, the core of the RSW progressively deepens to 1100m south of 10°S with increased density. To the southeast, this water mass is seen between the depths of 500 and 800m; between 15°N and 5°N; the average depth of the core layer is around 600m. The southern boundary of this water mass extends to 10°S where it mixes with the core of the Antarctic Intermediate Water (AAIW) (see Figure 3.10 for schematic salinity section).

Like the Red Sea, the Persian Gulf lies in an arid zone where evaporation exceeds precipitation and, as a result, warm and highly saline waters are formed in the Gulf, which has an average depth of 25m and opens into the Gulf of Oman through the Strait of Hormuz (Premachand *et al.*, 1968b). Its sea-surface salinity is between 36.36 and 42 and its temperature ranges between 30°and 35°C in summer and 14° and 15°C in winter, in the north. At the bottom, in the south, the water has a temperature close to 23°C and a salinity of 40. Since the monsoon circulation is weak and unsettled, the flow of this highly saline Persian Gulf Water (PGW) into the Gulf of Oman is

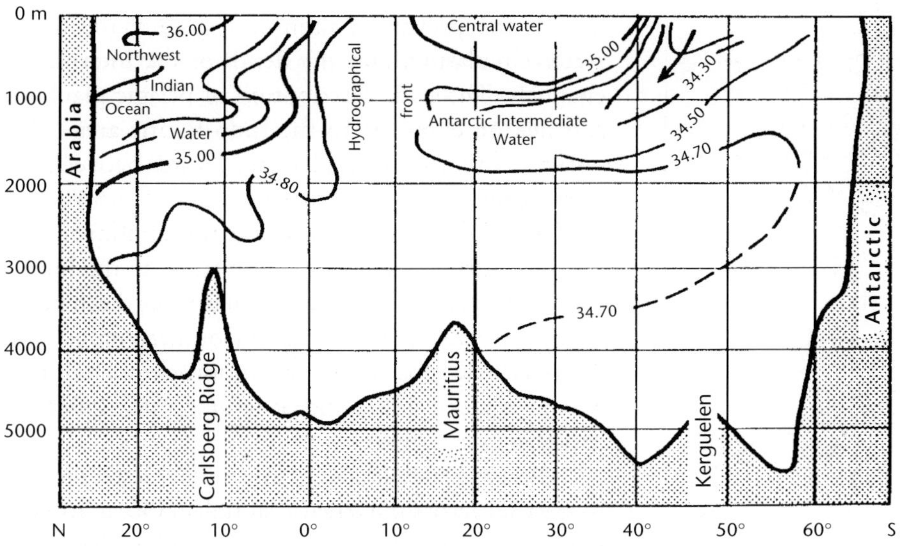

Figure 3.10
A schematic salinity section from Arabia to the Antarctic. (After Tchernia, 1980)

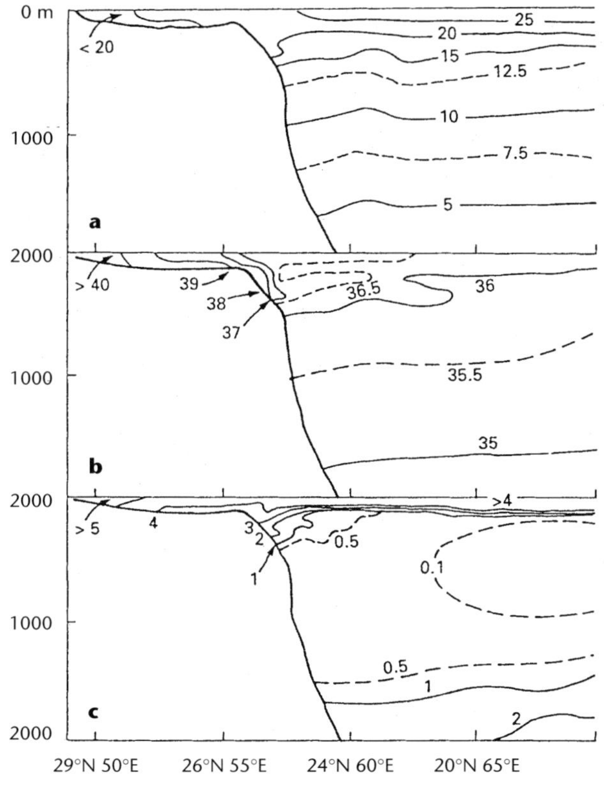

Figure 3.11
Hydrographic section along the axis of the Persian Gulf in winter. (a) = potential temperature in °C; (b) = salinity; (c) = dissolved oxygen in ml/l. (From Wyrtki, 1971)

small and intermittent; however, on entering the Gulf at depths of 25-70m, this water mass sinks to a depth of 200-250m as it spreads in the Arabian Sea and exhibits a tendency to deepen southwards (Figure 3.11).

In the Gulf of Oman, the core layer of Persian Gulf Water has a salinity of 37.9 and, as it spreads in the Arabian Sea, the salinity rapidly decreases to 35.5. In the Gulf of Oman, the temperature of this water exceeds 22°C, but southwards to the Laccadive Sea, the temperature decreases to about 12°C. Rochford (1964) traces the PGW in the Bay of Bengal and then into the Indian Ocean down to about 10°S. Wyrtki (1971) has identified this water mass along the Somali coast, throughout the central Arabian Sea and off the southwest coast of India down to 10°S and 100°E.

Equatorial surface water masses (0-500m): In the equatorial region, there are three water masses in the upper 500m of the Indian Ocean. They are usually called the Indonesian Upper Water (IUW) or Australasian Mediterranean Water (AAMW), the Indian Equatorial Water (IEW) and the south Indian Central Water (ICW).

Indian Central Water is formed and subducted at the Subtropical Convergence; its T-S characteristics are the same as those of the South Atlantic and the Western South Pacific Central Waters.

The Australasian Mediterranean Water is a tropical water mass derived from the North Pacific Ocean Central Water and subtropical water, and acquires its specific properties during its throughflow from the Pacific to the Indian Ocean, which occurs chiefly between Timor and the northwest Australian shelf and between the Indonesian islands east of Bali (Figures 3.12a, b). There are many channels separating the southeast Asian peninsula, the intervening Indonesian archipelago and Australia which allow

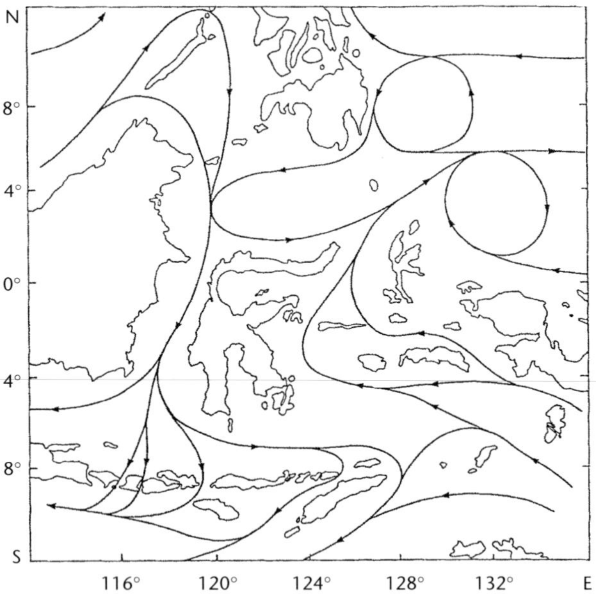

Figure 3.12a
Surface currents in the Australian Mediterranean Sea (the site of the Indonesian Throughflow) in August (maximum westward flow). (From Tomczak and Godfrey, 1994)

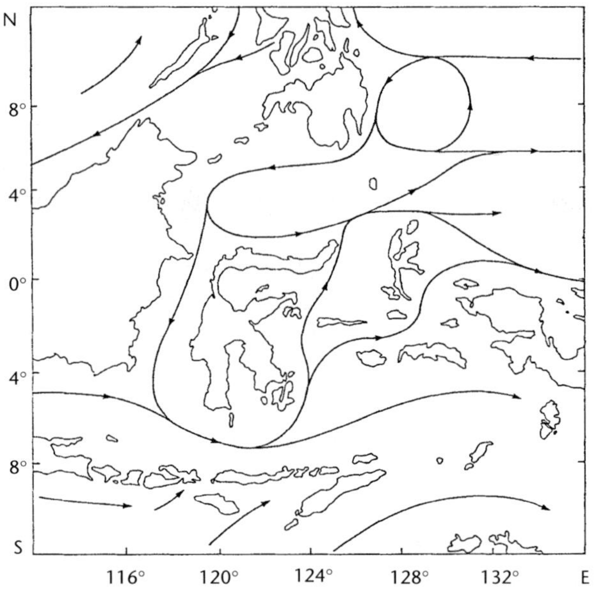

Figure 3.12b
Surface currents in the Australian Mediterranean Sea (the site of the Indonesian Throughflow) in February (maximum eastward flow). (From Tomczak and Godfrey, 1994)

the free exchange of surface water between the western Pacific and the eastern Indian Ocean. At the surface, water may flow from the Pacific through the Lombok, Sunda, Ombai and Timor passages into the Indian Ocean, or in the opposite direction, depending on the prevailing monsoon system, the deep-water circulation, and the depth of the inter-island channels. The wide distribution of pelagic plankton and nekton common to the Indo-Pacific area bears testimony to the free exchange of water between the two oceans.

The amount and timing of the throughflow is still poorly known. It is mainly westwards (to the Indian Ocean) and is minimum in February (northern monsoon) and maximum in August (southern monsoon). There is little geostrophic transport below a depth of 500m (Wyrtki, 1987; Fieux *et al.*, 1994, 1996a, b; Molcard *et al.*, 1996).

AAMW's comparatively low salinity makes it easy to trace all across the equatorial Indian Ocean to eastern Madagascar around 10°S down to 600m depth; the silicate distribution marks the IIW below 600m down to 1000m. The AAMW forms a strong salinity gradient with the ICW, although both these water masses have similar temperatures, 8.0-25°C (in the upper 600m), but the AAMW has a lower salinity range, from 34.6 to 34.7, whereas the ICW ranges from about 35.0 to 35.2, in the northern hemisphere, and from about 35.2 to 35.75 in the southern hemisphere (Figure 3.13). Some researchers divide the AAMW, in the east, into Indonesian Upper Water and Indonesian Intermediate Water.

The wind-driven throughflow, in effect, forces the ICW in the southern hemisphere to circulate from east to west in the South Equatorial Current, past the northern end of Madagascar into the Arabian Sea via the Somali Current (during the SW monsoon) and then into the Bay of Bengal, but its fate thereafter remains unknown; this is the only path by which southern hemisphere ICW reaches the northern hemisphere.

The dissolved-oxygen values of the ICW fall rapidly during this transit, indicating rapid aging (i.e., slow rate of renewal) of the water, and this continues into the Bay of Bengal which has the oldest Indian Ocean Central Water.

Some Australasian Mediterranean (or Indonesian Upper) Water contributes to the Agulhas Current, but some also spreads into the Bay of Bengal below the near-surface water (marked by the low salinities – <34.5 – of the Bay of Bengal). However, the fate of the AAMW, like that of the ICW, is poorly known.

Some workers also distinguish a Bay of Bengal Water (BBW); it is about 100m thick, at the sea surface, and is marked by a low salinity (<33 throughout the year), due largely to dilution by monsoonal river-

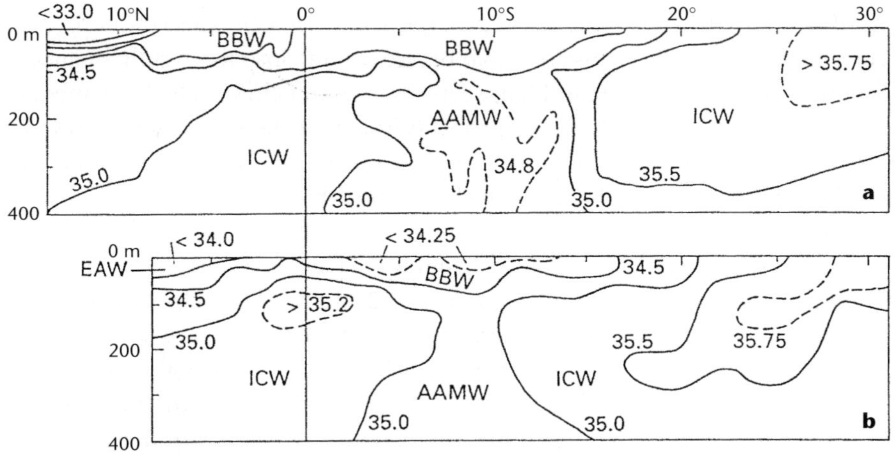

Figure 3.13
Meridional salinity section showing the relationship between Bay of Bengal Water (BBW), Indian Central Water (ICW) and Australasian Mediterranean Water (AAMW) – (a) for the BBW component originating in the inner Bay, along 92°E; (b) for the BBW component found along the west coast of India, along 75°E (note the East Arabian Sea Water (EAW). (From Wyrtki, 1971)

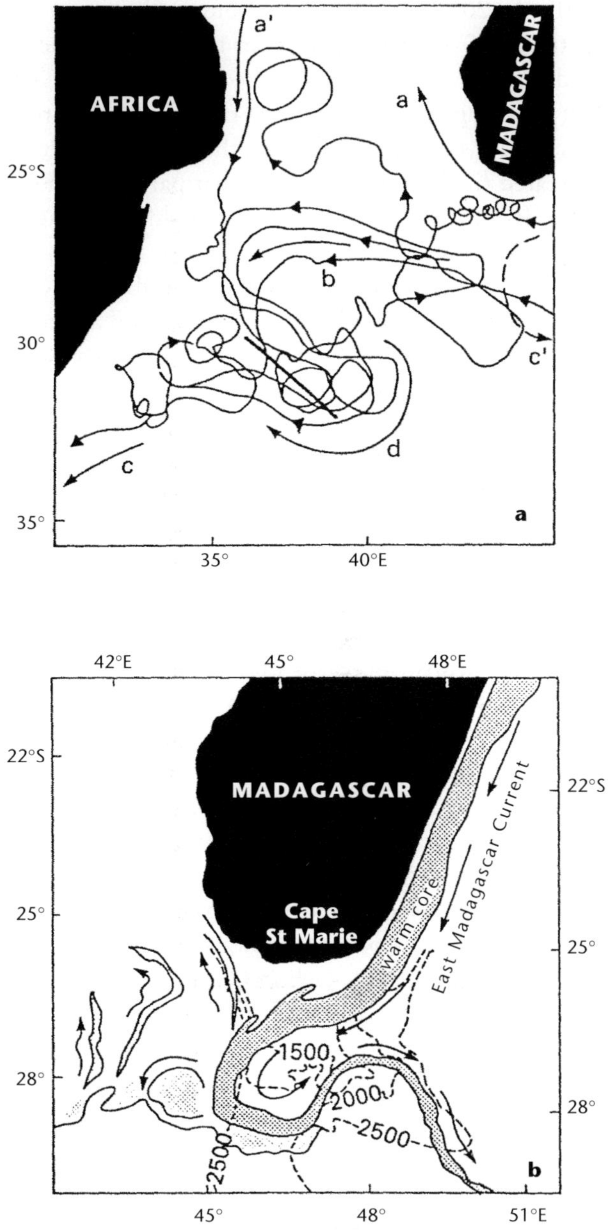

Figure 3.14
Paths of the East Madagascar Current extension: (a) from tracks of drifters remotely sensed by satellite; (b) from a satellite image of sea-surface temperature (13.8.84). The warm core of the EMC is dark-shaded and the intermediate water temperatures are light-shaded. The various letters in (a) represent the various parts of the EMC extension described in the text; bathymetric contours in metres, dashed lines. (From Lutjeharms *et al.*, 1981, Gründlingh, 1987, and Lutjeharms, 1988)

water input. Nevertheless, BBW's influence extends well into the tropics and, in October-December, it reaches the west coast of India (where it is sometimes identified as the East Arabian Sea Current). Oceanic salinities (≈ 35) reoccur in April-June but fall again as the SW monsoon rains start.

The heat flux in the Bay of Bengal is small and positive from the air to the sea; no satisfactory heat sink has been proposed to equilibrate the heat balance, except dispersion into the Indian Ocean proper.

The anticyclonic circulation in the subequatorial zone: This circulation is simpler than that in the northern Indian Ocean. The three components of the subequatorial circulation are the South Equatorial Current, flowing from east to west, the Agulhas/Mozambique/Madagascar Current system, flowing southwestwards along the eastern coast of Africa, and the South Indian Ocean Current, flowing eastwards just north of the Subtropical Convergence.

The South Equatorial Current originates south of Java and is fed by Pacific water flowing through the Indonesian straits, the Timor Sea and the southeastern Indian Ocean. This current, after reaching Madagascar, flows partly south along the east coast of Madagascar, as explained earlier.

The East Madagascar Current (EMC) is a narrow but well defined western boundary current between the surface and about 2000m depth (Schott *et al.*, 1989) (below 3100m depth, there is a northward-flowing countercurrent). The boundary current is derived from the South Equatorial Current which splits on arrival at the northeast coast of Madagascar, about 40% turning southwards as the EMC. On arriving at the southern end of the island, this Current undergoes retroflection through an anticyclonic loop, eastwards back towards the southern central Indian Ocean (c' in Figure 3.14a); in this sense, it resembles the Agulhas Current retroflection, described below (Lutjeharms *et al.*, 1981; Figure 3.17). Some water entrained by the EMC passes round the southern end of Madagascar (a in Figure 3.14a) and about half-way up the west coast till it is entrained by the southward-flowing Mozambique Current on the western side of the Mozambique Channel. Some associated water also flows westwards from the southern tip of the island (b in Figure 3.14a) and, together with the first part, gets into a cyclonic (clockwise, in the southern hemisphere) loop (d in Figure 3.14a) before feeding into the Agulhas Current (c in Figure 3.14a) (see also below). The Mozambique Ridge is probably the main factor in the creation of this loop/gyre since the Ridge rises to a depth of about 1500m which is enough to influence the current (Figure 3.15).

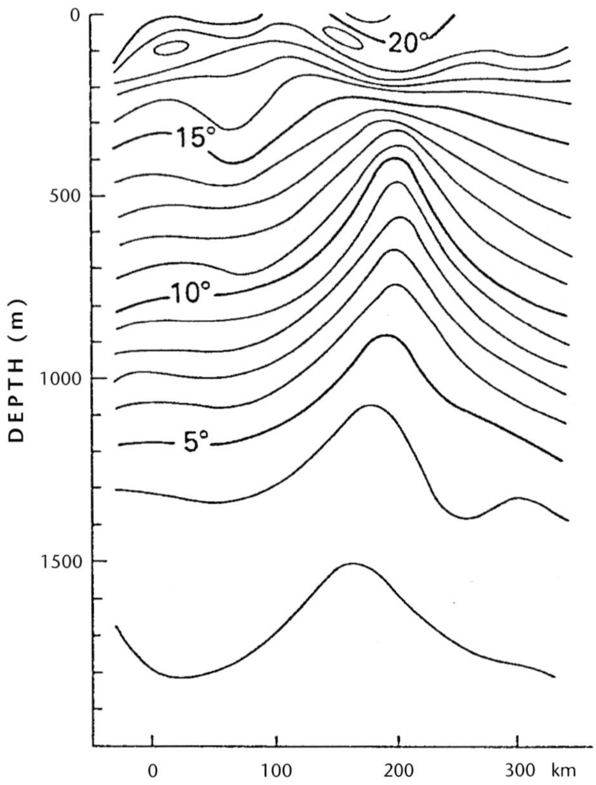

Figure 3.15
A temperature section through a cyclonic eddy near the Mozambique Ridge; the form of the 3°C isotherm indicates that the eddy reaches deeper than 2000m. (From Gründlingh, 1985)

The Mozambique Current starts near the East African coast at 10°-12°S (Menaché, 1955, 1958; Donguy and Piton, 1991) mainly as a contribution from that part of the South Equatorial Current that passed north of Madagascar (the rest turning northwards into the Zanzibar Current (or East African Coastal Current) then the Somali Current during the SW monsoon (Luyten *et al.*, 1980; Swallow *et al.*, 1991). At 20°-22°S, the Mozambique Current picks up the EMC contribution from the western side of Madagascar. At about 30°S, the Mozambique Current picks up the third part of the EMC; it is from these various contributions that the Agulhas Current is created (it should be recalled that part of the SEC came from the Indonesian Throughflow, briefly described above).

The Agulhas Current was first described in comparative detail by Rennell in the 1770s and again by the oceanographers aboard the RRS *Discovery* (UK) in the 1930s. Its oceanography has recently been reviewed by Lutjeharms (1994). It is a very narrow high-speed flow close to the southeast coast of Africa, particularly between Durban and Port Elizabeth. It is one of the strongest currents in the world ocean and shows comparatively little seasonal variation. The amount of water transported (expressed in millions of cubic metres per second, sometimes called Sverdrups) increases rapidly southwards. On reaching the Agulhas Bank (≈ 35°S), the transport is well over 100 × $10^6 m^3/s$. This fast current creates intense upwelling cells of much colder water on its inshore border, but only at distinct locations where the shelf topography induces such upwelling (Lutjeharms *et al.*, 1989). Farther south, towards 20°E, the Agulhas Current, owing to an imbalance in its planetary vorticity, retroflects eastwards as the Agulhas Return Current/South Indian Ocean Current flowing eastwards (Lutjeharms and Ansorge, 1997). The tight retroflection loop is unstable and sheds Agulhas rings by the occlusion of the loop (Figures 3.16 and 3.17). In effect, the Subtropical Convergence appears to pinch off rings as colder Antarctic water penetrates northwards towards the African continent (Figure 3.18a, b). The rings carry large amounts of

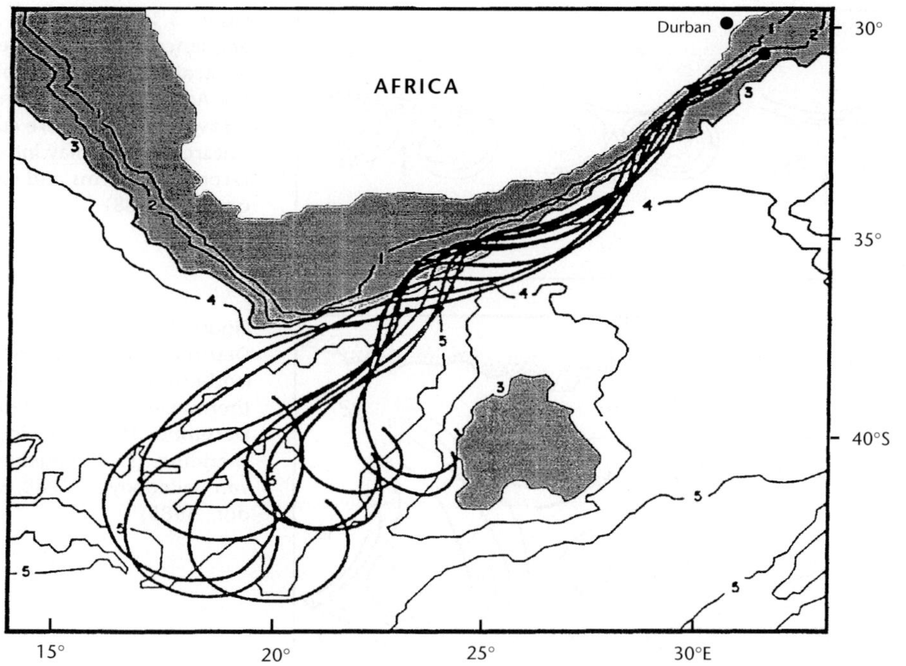

Figure 3.16
Agulhas Current trajectories modelled as a free inertial jet starting off Durban with a range of velocity profiles; the depth contours are labelled in units of × 10³m and areas shallower than 3000m are shaded. (After Lutjeharms and van Ballegooyen, 1984)

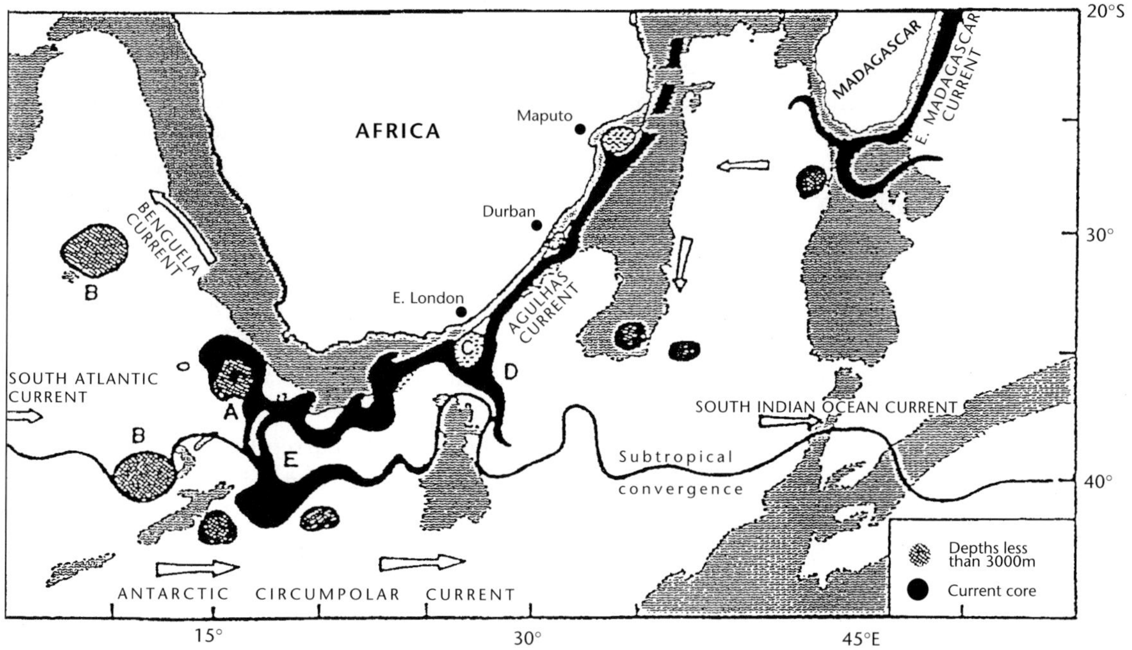

Figure 3.17
The main features of the surface circulation in the southwestern Indian Ocean and the southeastern Atlantic Ocean, showing the Agulhas retroflection (E), a new ring (A) and two older rings (B) entering the South Atlantic Ocean. The new ring (A) is encircled by an Agulhas Current filament. (C) is a coastal cyclonic eddy off East London which produces an upstream retroflexion (D); other trapped eddies off Durban and Maputo are also shown. The large-scale current circulation of the region is shown by open arrows. Depths of 3000m or less are shaded. (After van Ballegooyen *et al.*, 1991)

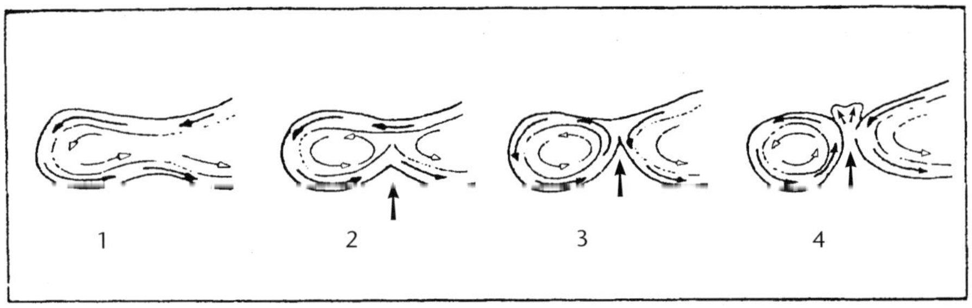

Figure 3.18a
Sequence of four sketches illustrating ring formation from the Agulhas retroflection. Open arrows indicate where colder Antarctic water may intrude. (From Lutjeharms and van Ballegooyen, 1988)

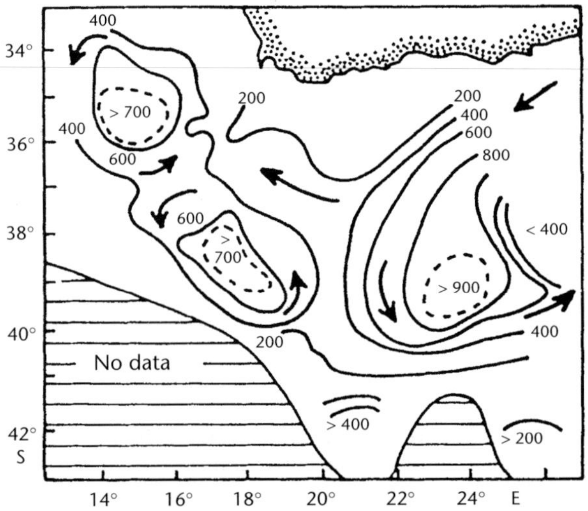

Figure 3.18b
Depth (in metres) of the 10°C isotherm (representing the thermocline), in November-December 1983, showing ring-shedding. (From Lutjeharms and van Ballegooyen, 1988, and Gordon, 1985)

heat, salt and organisms from the Indian Ocean and may have a life of several months or even years, since they are still detectable on the western side of the South Atlantic.

The retroflection, which has the form of a large loop stretching for nearly 300km parallel to the South African coast, seems to be a permanent feature of the circulation in the Agulhas area. The retroflected water is carried off into the South Indian Ocean Current, although some of the water may recycle back to the Agulhas Current itself via the western part of the subtropical gyre.

Agulhas Current water leaks into the Atlantic predominantly as rings, but shallow filaments of this water also make a contribution of about 10% to the salt exchange (Lutjeharms and Cooper, 1996).

Needless to say, the warm Agulhas Current is easily detectable by satellite radiometers (heat) and altimeters (sea surface) and its variability, although not great, is now relatively well known, so that interest in this variability is centred mainly on the ring generation and retroflection (the failed ring generation, so to say). The Current's role in the global heat and salt balance – that is, as part of the so-called global ocean 'conveyor belt' – is significant.

There is no evidence for the occurrence of an eastern boundary current in the southern Indian Ocean. However, along the western Australian coast, close to the continent, a warm-water current originating in the northeastern part of the Indian Ocean is found flowing south and is called the Leeuwin Current (Thompson, 1984; Cresswell, 1991). The mechanics of its formation and its dynamics are described by Tomczak and Godfrey (1994).

In the eastern boundary currents in the South Atlantic (i.e., Benguela Current) and the South Pacific (i.e., Peru Current), the predominantly equatorward winds drive the currents equatorwards, a compensatory undercurrent flowing polewards. In the southeastern Indian Ocean, the Leeuwin Current does the opposite, running against the predominantly equatorward winds. This is because, unlike the longshore winds, the longshore pressure gradient along the western Australian coast is quite different from, and larger than, those on the southeastern sides of the other two oceans: the geopotential anomaly (steric height) is $0.5 m^2/s^2$, compared to values of $\leq 0.1 m^2/s^2$. In the open ocean, this gradient drives an eastward geostrophic flow, but this becomes impossible close to the coast and the water accelerates down the gradient, the

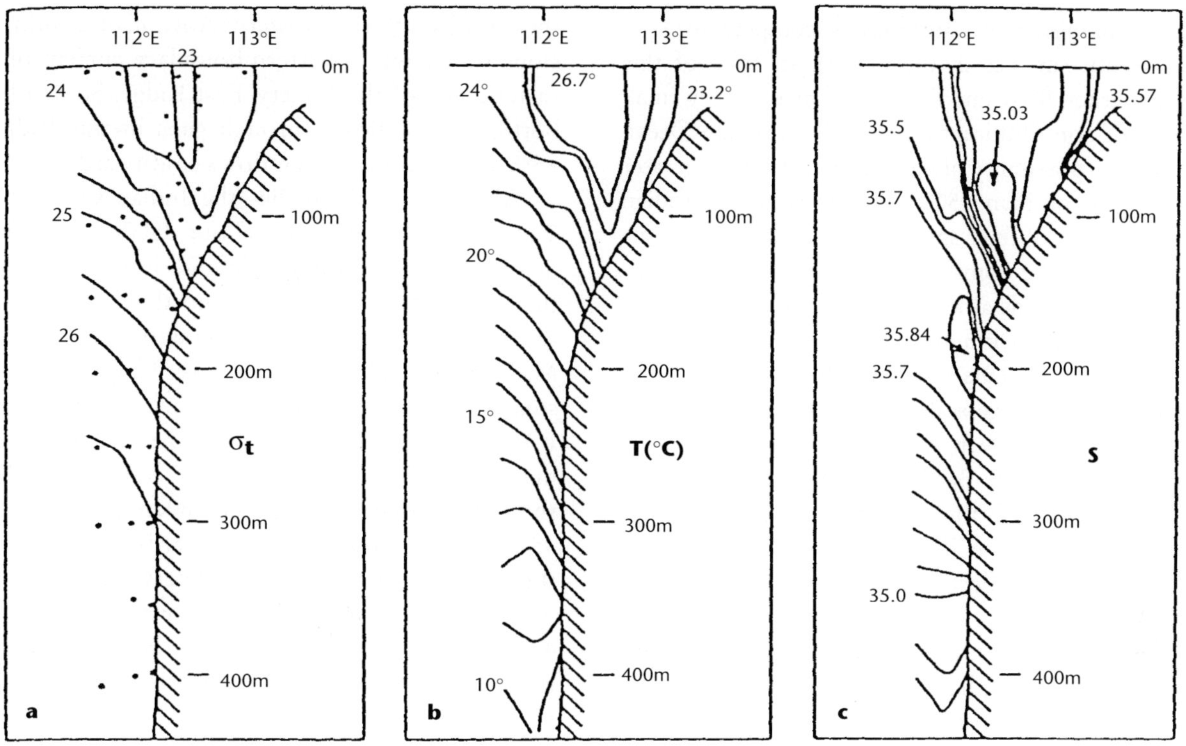

Figure 3.19
Hydrographic properties of the Leeuwin Current near 25°S – (a) density (σ_t); (b) temperature in °C; (c) salinity. Note the deep mixed layer, due to convective cooling between 112°E and 113°E. (From Thompson, 1984)

resulting poleward flow overriding the wind-driven equatorward flow. The difference between the geopotential anomalies is generally considered to be an effect of the Indonesian Throughflow. The large anomaly in the western Pacific (a product of the El Niño-Southern Oscillation phenomenon) can be 'extended' into the eastern Indian Ocean. The Atlantic Ocean and the Panama Canal are not so obliging.

The Leeuwin Current takes the form of a wedge some 200m in depth and about as wide at the surface (Figure 3.19). It is quite variable in its intensity (mean 0.1-0.2m/s) and southward extension, interseasonally, but may reach 1.5m/s on rounding Cape Leeuwin before entering the Great Australian Bight. As Figure 3.19 shows, the Current forms strong fronts on its offshore and inshore sides; eddies are formed at these frontal edges.

Intermediate layer (500-1500m)

The marked salinity minimum (34.5) south of 10°S indicates Antarctic Intermediate Water (AAIW) which is at the surface at the Antarctic Convergence (or Polar Front), deepening to a depth of about 1000m northwards. However, it is prevented from entering the northern hemisphere by the equatorial current system north of 10°S. At its source, in the Atlantic, it has a temperature of 2.0°-2.5°C and a salinity of about 33.8, but when it enters the subtropical (subequatorial) gyre its temperature increases to 3°-4°C and its salinity is 34.3. The final fate of the AAIW is still unknown; some of it may leak from the southern Indian Ocean in Agulhas rings at the Agulhas Current retroflection and/or possibly into the Pacific across the Great Australian Bight, though not unrestrictedly so.

Only Red Sea and Persian Gulf Waters are able, because of their considerable density, to supply water to intermediate depth ranges; they can be detected all over the northern Indian Ocean (they are often called High Salinity Maximum Waters). Red Sea Water enters the Indian Ocean with $T \approx 22°C$ and $S \approx 39$, with a resultant density (σ_t) of about 27.25. Such a density is found at 600-800m depth in the Arabian Sea, deepening to about 1000m depth at 30°S.

Persian Gulf Water has a similar temperature and salinity but a slightly lower density ($\sigma_t \geq 26.7$); such a density is found at 250-300m depth in the Arabian Sea and at 500-600m depth south of Madagascar (it should be borne in mind that water masses 'compete' for specific density surfaces in the ocean). In spite of this strong marker, RSW and PGW only make a very small volumetric contribution to the northern Indian Ocean and have no possibility of reaching depths there greater than 1000m or than 1500m in the southern Indian Ocean. Details of their spreading are only known satisfactorily for the Somali Current region where they are both identifiable as lenses and intrusions 200-250m thick and about 100km in diameter. They are mixed in the boundary currents to create the smooth salinity maximum in the large-scale salinity distribution.

Indian Ocean Deep Water, as it spreads into the Arabian Sea and the Bay of Bengal, becomes modified by mixing with the surface (thermocline) water above it, with upwelling Antarctic Bottom Water below it, and by injections of Red Sea and Persian Gulf Waters.

The Indonesian Intermediate Water, which comes from the Timor Sea, has a temperature range of 3.5-5.5°C and salinities from 34.6 to 34.7.

Deep layer (below 1500m.)

Below 1500m, the water mass is mainly identified as Circumpolar Deep Water (CDW), with a temperature range of 0.1 to 2.0°C and salinities from 34.62 to 34.73. Since these waters practically fill all the world oceans, it was also called the 'common water' by Montgomery (1958). These waters are a mixture of deep and bottom waters mainly contributed by the Weddell and Ross Sea areas. According to Broecker and Takahashi (1985), the common water is composed of 45% Weddell Sea water, 30% Pacific and Indian Ocean Intermediate Water and 25% water from the North Atlantic Deep Water (NADW). It fills the Indian Ocean below a depth of 3800m. It enters on the western side into the Madagascar Basin through gaps in the Southwestern Indian Ocean Ridge near 30°S, 56°-59°E and then moves towards the continental slope off eastern Madagascar, forming a deep western boundary current flowing northwards to the Arabian Basin (Tchernia, 1957b; Fieux *et al.*, 1986; Fieux and Swallow, 1988) before returning southwards at a shallower depth. In this progress it is subject to deep upwelling to mix with overlying deep water. On the eastern side of the Ocean, entry is via the Australian-Antarctic Discordance (a gap in the Southeast Indian Ocean Ridge south of Australia at 120°-125°E) into the South Australian Basin, filling the Great Australian Bight, then moving west then north off the western Australian continental slope to become a western boundary current on the eastern side of the Ninety-East Ridge. Some of this bottom water flows through gaps in this Ridge at ≈5°S and ≈10°S and then flows southwards.

Above the AABW lies the Indian Ocean Deep Water (IODW) between about 3800m and 1500m depth, but there is not a definite interface between the two; although both move northwards, mainly in the western boundary currents (along the east side of Madagascar and of Ninety-East Ridge), there are small but definite differences in their parameters. The IODW in the southern hemisphere has a salinity maximum (from $S = 34.75$ in the east to $S > 34.8$ in the west) between 2000m and 3800m depth north of 45°S; this maximum shoals towards 500m depth in the south. The marker parameters ($T = 2°C$, $S = 34.75$, $DO = 4.7ml/l$) of the salinity maximum near 40°E (roughly, the Mozambique Basin) are exactly those of North Atlantic Deep Water (NADW), suggesting the origin of the IODW. The NADW carried by the upper Circumpolar Current becomes IODW. A third western boundary current occurs along the eastern side of the Mid-Indian Ocean Ridge because depths south of this ridge are sufficient to allow direct advection of IODW from the Southern Ocean.

Surrounding the Antarctic Continent, the Southern Ocean forms a large mass of cold water whose hydrographic structure is very interesting. It is the only ocean where currents flow uninterruptedly round and round the globe south of 40-45°S. Close to the continent there is an East Wind Drift which is a westward-flowing current all round the continent. North of this there is throughout the year a strong wind system called the West Wind Drift which drives the Antarctic waters eastwards. This is known as the Antarctic Circumpolar Current (ACC). Its northern boundary is marked by the Subtropical Convergence.

In the Antarctic zone, the Antarctic Surface Water (AASW) is greatly influenced by summer melting of ice, resulting in lowering of salinities, and in winter, when ice is formed, there would be a slight increase in salinity. This water, with a thickness of 100-250m, has a salinity of less than 34.5 and a temperature close to the freezing point of sea water, about -1.9°C.

Below the AASW and the NADW/IODW, the Antarctic Bottom Water exists right down to the bottom at about 5000m. AABW is formed in the Weddell and Ross Seas off the Antarctic continent. This water mass has the highest density and therefore sinks and flows down the shelf slope into the deep

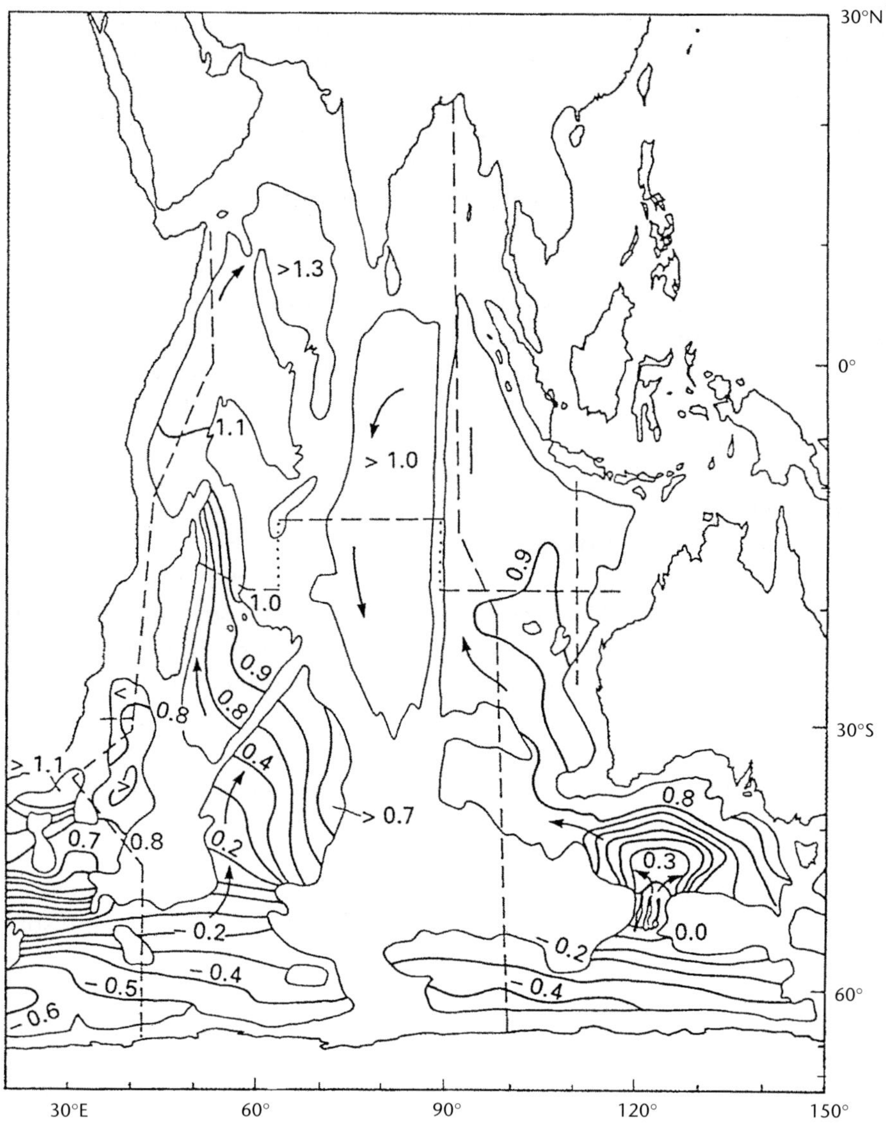

Figure 3.20
Potential temperature at 4000m depth (labelled isotherms); arrows show the likely path taken by Antarctic Bottom Water; the dashed lines indicate key hydrographic sections. (After Wyrtki, 1971, and Rodman and Gordon, 1982)

basins of the South Atlantic Ocean and from there to the Indian and Pacific Oceans (Warren, 1981). The Antarctic Intermediate Water (AAIW) is formed in the sub-Antarctic zone and has a thickness of about 500m under the Subtropical Convergence and flows below the sub-Antarctic upper water. It has a temperature range of 2°-3°C and a salinity of 34.2 and spreads northwards into the Indian Ocean at depths ranging from 800 to 1000m.

The deep-water circulation in the Indian Ocean is, according to Warren (1978), more complicated than that in the other two major oceans because of the multiplicity of basins separated by many ridges (Figure 3.20). The fact that water enters the western Indian Ocean basins is demonstrated by potential temperatures at 4000m depth through the Crozet Basin; this flow has been observed to pass over the Southwest Indian Ocean Ridge via the Atlantis II and Melville Fracture Zones and then into the Madagascar Basin through Amirante passage near 9°S, 52°E (Fieux and Swallow, 1988). There is some evidence, based on salinity and potential temperature values, that all this flow, from the Crozet Basin to the Somali Basin, is part of a northward-flowing western boundary current (Warren *et al.*, 1966). On the eastern side, the bottom water may be entering through the South Australian Basin around 120°E and flowing northwards till it reaches the eastern basins, namely, the West Australian, Wharton and Central Indian Ocean Basins, as indicated by the evolution of potential temperatures (Tchernia, 1980). In a recent paper, You and Tomczak (1993) have discussed the thermocline circulation between the depths 150-800m (the zone of the permanent thermocline) and ventilation in the Indian Ocean by means of water-mass analysis using cluster analysis and optimum multiparameter analysis. The mixing model presented focuses on ventilation, covering five isopycnal surfaces and two meridional sections, along 60° and 90°E and a zonal section along 10°S.

Their conclusions are that the input of the Red Sea Water is insufficient to renew the thermocline waters of the northern Indian Ocean which therefore has to be ventilated by advection from the south. More importantly, the jet-like inflow of Australasian Mediterranean Water produces one of the strongest frontal systems in the world oceans' thermocline, suppressing meridional motion across the 10-15°S latitude band southeast of 50°E. As a result, advection and transfer of thermocline water northwards and across the equator is possible only in the region of the western boundary current. The Indian Central Water is shown to flow along this path in the depth range of 300-400m and, as it crosses the equator, it ages rapidly producing very little mean transport. As a result, the Indian Central Water in the northern hemisphere is extremely low in oxygen and high in nutrients. These results fully corroborate the observations made by Wyrtki (1973).

Metzl *et al.* (1990) have attempted to resolve the intermediate and deep-water advective flow in the Indian Ocean through the analysis of temperature, salinity, oxygen and phosphate data, stressing the interplay of biogeochemical and geophysical tracers.

In the southwestern Indian Ocean, a salinity maximum is found at about 2500m depth and this is attributed to northward flow of North Atlantic Deep Water brought to the Indian Ocean through the Antarctic Circumpolar Current. As these waters flow north, their salinity decreases up to about 15°S, and, farther to the north, in the Arabian Sea, the salinity increases; this is attributed to the influence of Red Sea Water. Perhaps the most important aspect is the relatively low concentration of oxygen in the Arabian Basin (3.6ml/l at 4000m depth, in contrast to 4.0ml/l in the Somali Basin), an indication of the slow renewal of the deep waters (Fieux *et al.*, 1986; Warren and Johnson, 1992).

The deep Indian Ocean is open to the Antarctic; its western basins are separated from the Central Basin and the West Australian Basin by the Mid-Indian Ocean Ridge. The temperatures at 4000m depth here are 0.2-0.3°C lower than in the Somali Basin, indicating that the deep waters in these basins originate in the Antarctic Circumpolar waters through the eastern passage as a deep western boundary current along the eastern flank of the Ninety-East Ridge. In the absence of a sufficient number of observations, the foregoing observations are tentative as far as the deep-water circulation is concerned.

The Indian Ocean bottom water and the deep water are derived from the Atlantic and Southern Oceans, but the deep circulation in the Indian Ocean is weaker than that in the Atlantic Ocean.

The age of the deep waters (below 1500m depth) was worked out by Stüiver *et al.* (1983) using GEOSECS data, and the results indicated that replacement times for Pacific, Indian and Atlantic Ocean waters are approximately 510, 250 and 275 years, respectively. The deep waters of the entire world ocean are replaced on average every 500 years.

Conclusion

The exclusion of the Indian Ocean from the north polar region has a great influence on the meteorology and hydrography of the northern Indian Ocean. There is an asymmetrical development of its structure and circulation, compared to the Atlantic and Pacific Oceans. This is borne out by the huge oxygen-minimum layer which pervades the whole of the Indian Ocean, particularly the Arabian Sea and the Bay of Bengal and by the distribution of nutrients (the following chapter highlights the nutrient distribution in the Indian Ocean).

The arid climatic conditions of the northwestern Indian Ocean have also contributed to the formation of the high-salinity waters of the Red Sea and the Persian Gulf which spread in the Arabian Sea to affect the subsurface circulation by preventing vertical mixing of southern hemisphere waters into the northern Indian Ocean.

Madagascar, with its physical relation to the East African coast, also has a specific effect on the western boundary current anatomy which is peculiar to this Ocean. The region between South Africa and the Subtropical Convergence allows the Agulhas Current to 'leak' water (mass, heat and salt) into the Atlantic Ocean. The potentially analogous currents, at the southern end of South America (the Brazil Current) and the southeast corner of Australia (the East Australian Current) are blocked by the proximity of the continent to the Circumpolar Current (i.e., about 55°-60°S), in the first case, and by Tasmania (and the Subtropical Convergence), around 45°S, in the second case. Thus, the Indian Ocean western boundary current (Agulhas) uniquely helps to complete the global thermohaline circulation (the so-called 'conveyor belt').

The Oceans ~ coast to coast

South coast of Java

Complementing the contributions of T.S.S. Rao and Ray Griffiths, this special section contains information about IOC and some of its major programmes, both global and those in the Indian Ocean, as well as brief discussions provided by the other sponsors of the book. These partners are: Global Environmental and Oceanic Sciences Ltd. (GEOS), US Office of Naval Research (ONR) and France's ORSTOM (Institut Français de Recherche Scientifique pour le Développement en Coopération). The British Oceanographic Data Centre (BODC) assisted in the inclusion of the Indian Ocean bathymetric chart.

AN IOC ROLE:
SCIENCE FOR OCEAN MANAGEMENT

Three decades have transpired since the completion of the historic International Indian Ocean Expedition, a multi-national effort the coordination of which became a cornerstone task of the Intergovernmental Oceanographic Commission (IOC) of UNESCO during its early developmental stages. Basically performing a coordinating function, at least at the outset, the IOC, on behalf of its Member States and other partners, continues to show considerable interest in the Indian Ocean – as an essential component in the greater World Ocean. Scientific discoveries in one ocean often have inescapable impacts on the advancement of knowledge of the other oceans, which are all parts of a single interconnected planetary system.

In 1994, the United Nations General Assembly formally adopted a proposal, originated by IOC and transmitted via the UNESCO General Conference, to declare 1998 as the International Year of the Ocean (IYO). This decision was a catalyst to foster a worldwide campaign to raise public awareness as to the importance of the oceans in general, and specifically to enhance the establishment and execution of programmes to better understand and protect the marine environment.

IOC Member States around the world proposed and launched a number of activities focusing the attention of governments, decision-makers and the general public on the need for healthy oceans, and for sustainable development of the marine environment. Information on IYO activities in which IOC is directly involved can be found on the Commission's web site (*http://www.unesco.org/ioc)*.

The publication of *Understanding the Indian Ocean – Perspectives on Oceanography* by UNESCO is offered as a contribution to the IYO. Here the reader can find information about this ocean's biological, physico-chemical and geological characteristics. Included are future research directions and the possible role of IOC (which of late increasingly involves related socio-economic considerations) in the continuing study of this ocean.

Albeit basically scientific in its approach, the book also takes into account relevant cultural and other dimensions. Oceanographic knowledge should be optimally utilized, and this is an underlying premise for the IYO. Yet it is only through the concerted efforts of the various research and user communities that effective integrated ocean and coastal management can be achieved for the common benefit of mankind.

IOC's wide-ranging global programmes

The Commission operates both on global and regional scales, and in various ocean-related domains. A major IOC initiative is the Global Ocean Observing System (GOOS), a comprehensive system to provide information needed for oceanic and atmospheric forecasting, for ocean managment by coastal nations, for the needs of global environmental change research, with education and training programmes and technical assistance essential to ensure that all countries can participate.

Still in its planning phase, through IOC's efforts GOOS has secured the additional international co-sponsorship of the World Meteorological Organization, the United Nations Environment Programme and the International Council of Scientific Unions. National and regional GOOS projects also have been launched. Already, Indian Ocean interests have been addressed through capacity-building workshops, e.g. in Goa and Mombasa, with plans for growth in this ocean over the next few years.

Several disciplines come together also in areas of research and services supportive to integrated coastal management. The assessment of geological and geomorphological characteristics is very important in helping to determine what controls coastal changes. Thus marine geoscience can provide strategic information on the nature and evolution of the coastal zone. Such work is a major factor in IOC's programmes on Ocean Science in Relation to Non-Living Resources.

A wide variety of Member State activities and platforms provide input for GOOS. Example: the US-French TOPEX-POSEIDON programme (design below), where on-site observations are obtained by: (1) instrument-carrying buoys, localized by satellite, (2) current meters, (3) hydrological probes and (4) subsurface floats.
© Patrick Buat-Ménard, French Ministry of Research and Technology

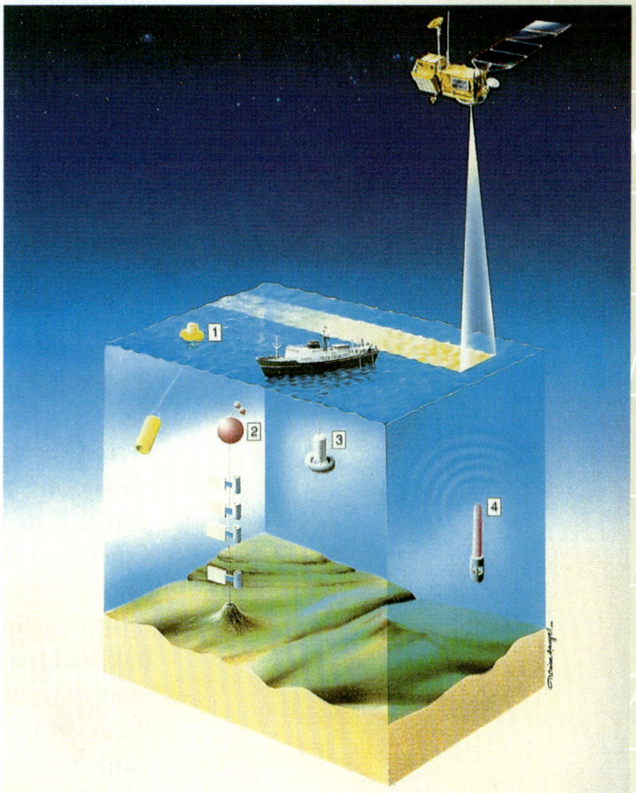

The IOC provides responses to sensitive pollution issues through the Committee for the Global Investigation of Pollution in the Marine Environment. This committee is also sponsored by the United Nations Environment Programme and the International Maritime Organization. The programme provides scientific information concerning the detection and analysis of contaminants, and documents concentration levels of pollutants from countries throughout the world. Thus more rational management strategies can be formulated.

Climate and remote sensing

In climate research, the IOC acts in partnership with the International Council of Scientific Unions and the World Meteorological Organization within the framework of the World Climate Research Programme. The most important components of this programme are the Tropical Oceans and Global Atmosphere study (completed), the World Ocean

Circulation Experiment and the more recent effort on Climate Variability and Predictability. Today, these activities collectively provide additional ocean information which is an order of magnitude greater than that produced a decade ago. The programmes in turn generate modelling activities at regional and global scales, all of which confirm the ocean's role in influencing climate. The ultimate goal is improved climate prediction.

The IOC and, in particular, its Global Ocean Observing System are affiliated with the Committee on Earth Observing Satellites. This committee (i) optimizes the benefits of satellites through mission planning and development of data, formats and services, (ii) aids its members and the international user community in general by serving as a focal point on remote-sensing activities and (iii) exchanges policy and technical information to encourage complementarity and compatibility among space-borne Earth observing systems.

Data and information

In addition to GEBCO (see following pages), IOC and its partners, as a service to the global scientific community, are active in projects that produce data and information products for researchers of the Indian Ocean and other marine regions. These projects are co-ordinated through IOC's International Oceanographic Data and Information Exchange (IODE) programme. Enquiries on IODE can be directed to the IOC address listed on the following page.

An exemplary result of IODE's activities is the *World Ocean Atlas 1994* (currently being updated), which provides considerable data on various global ocean parameters, such as: ocean temperature, salinity, oxygen saturation, apparent oxygen utilization, and the concentrations of dissolved oxygen, phosphate, nitrate and silicate. For more information on the Atlas, contact: US NODC/NOAA, E/OC5, 1315 East West Highway, Room 4362, Silver Spring, Maryland 20910-3282, USA; fax: (+1-301) 713 3303; e-mail: services@nodc.noaa.gov. The NODC Home Page (http://www.nodc.noaa.gov) also contains pertinent information.

Specific IOC Indian Ocean activities

In this broad geographical area, the IOC operates through its Regional Committee for the Cooperative Investigation in the North and Central Western Indian Ocean (IOCINCWIO) and its Regional Committee for the Central Western Indian Ocean (IOCINDIO). During the last decade, the regional programmes have concentrated on enhancing marine science capacity. In ten years the countries involved, as a whole, have tripled their human capacity and significantly upgraded their research infrastructure in this field. Not only is the level of participation higher, but also the type of activities has matured from mainly educational to more operational.

Major research and monitoring programmes concern natural and human-induced processes that contribute to coastal erosion and the impact of the oceans on climate change. Ocean sciences, as well, should focus on living and non-living resources as keys to sustainable development. Efforts are underway to develop and help implement national and regional plans for Integrated Coastal Area Management. Regional networks have been set up for marine pollution monitoring and for data management.

With regard to living resources, the Global Coral Reef Monitoring Network (being set up by IOC and its partners) will provide information on the status and trends in reefs, many of which are found in the Indian Ocean. This will lead to improved management and sustainable conservation of these highly diverse and productive communities. As well, the growing interest in shellfish and fish farming in the Indian Ocean area makes monitoring of harmful algae species essential because of possible contamination of seafood products destined for local consumption and export. To help establish monitoring systems, the IOC's Harmful Algal Bloom programme trains scientists of the region in the identification of phytoplankton species causative of harmful algal events, and a guide is being prepared for the identification of harmful algae in the Western Indian Ocean. Valuable to scientists worldwide in this regard, UNESCO recently published *Phytoplankton Pigments in Oceanography*, a monograph containing guidelines to modern methods, produced as a joint effort with the Scientific Committee on Oceanic Research (of ICSU) and supported by IOC.

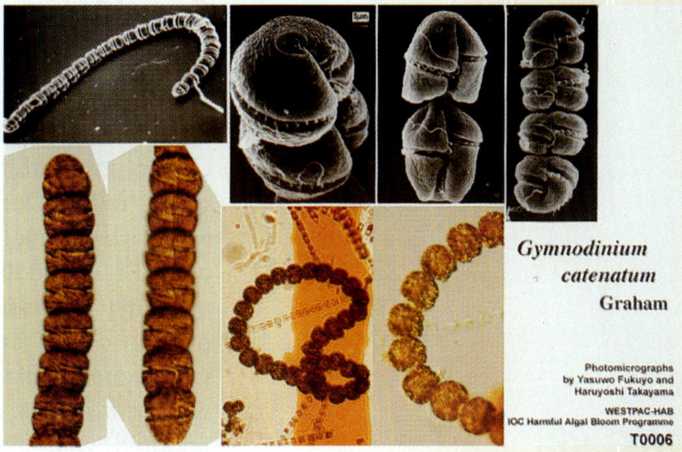

Gymnodinium catenatum*, a toxic dinoflagellate recently found on India's west coast. Source: IOC's Harmful Algae News, No. 15, 1996. © WESTPAC HAB IOC Harmful Algal Bloom Programme.*

In various ways, all these activities contribute to the Global Ocean Observing System and strengthen capacities for management of the coastal zone as well, from the perspectives of both natural and social sciences. This is consistent with IOC's support of multi-disciplinary approaches to the problems of the oceans and, ultimately, to the major issues facing society.

For more information, contact:
Executive Secretary, IOC,
or IOC Coordinator for the IYO,
UNESCO, 1 rue Miollis,
75732 Paris cedex 15, France.
Fax: (+33-1) 45 68 58 12
E-mail: i.oliounine@unesco.org
Web site: http://www.unesco.org/ioc

INDIAN OCEAN BOTTOM:
A DIGITIZED VIEW

From an ongoing project contributing to the General Bathymetric Chart of the Oceans (GEBCO)

Shown on the following page is an extract, covering the Indian Ocean, taken from the *GEBCO Digital Atlas*. This Atlas is maintained by the British Oceanographic Data Centre (BODC) on behalf of the Intergovernmental Oceanographic Commission (IOC) of UNESCO and the International Hydrographic Organisation (IHO). When originally published in 1994 the Atlas was the first seamless, high-quality, digital bathymetric chart of the world's oceans and contained the digitized bathymetric contours, coastlines and trackline control from the printed Fifth Edition of GEBCO. It is now being updated with new bathymetric compilations as and when they become available. For example, the area south of 30° S on the following page is based on a recent compilation (1997) by Dr. Robert L. Fisher of the Scripps Institution of Oceanography. Dr. Fisher is currently working in collaboration with BODC to revise the bathymetry of the whole of the Indian Ocean for inclusion in the next release of this Atlas.

The *GEBCO Digital Atlas* is published by BODC as a CD-ROM together with a sophisticated but easy-to-use software interface for PC use. A coloured brochure containing more details can be obtained from the address below, or viewed on the web site.

BODC (GEBCO),
Proudman Oceanographic Laboratory,
Bidston Observatory,
Birkenhead, Merseyside L43 7RA,
United Kingdom.
Fax: (+44-151) 652 3950
E-mail: bodcmail@pol.ac.uk
Web site: http://www.nbi.ac.uk/bodc/gebco.html

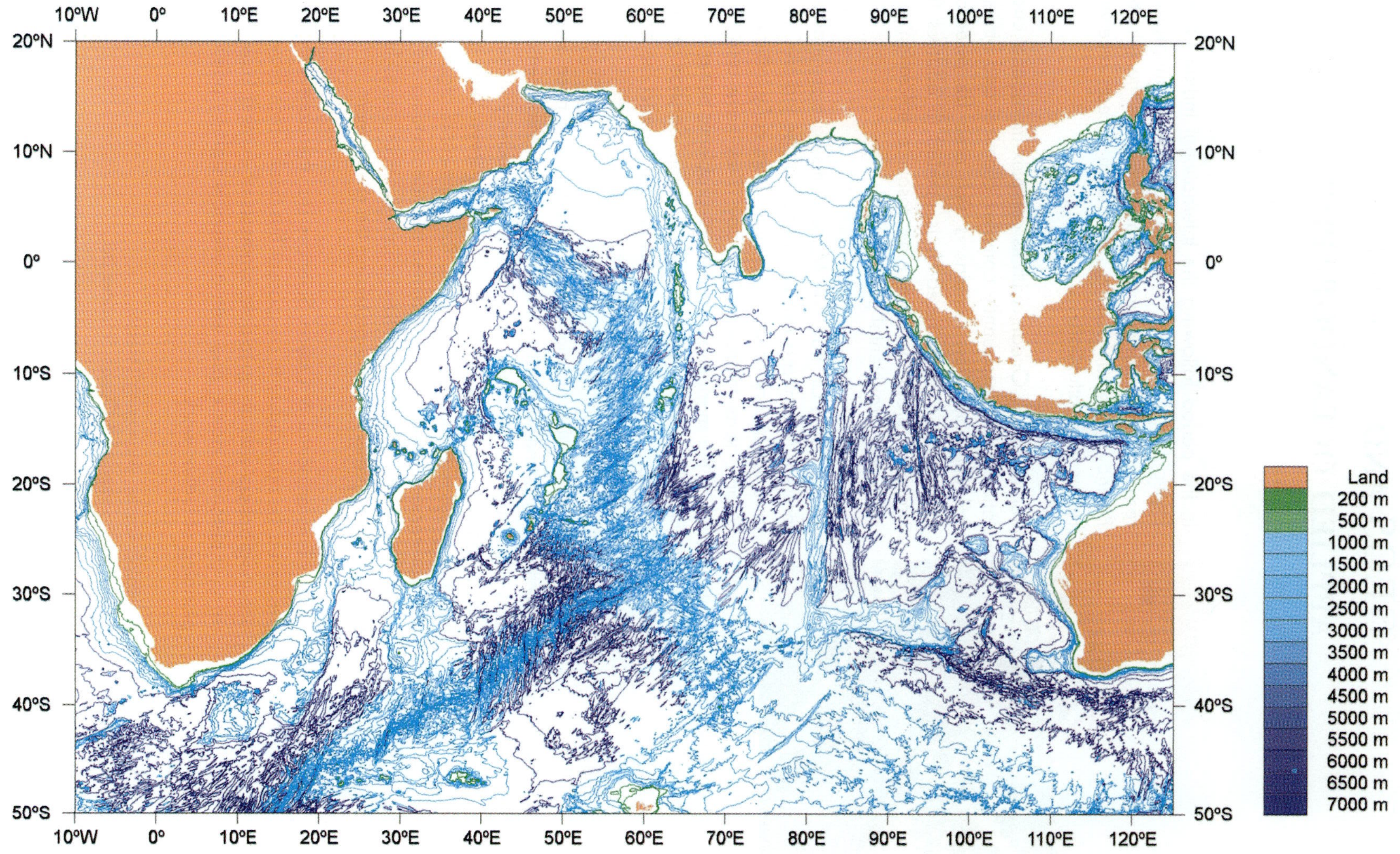
Land
200 m
500 m
1000 m
1500 m
2000 m
2500 m
3000 m
3500 m
4000 m
4500 m
5000 m
5500 m
6000 m
6500 m
7000 m
20°N
10°N
0°
10°S
20°S
30°S
40°S
50°S
10°W
0°
10°E
20°E
30°E
40°E
50°E
60°E
70°E
80°E
90°E
100°E
110°E
120°E

GEOS AND APPLIED OCEANOGRAPHY

Preparation for a deep water measurement programme in the Arabian Sea. © GEOS

In the period following the International Indian Ocean Expedition (1959-65), the emphasis of oceanographic research has been strongly biased towards the Atlantic, Pacific and polar oceans, the Indian Ocean receiving rather scant attention.

Yet the Indian Ocean, with its complex monsoonal circulation and consequently complex chemical and biological characteristics, remains a rich area for research. Furthermore, much research is needed to address the practical problems of marine resource utilization and marine environmental protection.

In this book, Dr. Rao and Mr. Griffiths have summarized present oceanographic knowledge of the Indian Ocean and the history of its exploration, from early Indian and Arabic texts to the present day. They have also highlighted some of the marine environmental problems being experienced by the nations of the Indian Ocean region. Solution of these problems relies on our ability to understand and predict the behaviour of the marine environment.

Global Environmental & Ocean Sciences Ltd. has long been associated with efforts to find solutions to practical marine environmental problems in this region. It is a measure of the level of activity in the region that, at the time of writing, GEOS is conducting seven large applied-research projects in the Indian Ocean, for commercial organizations and governmental bodies; all are aimed at understanding the marine environment better in order to utilize ocean resources successfully and sensibly, while preserving and protecting the environment.

Global Environmental & Ocean Sciences Ltd. therefore very much welcomed the opportunity to co-sponsor this book which provides all oceanographers with an up-to-date synthesis of existing knowledge, set in a historical matrix, and with a valuable starting point for the planning of further work.

Dr. Ralph Rayner, Managing Director,
Global Environmental & Ocean Sciences Ltd.,
Hargreaves Road,
Swindon, Wiltshire SN2 5AZ, UK.
Tel: (44-1793) 725766
Fax: (44-1793) 706604
E-mail: GEOSUK@geos.co.uk

ONR AND THE ARABIAN SEA

The US Office of Naval Research (ONR) recently sponsored (1992-1997) a research initiative in the Arabian Sea (NW Indian Ocean). The region experiences two monsoon seasons yearly, with reversals in wind direction as well as alterations of wind strength and duration associated with each monsoon. The research initiative was undertaken to study the response of the upper ocean to strong monsoonal forcing, and the impact of the physics on the biological and optical properties of the upper water column. An intensive interdisciplinary field effort consisted of multi-sensor moorings left in place for over one year, research cruises with SEASOAR surveys and CTD casts during different seasons, plus modelling efforts to analyse and synthesize the data collected. Fifteen US institutions and 25 principal investigators participated in the ONR Arabian Sea research programme. This programme complemented research efforts sponsored by the US National Science Foundation (NSF) under the auspices of the Joint Global Ocean Flux Study (JGOFS) and the World Ocean Circulation Experiment (WOCE). Participants in those studies included researchers from The Netherlands, Germany, United Kingdom, India and Pakistan. Oman assisted in regard to the logistics of shipping science equipment.

Moorings were deployed at 15°30'N, 61°30'E from October 1994 through October 1995. At the sea surface, they collected data on wind velocity, incoming short-wave and long-wave radiation, barometric pressure, air temperature, sea temperature, relative humidity, and precipitation. Subsurface, the moorings measured current speed and direction, temperature, salinity, fluorescence, dissolved oxygen, photosynthetically available radiation, and light transmission, with a vertical resolution of less than 5 m near the surface. This data set is unprecedented in respect of the suite of variables measured and the duration of the time series.

Fig. 1. Temperature contour and 0.1 degC MLD at the central mooring Arabian Sea 15°30'N, 61°30'E. © R. Weller, T. Dickey and J. Marra

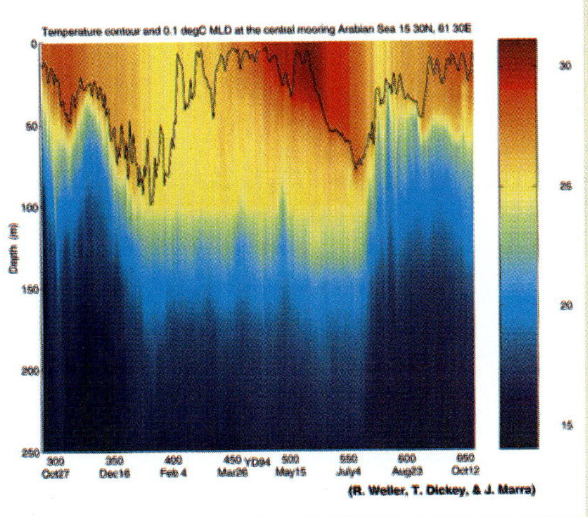

Results from the mooring data (Rudnick et al., 1997) show clear signals of the monsoon cycles in temperature (Fig. 1) and chlorophyll (Fig. 2: dotted lines indicate locations where moored equipment was fouled, and shipboard measurements replaced moored measurements). Winds averaged 5 m/s during the NE monsoon in winter, but were much stronger, sometimes exceeding 15 m/s during the SW monsoon in spring. The mixed layer was deepest during the NE monsoon (Fig. 1; mid-November 1994 to mid-February 1995) due to intense surface cooling from the low air temperatures combined with wind-driven mixing. Although winds were stronger in the SW monsoon, the mixed layer was not as deep because surface waters were warmer at the initiation of the increase in wind mixing (Fig 1; June to September 1995). Elevated chlorophyll concentrations were observed in the mixed layer as mesoscale features (possibly eddies) early in both monsoonal seasons (Fig 2; December 1994 and July 1995). These were followed by blooms with deeper extent and longer duration at the end of each monsoon (Fig 2; February to April 1995 and September to October 1995). The blooms at the end of each monsoon were associated with shoaling of the isotherms and mixed layer (Fig. 1; February to April 1995 and September to October 1995).

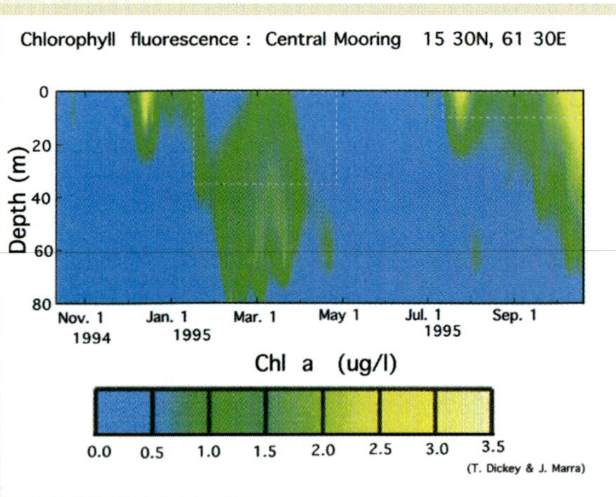

Fig. 2. Chlorophyll fluorescence: central mooring 15°30'N, 61°30'E. © T. Dickey and J. Marra

The Arabian Sea research initiative has provided new insights into the response of ocean physical, biological and optical properties to monsoon forcing. The first stage of data interpretation has shown the strong effects of wind stress and surface temperature on latent heat flux, shear, mixed-layer depth, chlorophyll concentration, and oxygen dynamics. Further treatment of the data will include more detailed modeling efforts and correlation of the mooring data to cruise measurements. The year-long time series from the moorings has allowed an unprecedented look at an ocean system responding to strong periodic wind forcing.

US PARTICIPANTS

B. Arnone, K. Brink, O. Brown, T. Dickey, C. Erickson, R. Evans, C. Flagg, G. Hitchcock, V. Holliday, B. Jones, J. Kindle, M. Luther, J. Luyten, J. Marra, J. McCreary, D. Olson, P. Ortner, D. Phinney, D. Rudnick, S. Smith, B. Weller, A. Wiedemann, M. Wood, C. Yentsch and D. Young

INTERNATIONAL COOPERATION

The Netherlands (1992-1993) RV Tyro
Pakistan (1993-1996) NASEER Programme (Northern Arabian Sea Environmental and Ecological Research)
UK (1994-1995) RRS Discovery
India (1994-1995) RV Sagar Kanya
Germany (1995) RV Meteor and (1997) RV Sonne

REFERENCE

Rudnick, D.L., R.A. Weller, C.C. Erickson, T.D. Dickey, J. Marra and C. Langdon, 1997. *EOS* Vol 78, No. 11, March 1997, pp. 117, 120-121.

FOR FURTHER INFORMATION:

Dr. Elizabeth Turner,
Biological Oceanography Program, Code 322BC,
Office of Naval Research,
800 N. Quincy Street,
Arlington, VA 22217, USA.
Tel: (+1-703) 696 2495
Fax: (+1-703) 696 3390
E-mail: TURNERE@ONR.NAVY.MIL

ORSTOM and Indian Ocean Development

Programmes undertaken in co-operation with the Institut Français de Recherche Scientifique pour le Développement en Coopération (ORSTOM)

A word on ORSTOM

Research undertaken by ORSTOM (France's national institute of cooperative research for development) covers the tropical zone and therefore includes oceanography.

ORSTOM brings together a large number of researchers from all the disciplines concerned with Man and his environment. The main objective is to contribute to the sustainable development of the countries with which the Institute co-operates. This co-operation is mainly carried out through the relevant organisms and researchers of the partner countries, to strengthen their own research capacity and thus promote the emergence of a scientific community that is widely acknowledged in each of these countries. Research is undertaken in partnership by favouring training in, and transfer of, the relevant competences while supporting the execution of well defined programmes carried out in close collaboration. The principal themes and programmes of research undertaken with ORSTOM in the oceanography of the Indian Ocean are briefly described herebelow.

Regional tuna project

The 'raison d'être' of this Project lies in the rapid growth in the industrial exploitation of tunas in the Indian Ocean. Started in 1987 in the framework of the Indian Ocean Fisheries Commission (of FAO), it is financed by the European Union and its European Development Fund, as well as by contributions from the Member States of the Commission participating in the Project (Comoros, France/La Réunion, Madagascar, Mauritius, and Seychelles). The first phase of the Project was completed in 1992 and has been followed by a second phase which was started in 1993 and completed in 1996.

Tuna purse-seiner awaiting a berth in the port of Victoria in Mahé, the Seychelles.
© ORSTOM/Patrice Cayré

Live-bait fishing for tuna by pole and line in the Indian Ocean.
© ORSTOM/Patrice Cayré

The objective of this Project, besides the establishment of a regional system for fishery monitoring and the acquisition of knowledge on the biology and the dynamics of exploited fish populations, is to provide the participating countries with the means for managing these important resources. The programme, which is administered by the Association Thonière (Tunafishing Association) from Madagascar, covers three fields: scientific research, training, and development of the national fishery capabilities.

ORSTOM co-ordinates research under this programme, in which the researchers in the countries concerned participate actively. The principal actions and results obtained so far may be summarized as follows:

- establishment and operation of a regional network for the collection and analysis of fishery statistics (artisanal and industrial),
- analysis of the state of the exploited stocks and the related biological and economic factors,
- modelling of tuna behaviour on small and large spatial scales (acoustic and classical tagging, remote sensing) and the relation to the environment, with application to the forecasting of favourable fishing areas,
- reproduction and growth biology of the main exploited species,
- analysis of the food chain leading to tunas and the forecasting of favourable fishing areas,
- development of a regional oceanographic data base, and
- analysis of economically efficient fishing aids (fish-aggregating devices).

Transfer of large albacore tuna from fishing vessel to refrigerator ship in Port Victoria for export to a distant cannery.
© ORSTOM/Patrice Cayré

The results are presented at international meetings (in the framework of the FAO's Indo-Pacific Tuna Development and Management Programme) and published in the reports of these meetings. They also appear in international publications and in communications of the Association Thonière. The summarized results (Proceedings of the Tuna Conference) of the first phase (1987-1991) of the Project are available from the Association Thonière or ORSTOM.

IMPACT OF THE INDUSTRIAL TUNA FISHERY ON MARINE MAMMALS AND OTHER SPECIES INCIDENTAL TO THE TUNA FISHERY

This programme, which is financed by the European Union, involved the Instituto Español de Oceanografía (Spanish Oceanographic Institute) and ORSTOM. It was started in 1994 and completed in 1996. Its purpose was to evaluate the precise impact of the industrial tuna purse-seine fishery on other species, some of which occur in the by-catch (marine mammals, swordfishes, marine turtles etc.). This programme covered the Atlantic as well as the Indian Ocean.

The results were obtained through the analysis of existing data as well as of data specially collected during the programme by scientific observers on board vessels. The results are throwing light on the specific association of tunas with other species (sharks, marlins, marine mammals etc.) which are abundant in these two oceans.

Preliminary analysis suggests that the association of tunas and marine mammals is much rarer in the Indian Ocean and, in any case, is not systematic, in contrast to the situation in the western Pacific.

Remote-sensing station in La Réunion

A station for receiving satellite images and for their data processing was established by ORSTOM on La Réunion in 1989. The station is designed to receive very-high-resolution radar/radiometer images (VHRR) and can receive, store or analyse data from various satellites (NOAA, ERS1, SPOT etc.). The archived data are downloaded onto the EARTHNET network. The station will shortly be equipped to handle ocean colour data from SeaWIFS (Sea-viewing Wide-field-of-view Sensor).

Sea-surface temperatures in the southwestern Indian Ocean a) at the end of the southern-hemisphere summer (March 1995): the pre-cyclonic situation; b) during passage of cyclone Marlene (6 April 1995); c) after passage of cyclone Marlene (12 April 1995), showing evidence of upwelling due to 'pumping'; and d) in the southern-hemisphere winter (July 1995), showing evidence of the convergence front, eddy and upwelling off Somalia.
© ORSTOM/Michel Petit

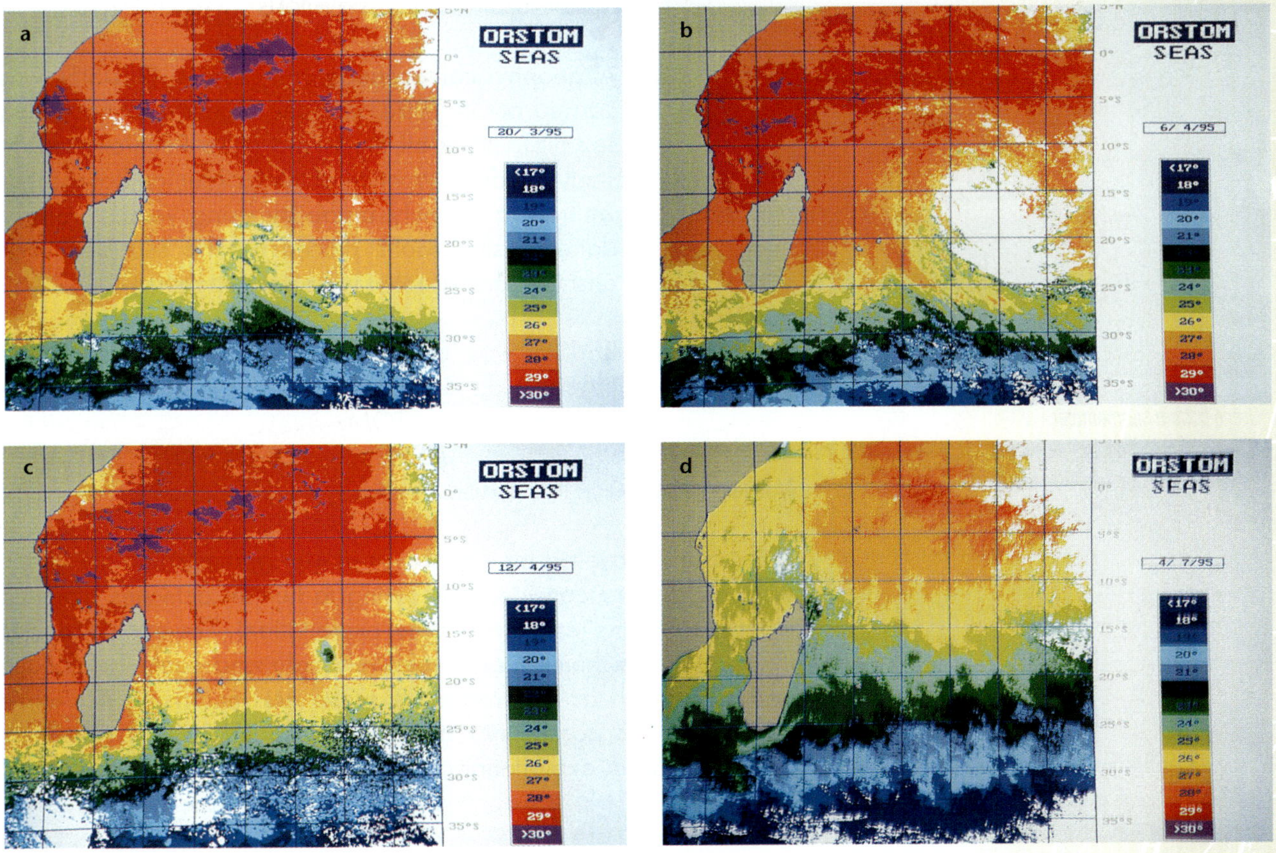

The geographical zone covered is from 5°N to 40°S and from the great African lakes to the middle of the Indian Ocean. The station's activities are linked to those of the European Space Agency (ESA), the US National Oceanic and Atmospheric Administration (NOAA), and the US National Aeronautics and Space Administration (NASA). This station may be seen as being a centre for the provision of high technology to the technical and scientific experts of all the coastal countries of the Indian Ocean. Training sessions and courses are organized to this end.

Some of the activities of the station are:

- vegetal-cover (vegetation index) mapping, in the framework of the ESA programme,
- processing and analysis of SPOT data, in the framework of the Mangrove Research Programme on-going in Madagasar (see below),
- periodic issuance of ten-day sea-surface temperature charts (Regional Tuna Project; see above), and
- pilot analysis of capacity for ocean fishery surveillance by means of satellite remote sensing (in co-operation with the Canadian Centre for Remote Sensing).

Ocean-bottom seismometer, capable of recording seismic waves down to depths of 10,000 metres. © ORSTOM

Marine geophysical observations

A geophysical oceanography cruise (REUSIS) was carried out in 1993, with the objective of recording seismic activity off the island of La Réunion, and to study the deep structure of the La Réunion hot-spot, as well as to quantify instability using vertical seismic data. Seismographs placed on the sea bed, and onshore tomography, were used together.

REUSIS was undertaken jointly by the Institut de Physique du Globe de Paris (the Paris Earth Physics Institute) and ORSTOM.

Mangrove research programme

The surfaces covered by mangroves in Madagascar (mainly on the west coast) vary considerably, owing to human activities affecting other ecosystems (especially forest exploitation). The exploitation of the mangroves themselves is relatively light and could be encouraged inasmuch as the mangrove areas are increasing in certain places, but this possibility cannot be envisaged before a development plan and exploitation procedures have been defined and developed.

A joint research programme (involving the University of Madagascar, the National Environmental Research Centre, and ORSTOM) was initiated in 1991. It comprises two main lines of research: the first is to undertake an inventory of the mangrove areas and their changes, as well as to understand the mechanisms and parameters underlying these changes (the dynamics of the vegetation, the geology and sedimentation, and the hydrology); the second is to draw up an inventory of the human activities that affect, or could affect, the dynamics of mangroves (forest exploitation, rice-growing, fishing, extensive aquaculture etc.).

View of a mangrove at the ORSTOM Centre at M'Bour.
© ORSTOM/Yves Paris

This essentially multidisciplinary programme depends appreciably on remote sensing which relies on the ORSTOM remote-sensing station on La Réunion, and on local image-analysis capability.

Active marine substances and biodiversity

A joint research programme (between the University of Madagascar, the National Oceanographic Research Centre, and ORSTOM) has been initiated to draw up an inventory, extract and test (mainly for pharmaceutical purposes) the activity of molecules from marine organisms. Some results have been obtained, notably relative to the treatment of malaria, but are still preliminary. The programme is now being carried out entirely by researchers at the University of Madagascar.

View of mangrove 'roots' to which bivalve molluscs are attached.
© ORSTOM/Jean-Pierre Montoroi

A significant quantity of organisms have been collected and photographed, constituting a broad inventory of the fauna (biodiversity) of the north-west part of Madagascar. This information will be made available to the general public in the form of a review prepared jointly by ORSTOM and the National Oceanographic Research Centre (Centre National de Recherche Océanographique – CNRO) at Nosy-Bé.

Support for the activities of the Intergovernmental Oceanographic Commission

ORSTOM participates in the meetings of UNESCO's IOC relative to the western Indian Ocean (IOC Regional Committee for the Co-operative Investigation in the North and Central Western Indian Ocean – IOCINCWIO) and in the definition of research activities and programmes to be carried out at the regional level. ORSTOM supports and participates particularly in the IOC Group of Experts on Ocean Science in relation to Living Resources (OSLR).

ORSTOM has recently offered its support to the organization of specific training seminars.

Upcoming programmes

Several research projects that ORSTOM could support are being developed for the Indian Ocean region. ORSTOM could contribute its expertise to the major programme entitled 'Support for Environmental Programmes in the Member States of the Indian Ocean Commission' which is just beginning.

Another example is the involvement of ORSTOM in the definition of the research project of the shrimp fisheries of Madagascar; this project should also involve the University, the Ministry of Research (National Oceanographic Research Centre) and the Ministry of Rural Development (Marine Resources Department). Its overall objective is to provide the Marine Resources Department with advice on the development and conduct of marine fishery management. The research aspect of the project deals with resource evaluation, the identification of the exploited stocks, determination of stock structure, interaction among different stocks, competition among fisheries (traditional, artisanal and industrial), with respect to the resource as well as to the socio-economic aspects, and the evaluation of various management methods. Industrial fishing-boat operators are involved in the design of the project and will be involved in its execution.

Conclusion

The programmes outlined above and the research work to which ORSTOM is contributing in the Indian Ocean illustrate the means and the objectives of ORSTOM's involvement. This brief presentation shows the possibility for carrying out high-level research in partnership while ensuring the transfer of knowledge to the scientific community and the decision-making entities of the developing counterpart countries. This should be the point of departure for further collaboration and to ensure progress in achieving the common wish for better management and conservation of natural resources and the environment of all the countries concerned and involved in a joint research effort.

Patrice Cayré,
Département RED (Ressources, Environnement, Développement),
Ecosystèmes marins et littoraux: Ressources et usages,
ORSTOM,
Institut Français de Recherche Scientifique pour
le Développement en Coopération,
209-213, rue La Fayette,
75480 Paris cedex 10, France.
Fax: (+33-1) 48 03 76 81
E-mail: cayre@orstom.fr

CHAPTER 4

NUTRIENT DISTRIBUTION IN THE INDIAN OCEAN

Introduction

It is now well established that a constant supply of nutrient salts, particularly nitrates, phosphates and silicates, is a prerequisite for primary production in the oceans. This apart, the origin and source of the nutrients, their interactions and interrelations, their residence times and fates reflect the history of the oceans and make an extremely interesting study.

Such studies in the Indian Ocean are inadequate, considering the information available for the Atlantic and the Pacific Oceans. However, the pioneering work done during the IIOE and subsequently by Qasim and his colleagues at the National Institute of Oceanography, Goa, since the 1960s, has revealed some unique features in the chemistry and distribution of nutrients in the Indian Ocean.

Prior to the time of the International Indian Ocean Expedition (1959-65), neither India nor any other country around the Indian Ocean had done any work on this Ocean's marine chemistry. Some of the expeditions, such as the *Valdivia* (1888-89), the *Dana* (1928-30), the John Murray (1933-34) and the *Galathea* (1950-52), did make some observations on the hydrography and nutrients in the Indian Ocean. The reports of these expeditions form a basic source for our understanding of the nutrient distribution in the Indian Ocean. During and after the IIOE, many ships criss-crossed the Indian Ocean and its adjacent seas and collected most valuable information on the concentration and distribution of the major nutrients (nitrates, phosphates and silicates) and of dissolved oxygen and carbon dioxide.

Earlier investigations in Europe and America, during the latter half of the 19th century, had posed two major questions to marine scientists: (1) why is productivity greater in coastal waters than offshore; and (2) what are the factors controlling the spring outburst of plankton populations. Practically all research in those days was directed towards finding answers to these questions. Particularly in England, Atkins and Harvey published a series of papers during the 1920s and 1930s explaining the role of nutrients in the chemistry of the sea water. So when the John Murray Expedition was launched in 1933-34, physical and chemical investigations of the north-western Indian Ocean formed an important part of the programme.

It was Brandt, in the years 1899-1920, who showed that the nitrogen and phosphorus compounds are often present in sea water in extremely minute quantities. He proposed a theory that the reason for the richness of certain areas in the oceans and at certain times of the year might be due to the presence of an abundant supply of these nutrient salts which, in other places and seasons, may not be available in sufficient quantities to support a rich phytoplankton growth. It was only in the years that followed the investigations of Atkins, Harvey and Cooper, mostly at the Plymouth Laboratories in England, that marine chemical studies became important.

General features of the nutrient distributions

The presence of the three primary nutrients is absolutely essential for the growth of the primary producers, namely the phytoplankton, and any variability in the distribution and concentration of these nutrients is therefore immediately reflected in the productivity of the waters. Bearing this in mind, the John Murray Expedition to the northwestern part of the Indian Ocean made a detailed study of the nutrient distribution in the Arabian Sea and adjacent areas for about nine months during 1933-34 (Gilson, 1937).

Gilson's report on the distribution of nitrates recognized that the Arabian Sea as a whole is dominated by the monsoon and therefore seasonal changes are marked. He noted upwelling off the Arabian coast as well as in the Gulf of Aden. However, the Expedition clearly missed the massive upwelling area off Somalia and the attendant increase in the organic production. He concluded that 'nitrate forms the main store of combined nitrogen in the sea. In algal cells, it is built up into organic compounds – principally the proteins of their protoplasm. If the algae are then eaten by animals, the plant proteins are broken down by the digestive juices and the amino-acids resynthesized into the proteins of the animal body. The animal phase may be prolonged through several stages, as small animals are eaten by large ones and they by larger ones again. Throughout, there is a continual return of a proportion of the nitrogen to the sea as urea or ammonium salts in their excreta. On the death of the uneaten plant cell or any member of the animal chain, a series of bacterial transformations take place'. These are made possible by a variety of bacteria, such as nitrogen-fixing bacteria, denitrifying bacteria, saprophytic bacteria and others. This is the nitrogen cycle in the sea (Figure 4.1).

Subsequent to Gilson's report on the chemistry of the northern Indian Ocean, Nielsen and Jensen (1957-59) reviewed the nutrient situation in the Indian Ocean, based on the data collected. The data from the second *Galathea* Expedition (1950-52) cover the Pacific, Atlantic and Indian Oceans.

In the western part of the Indian Ocean, in lower latitudes, the surface waters are poor in nutrients. However, water rich in nutrients is found here below the photosynthetic zone. Along the East African shelf, as also in the Agulhas Bank area, Clowes (1938) has reported upwelling or mixing of nutrient-rich waters in the surface layers, the nitrate values going up to 25µg atom/l. Here, even at a depth of 75m, 5µg atom NO_3-N/l and 0.3µg atom PO_4-P/l were reported by the Dana at its station 3971 (35°49'S, 23°09'E). Farther east, the Galathea found similar enrichment of the surface layers. The lower boundary of this layer was found at a depth of 81m, where 1.7µg atom PO_4- P/l was measured.

In the equatorial region of the Indian Ocean, nutrient-rich water was found in the lower part of the photosynthetic layer with a vertical extension down to 90m. The distribution of the inorganic phosphorus and nitrogen across three sections in this area (Figure 4.2 a, b) shows the upswing of the nutrients to occur between 40 and 100m depth with a value of at least 1.0µg atom PO_4-P/l and 6-12µg atom NO_3-N/l as measured.

In the Bay of Bengal, the *Galathea* occupied four stations outside the shelf area near the head of the Bay and found subsurface waters rich in nutrients near the surface, the thickness of the photosynthetic layer being 45-66m and the phosphate value being 1.3µg atom PO_4-P/l at 80m depth.

The two elements, nitrogen and phosphorus, often become limiting factors for the production of organic matter and, as such, it is important to study the source of the nutrient salts and their supply to the photosynthetic layers at the surface of the ocean. The major source of these nutrient salts is the deeper parts of the oceans, where, owing to bacterial action, all organic matter from living and dead organisms is broken down and returned to the surrounding water in simple chemical forms. The atmospheric source of nitrogen as

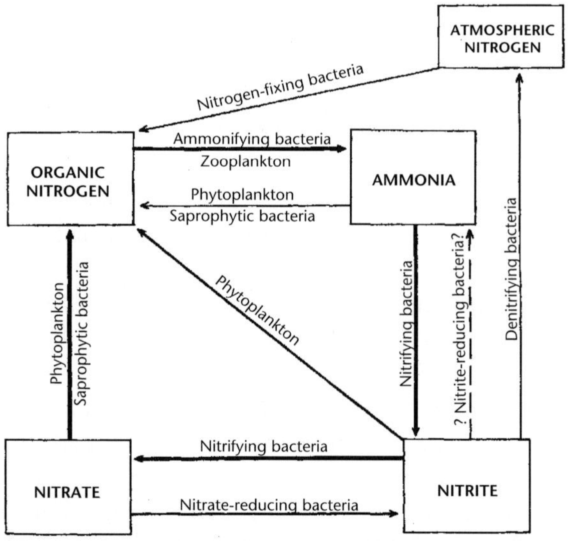

Figure 4.1
Diagram of the nitrogen cycle.

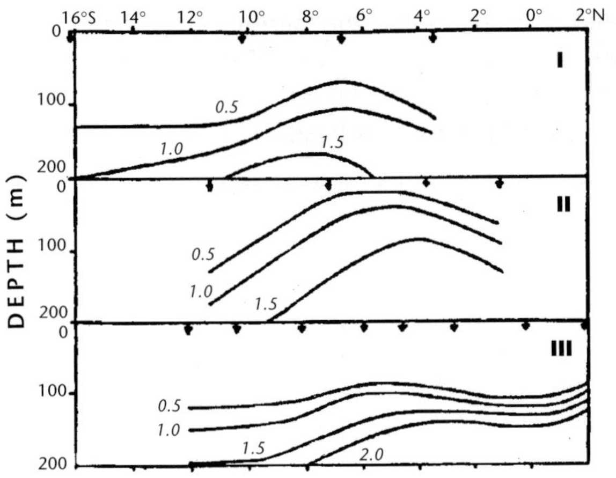

Figure 4.2a
The phosphate distribution in three Dana sections across the equator in the Indian Ocean; in μmol/l. Section I: at about 40°-45°E; section II: at about 50°-60°E; section III: near 95°E.

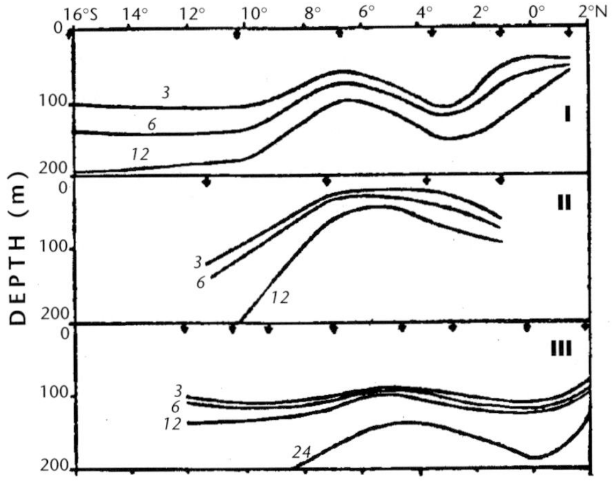

Figure 4.2b
The nitrate distribution in three Dana sections across the equator in the Indian Ocean; in μmol/l. Section I: at about 40°-45°E; section II: at about 50°-60°E; section III: near 95°E.

a result of electrical fixation in rain water is of very little significance in the sea, whereas the source from rivers may be significant. Nor is the input of nitrogen from the atmosphere in the form of oxides of nitrogen (notably nitrous oxide) from industrial sources to the Indian Ocean yet considered to be important.

It is already well established that there is some constancy in the proportion in which these nutrients occur in sea water and their utilization by organisms in the same proportions. Redfield (1934) and Cooper (1937) gave an ideal ratio of 1:15, when the concentrations are expressed as μg atom/l and the phosphate determinations are corrected for the salt error. Numerous workers have obtained similar ratios, though not always. Many expeditions have reported that, most of the time, the surface layers are generally poor in nitrate, though some small amounts of phosphate may be present in the photosynthetic layer or may even be totally absent, as at Dana station 3850 (Table 4.1).

In areas where surface water descends, as in convergences, the N:P ratio is very high in the photosynthetic layer. According to the Dana observations, positive anomalies were also found just beneath the photosynthetic layer near the divergence at the northern boundary of the South Equatorial Current in the Indian Ocean.

The Dana station 3925 (7°13'S, 52°22'E, in the Arabian Sea) presents the best illustration of N:P ratios (Table 4.2).

As to which one of the nutrients, the nitrate or the phosphate, is more important as a limiting factor, it is very difficult to answer, since it would mostly depend on the local conditions. Nitrogen becomes important if we consider its dependence on the upwelling of rich subsurface waters. Phosphates would be a limiting factor in coastal waters

Table 4.1 Nitrate and phosphate values with depth at Dana station 3850 (from Nielsen and Jensen, 1957-59)

Depth in metres	μg atom NO_3-N/l	μg atom P-PO_4/l
0	0.1	0.0
10	0.0	0.2
25	0.0	0.2
50	0.0	0.2
75	0.0	0.2
100	14.0	1.0

Table 4.2 Nitrate and phosphate values with depth at Dana station 3925 (from Nielsen and Jensen, 1957-59)

Depth in metres	µmol $N\text{-}NO_3$/l	µmol $P\text{-}PO_4$/l	N:P ratio
0	0.3	0.10	3.0
10	0.0	0.20	-
25	2.1	0.50	4.2
50	16.1	0.75	21.4
75	23.2	1.15	20.1
100	28.5	1.15	24.8
150	30.4	1.60	19.0
200	30.4	1.70	17.8
400	30.4	2.10	14.5
1000	43.0	3.15	13.7
2000	43.0	3.15	13.7

where their replenishment is due primarily to regeneration either in the photosynthetic layer or at the bottom.

It is well known that the rate of replenishment of the nutrient salts is much more important than their actual concentration in the water to promote organic production. In this, essentially two processes are involved: the regeneration by bacterial action; or upwelling and circulation. The concentration of inorganic phosphate at about 50m below the lower boundary of the photosynthetic zone, can generally be taken as indicative of the productivity of the area, but this may not always be the case.

Initial phosphate distribution – its significance

In the deeper layers of the ocean, in and below the thermocline, the concentration of certain nutrients and oxygen is made possible through the chemical processes active during remineralization. These facts can be used to derive truly conservative parameters for dissolved oxygen and nutrients by correcting for biochemical uptake and release. For this, it is assumed that a water mass is formed at the surface whose oxygen concentration is at saturation level and the nutrient concentrations are at their 'initial levels'. As the water mass moves away from the region of formation there will be oxygen loss and nutrient gain and it is possible to determine how much nutrient was remineralized since the water left the surface and to recover the initial nutrient level.

Redfield (1960) has commented on the significance of the distribution of phosphorus in the deep oceans of the world, where it was found that the Pacific and Indian Oceans contain much more phosphate-phosphorus than the North Atlantic; similarly, nitrate-nitrogen and phosphate-phosphorus showed concentrations of 3µg atom/l in the deep water of the North Pacific and Indian Oceans to less than 1.25µg atom/l in the North Atlantic and to less than 0.5µg atom/l in the Mediterranean, a total variation of 6:1. This magnitude of change clearly means that effective forces are at work to isolate the biologically active elements from the more conservative components of the waters. This is explained by the sinking of the organic matter formed at the surface to greater depths where it is decomposed, resulting in the increase of phosphorus and nitrogen and the redistribution of the deeper water by the oceanic circulation.

Accordingly, Redfield prepared two charts (Figure 4.3a, b) which show the possible pathways of deep-water circulation in the world oceans and the distribution of phosphorus at a depth of 2000m, since this depth is below the major influence of the zone of maximum phosphate and of the Antarctic Intermediate Water. For the interpretation of these charts, it is assumed: (1) that, in the absence of strong mixing processes, phosphorus concentration will increase with the time the water remains at depth and, in a current, will increase downstream; and (2) the mixing at depth of adjacent streams having different phosphorus values leads to intermediate concentrations. Therefore, the lowest phosphorus concentrations, and presumably the youngest deep water, are found north of 50°N in the Atlantic, where the surface water is denuded of phosphorus in its northward drift and is presumed to sink to form the North Atlantic Deep Water having a phosphorus

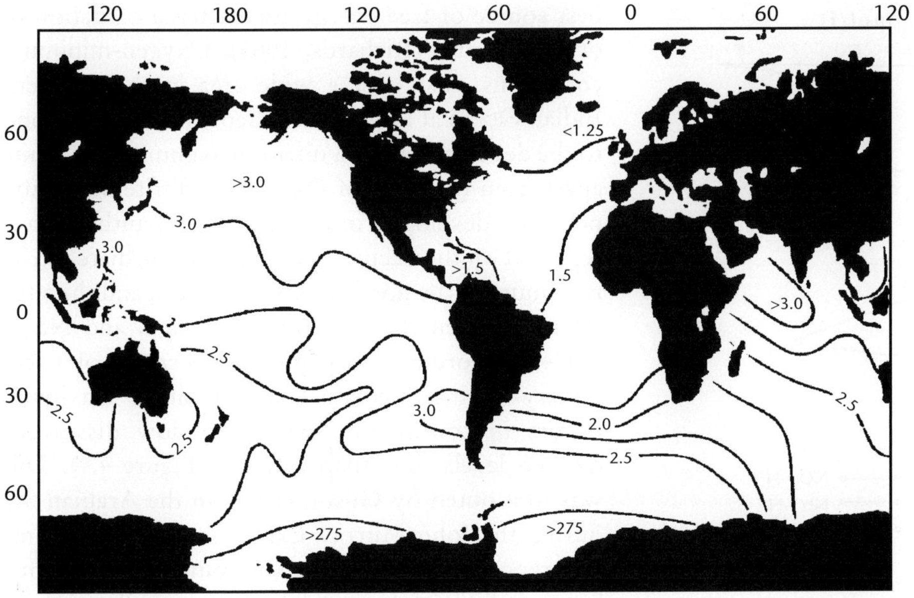

Figure 4.3a
Global oceanic distribution of phosphorus at 2000m depth (in millimoles/m³).

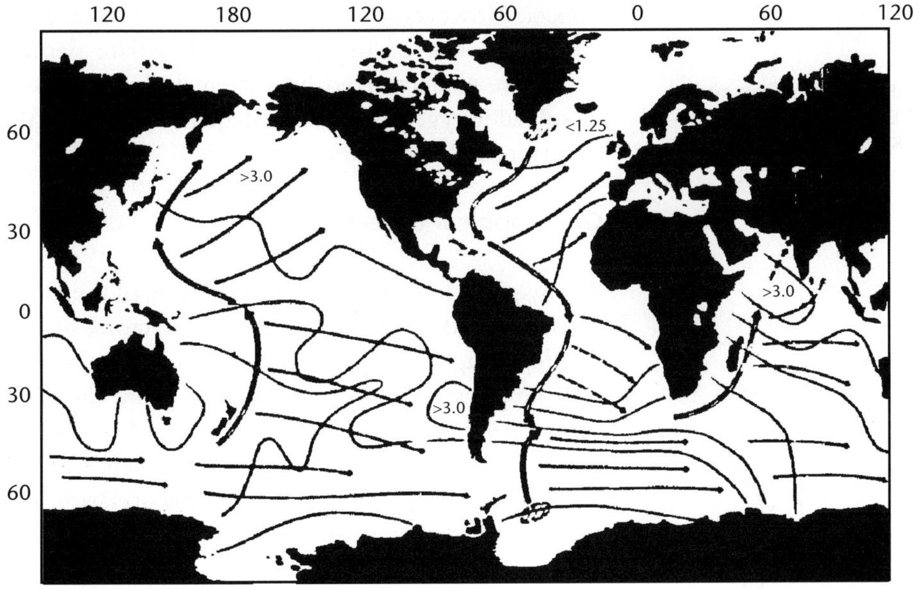

Figure 4.3b
The same distribution as in Figure 4.3a, with deep currents added (thick black arrows).

content of 1.25µg atom/l. As this water drifts southwards, the phosphorus concentration increases gradually and, beyond 35°S, there is a steeper gradient culminating in mixing with the southern waters whose phosphorus values vary from 2 to 2.5µg atom/l. Together, these mixed deep waters of the Atlantic and Southern Ocean move northwards into the Indian and Pacific Oceans, with the phosphorus values reaching above 3.0µg atom/l. In both cases, and in the North Atlantic, the phosphorus contours indicate higher values along the western side.

Subsequent surveys, including the Geochemical Ocean Sections Study (GEOSECS), support the elegant ideas of Redfield on the deep-water circulation at 2000m depth and possibly deeper, and the use of phosphate concentrations as a tracer.

Denitrification

It is believed that nitrate reduction to nitrite by bacteria is characteristic of areas that are deficient in oxygen. In such conditions, the nitrate ions are the next

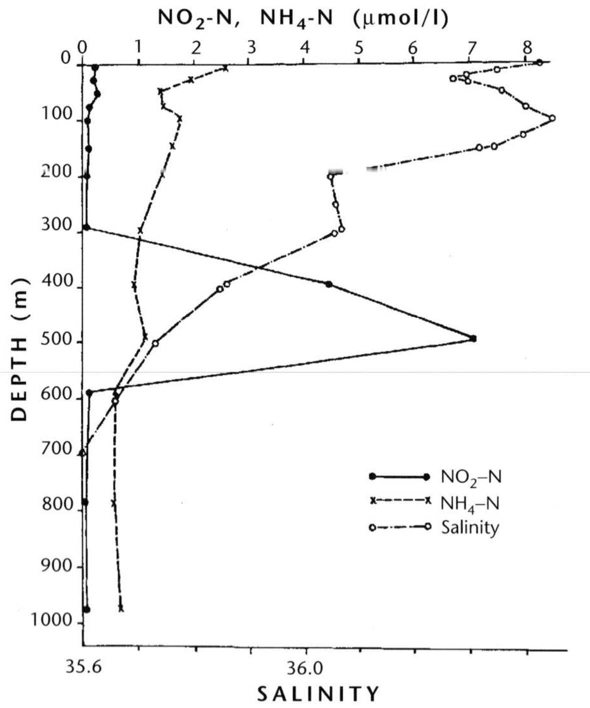

Figure 4.4
Vertical profiles of nitrite-nitrogen, ammonia-nitrogen and salinity at a station at 23°25′N, 65°53′E, in the northern Arabian Sea. (After Naqvi and Qasim, 1983)

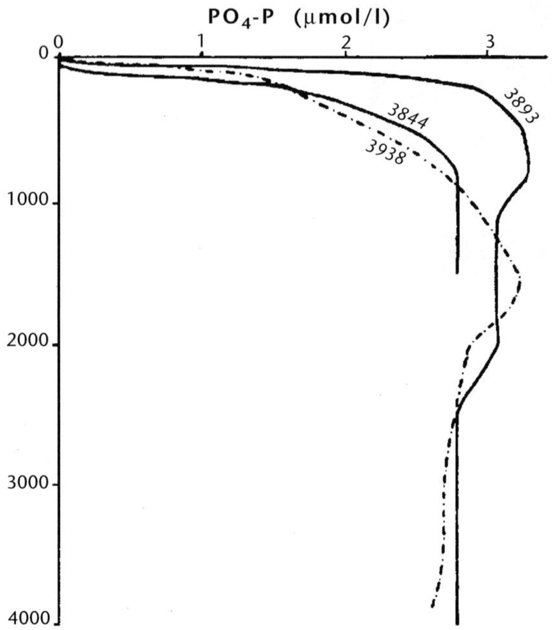

Figure 4.5
Vertical distribution of phosphate (µmol/l) at three Dana stations in the Indian Ocean: sta. 3844 (12°05′S, 96°44′E); sta. 3893 (5°59′N, 92°29′E); sta. 3938 (9°10′N, 45°17′E).

best source of free energy for bacterial oxidation of organic matter (Richards, 1965). Oxygen-minimum conditions exist over a wide area in the northern Indian Ocean at intermediate depths that correspond to the development of a nitrite-maximum zone, quite apart from the zone of the surface nitrite maximum normally developed in the sea. Several authors have reported on the occurrence of an intermediate nitrite maximum in the northern Indian Ocean and the situation has been reviewed by Sen Gupta and Naqvi (1984). According to them, the occurrence of a secondary or intermediate layer of nitrite maximum occurs in conjunction with very low dissolved-oxygen levels (less than 0.1ml/l) (Figure 4.4). This was first noted by Gilson (1937) in the Arabian Sea during the John Murray Expedition. There is some difference of opinion as to the source of the nitrite. Some believe that it results from nitrification of ammonia or amino-acids, whereas others say it is due to the reduction of the nitrates. Both processes may coexist. However, the apparent deficit in nitrate and the occurrence of nitrites at intermediate depths, which are characterized by extremely low levels of dissolved oxygen, provide adequate proof of the occurrence of such reduction.

Geographical aspects

At any given location, the concentration of the nutrients is dependent on a complex set of factors, such as the depth, age of the water, in situ biological events and the physical processes. The data from the Geochemical Ocean Sections Study (GEOSECS) stations 417 and 446 located in the Arabian Sea and the Bay of Bengal, respectively, show the relationship between the distributions of dissolved oxygen and of key nutrients (Tables 4.3 and 4.4).

Figure 4.5 shows the distribution of phosphate-phosphorus with depth at three Dana stations in the Indian Ocean. Further information on the distribution of nutrients and dissolved oxygen is available from two excellent reviews, one by McGill (1973) and the other by Sen Gupta and Naqvi (1984), as well as from the unique atlas of the hydrography of the Indian Ocean by Wyrtki (1971) which is based on the results of the International Indian Ocean Expedition (IIOE).

For marine chemistry, the most recent input has been GEOSECS, a global effort funded by the US National Science Foundation. The Indian Ocean portion of GEOSECS was carried out aboard the RV *Melville* as part of the Scripps Institute of Oceanography's INDOMED Expedition in 1977-1978, co-ordinated by Arnold E. Bainbridge (Spencer *et al.*, 1982; Weiss *et al.*, 1983). Previously, the RV *Knorr* had completed GEOSECS work in the Atlantic and the RV *Melville*, in the Pacific. The collection of sea-water samples and their analysis were meticulously planned and carried out by expert scientists drawn from many countries. The shipboard analytical programme included salinity, temperature, dissolved oxygen, nutrients (nitrate, phosphate and silicate), total dissolved inorganic carbon, radon, and measurements of atmospheric and surface-water CO_2 partial pressure, trace elements and particulates.

Western Indian Ocean and Arabian Sea nutrients

It has already been stated that the northern Indian Ocean is a seasonal water body greatly influenced by the prevailing monsoon. The complex water movements induced by wind forcing and density gradients affect the nutrient distribution and provide significant background to biological events.

Phosphate-phosphorus (Figure 4.6a): At the surface, oxygen and inorganic phosphates are fairly high in the entire area. Observations near Cape Guardafui and Socotra correspond to the peak upwelling season when the coastal areas close to Africa and Arabia

Table 4.3 Oceanographic data from GEOSECS station 417 (in the Arabian Sea, on 2 January 1978, at 12°58′N 64°28′E) showing, as an example, dissolved-oxygen and nutrient distributions. (After Weiss *et al.*, 1983)

Sample number	Pressure (decibars)	Depth metres	Temp. °C	Salinity	Oxygen µM/kg	Silicate µM/kg	P-PO_4 µM/kg	N-NO_3 µM/kg
301	9	9	26.618	36.410	195	2.8	0.45	1.8
302	34	34	26.605	36.408	200	2.8	0.45	1.8
303	54	54	26.610	36.406	195	2.6	0.45	1.9
304	86	86	23.371	35.810	120	7.9	1.29	16.0
305	104	104	21.854	36.161	40	11.4	1.79	22.8
306	146	146	17.459	35.647	3	23.5	2.25	23.1
307	166	166	16.566	35.622	2	25.2	2.29	24.2
308	207	206	14.935	35.601	3	28.0	2.32	25.3
309	236	235	13.752	35.483	3	30.2	2.36	27.4
310	307	306	12.638	35.430	9	32.9	2.37	30.0
311	389	387	11.813	35.410	17	36.9	2.40	31.0
312	468	466	11.545	35.452	18	41.5	2.47	31.2
313	570	567	11.039	35.448	17	47.3	2.54	32.3
314	625	622	10.627	35.410	15	51.5	2.61	33.4
315	702	698	10.232	35.422	16	56.8	2.65	34.0
316	801	796	9.636	35.388	23	62.9	2.70	34.9
317	901	895	8.990	35.332	19	69.4	2.75	35.9
318	1001	994	8.251	35.264	30	76.0	2.78	36.5
319	1104	1096	7.714	35.223	26	81.9	2.81	37.1
320	1254	1245	6.844	35.148	32	89.6	2.84	37.7
321	1405	1394	6.007	35.070	43	97.8	2.85	38.0
322	1554	1541	5.108	34.992	60	107.4	2.84	38.0
323	1704	1689	4.329	34.925	72	115.9	2.82	37.7
101	1845	1829	3.727	34.882	82	123.2	2.78	37.5
324	1955	1838	3.575	34.865	90	123.5	2.76	37.4
102	1995	1976	3.169	34.836	96	128.7	2.73	37.2
103	2145	2124	2.772	34.809	109	133.2	2.67	36.8
104	2296	2273	2.530	34.794	115	135.7	2.64	36.6
105	2448	2422	2.309	34.778	120	139.4	2.63	36.5
106	2600	2572	2.116	34.765	125	140.8	2.60	36.4
107	2750	2719	1.959	34.757	133	141.5	2.56	36.1
108	2903	2869	1.841	34.747	138	141.8	2.54	35.8
109	3055	3019	1.785	34.742	141	142.4	2.52	35.8
110	3207	3168	1.749	35.740	142	143.5	2.51	35.7
111	3358	3316	1.724	34.743	146	142.3	2.49	35.5

Table 4.4 Oceanographic data from GEOSECS station 446 (in the Bay of Bengal, on 28 March 1978, at 12°29'N 84°29'E) showing, as an example, dissolved-oxygen and nutrient distributions. (After Weiss *et al.*, 1983)

Sample number	Pressure decibars	Depth metres	Temp. °C	Salinity	Oxygen µM/kg	S-SiO_2 µM/kg	P-PO_4 µM/kg	N-NO_3 µM/kg
714	5H	5	28.242	33.331				
715	5H	5	28.242	33.341				
401	9	9	28.109	33.250	209	2.1	0.05	0.0
716	15H	15	28.460	33.665				
717	15H	15	28.460	33.673				
718	30H	30	28.043	33.717				
719	30H	30	28.043	33.717				
402	38	38	27.949	33.640	209	1.7	0.07	0.0
720	45H	45	27.995	33.711				
721	45H	45	27.995	33.710				
403	68	68	27.256	34.036	197	1.6	0.17	0.0
404	89	89	27.469	34.803	165	2.7	0.40	3.0
405	101	101	27.468	35.069	168	2.6	0.40	3.0
406	106	106	27.153	34.994	162	3.4	0.49	4.6
407	118	118	25.047	34.730	94	8.3	1.01	13.0
408	127	127	22.371	34.704	34	14.8	1.59	21.7
409	137	137	20.302	34.823	22	19.3	1.82	24.7
410	147	147	19.164	34.858	17	20.4	1.88	25.5
411	157	157	18.628	34.846	12	22.5	1.96	26.5
412	179	179	17.028	34.882	4	27.9	2.17	28.8
413	195	195	15.187	34.926	2	32.1	2.31	30.3
414	236	236	12.994	34.986	3	36.3	2.41	32.6
415	276	275	12.096	35.010	4	38.4	2.46	34.0
416	317	316	11.236	35.026	3	42.1	2.53	35.2
417	358	357	10.840	35.024	4	44.3	2.55	35.8
418	398	397	10.467	35.029	7	46.6	2.58	36.2
419	497	495	9.748	35.016	6	53.3	2.67	37.2

show an 18°C isotherm. Here PO_4-P values exceed 1µg atom/l. Oxygen values (Figure 4.6d) reach near saturation and exceed 5ml/l. Most indications of upwelling disappear by 200m depth and an excessive O_2 depletion to the level of 0.10-0.25ml/l is seen in the Arabian Sea. The phosphate maximum, with values as high as 2.70µg atom/l, begins at 600m depth; the maximum phosphate values, of 3.00µg atom/l, are, however, found in the Arabian Sea at 1200m depth.

Values for the central and equatorial regions are from 2.50 to 2.80µg atom/l. This is almost similar to the distribution found in the Atlantic.

At 2000m depth, the phosphate concentrations remain high in the northern areas (more than 2.60µg atom/l) but are slightly lower in the south. However, maximum concentration of phosphate has been reported at about 2000m in the western Indian Ocean. At 3000m, at the top of the Mid-Indian Ocean Ridge, deep-water circulation is affected and, as a result, only slight variations in phosphate and dissolved-oxygen values are noticed. In the Arabian Sea Basin, dissolved-oxygen levels range from 3 to 3.5ml/l and PO_4-P values are greater than 2.50µg atom/l, but they decrease to less than 2µg atom/l near South Africa. Dissolved-oxygen values are in excess of 5.0ml/l in more southerly locations.

Observations at 4000m depth are quite interesting. The Arabian Sea shows dissolved-oxygen concentrations of 3.75 to 3.80ml/l and phosphates from 2.4 to 2.5µg atom/l. The Southern Indian Ocean Bottom Water (temperature 1.1° to 1.5°C, salinity 34.72 to 34.73) has a dissolved-oxygen content of more than 4.00ml/l. South of the Mozambique Channel is a water mass with properties closely corresponding to those of Antarctic Bottom Water (temperature less than 1.0°C, salinity 34.73) with a dissolved-oxygen content of more than 5.0ml/l and a PO_4-P value of 2.0µg atom/l.

Organic phosphorus: During the IIOE (1959-65) and subsequently, some data on the distribution of particulate and dissolved organic phosphorus became

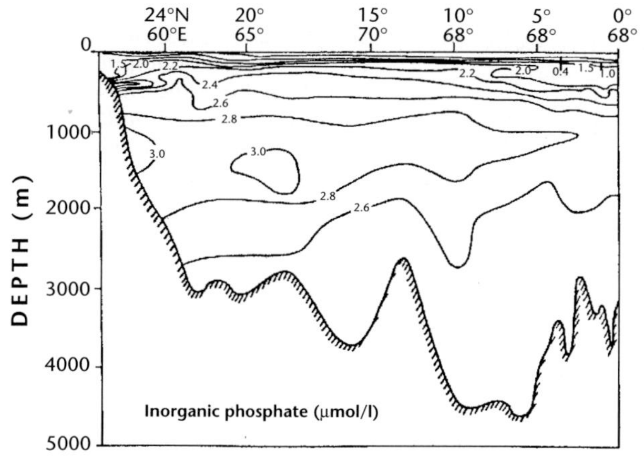

Figure 4.6a
Vertical distribution of inorganic phosphate (μmol/l) in a section from the Gulf of Oman to the equator. (From Sen Gupta and Naqvi, 1984)

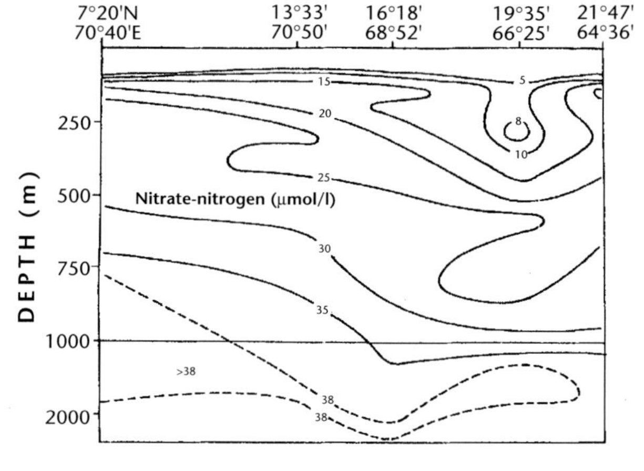

Figure 4.6b
Vertical distribution of nitrate-nitrogen (μmol/l) in an RV *Meteor* section parallel to the west coast of India. (From Sen Gupta and Naqvi, 1984)

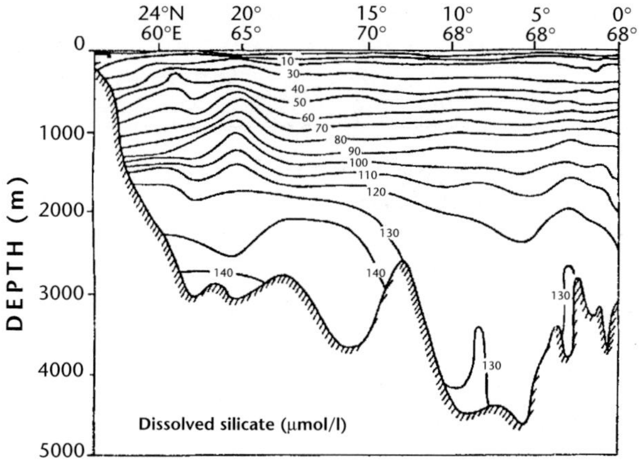

Figure 4.6c
Vertical distribution of reactive silicate (μmol/l) along the same section as in Figure 4.6a. (From Sen Gupta and Naqvi, 1984)

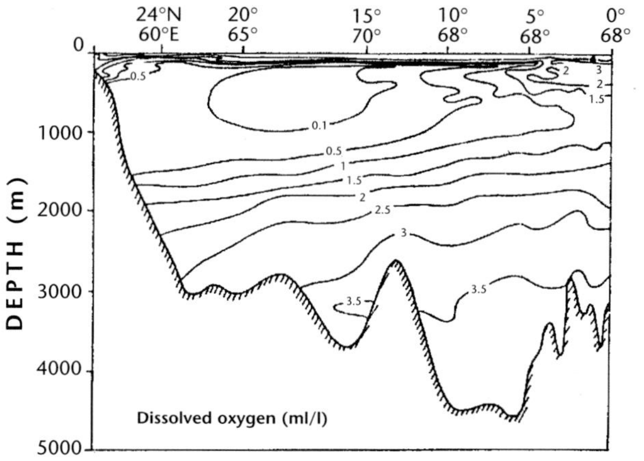

Figure 4.6d
Vertical distribution of dissolved oxygen (ml/l) along the same section as in Figure 4.6a. (From Sen Gupta and Naqvi, 1984)

available (Figure 4.7). From the surface to 1000m depth, the total organic phosphorus shows a high mean value ranging from 0.12 to 0.185μg atom/l, though very much less at greater depths (0.03μg atom/l or less).

The higher values noticed in the Arabian Sea can be attributed to the upwelling along the East African and southern Arabian coasts. Besides, all distributions for mean organic phosphorus concentration show a trend towards higher values north of the equator than to the south of it. There is no equatorial peak in the total organic phosphorus as was found in the Atlantic (McGill, 1973), and this may be due to the effect of the monsoon. The highest concentrations of organic phosphorus are encountered in the euphotic zone (0-200m), and the dissolved organic phosphorus constitutes two-thirds to three-fourths of the total organic phosphorus. The values for the aphotic zone for the organic phosphorus remain nearly constant at 0.03 to 0.04μg atom/l for the whole of the western Indian Ocean. This represents about one-fifth of the total organic phosphorus, while the remaining four-fifths occurs as dissolved organic phosphorus (0.15 to 0.18μg atom/l). Here also, there is no pronounced accumulation of organic phosphorus at the equator as is the case in the Atlantic. The decrease in their values southwards, compared to their increase in the Atlantic, may be attributed to the difference in the influence of the Antarctic waters in the two oceans.

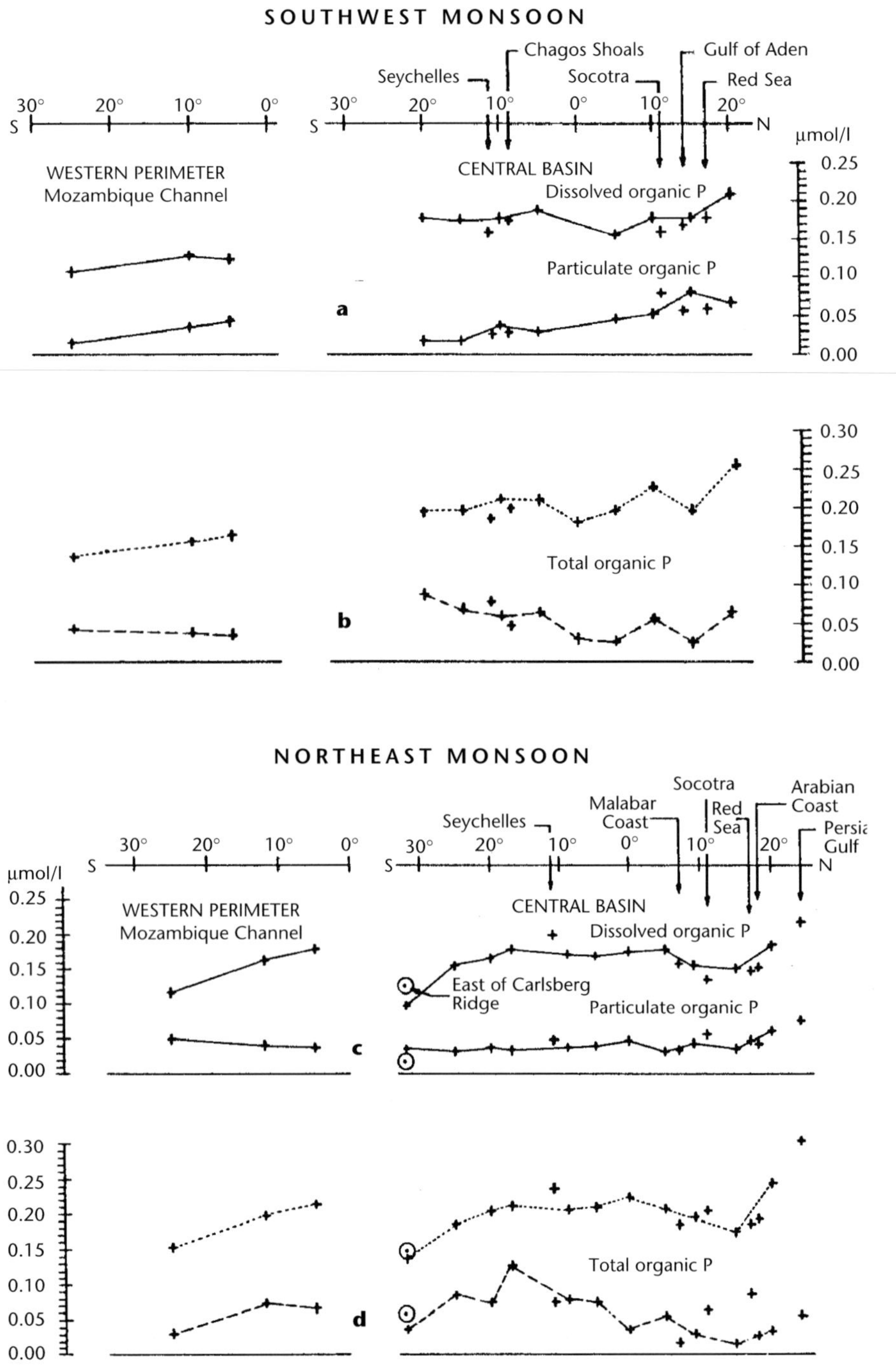

Figure 4.7
Distribution of particulate and dissolved organic phosphorus (a, c) and of total organic phosphorus (b, d) in the western Indian Ocean during the SW monsoon (a, b) and during the NE monsoon (c, d), based on data obtained by RV *Atlantis II*; dotted line = 0-200m depth, dashed line = 200-1000m depth. (After McGill, 1973)

Nitrate-nitrogen (Figure 4.6b): According to McGill (1973), the only significant nitrate concentration in the surface waters occurs near the upwelling zone off Socotra during the SW monsoon. Nitrate concentrations reaching 20μg atom/l have been observed by Smith and Codispoti (1980) in the Somali upwelling region. Values from 5 to 15μg atom/l are seen in this area whereas negligible amounts were detected elsewhere. The western Arabian Sea shows the highest nitrate levels at 200m (over 25μg atom/l) and a maximum of 30μg atom/l is found near southern India and Sri Lanka, probably due to an input from the coastal zone. The equatorial areas have less than 20μg atom/l, decreasing to less than 5μg atom/l near Madagascar.

At 600m depth, there is a small nitrate maximum in the East African coastal waters, also near the Seychelles bank and in southern Indian coastal areas. At 1200m depth, the whole of the Arabian Sea shows nitrate levels of about 35μg atom/l; lower concentrations are seen in the intrusive Antarctic Intermediate Water east of the Mascarene Ridge. Below this depth, most of the deeper waters in the western Indian Ocean show nitrate values between 30 and 35μg atom/l.

High dissolved and particulate organic nitrogen fractions are found in the surface layers of the equatorial region from 0° to 8°N and extend to deeper layers. Particulate nitrogen shows a maximum value at 10m and a very clear diurnal variation extending to a depth of 340m. During the night, nitrogen increases in the depth range 0-60m and decreases in the depth range 60-340m, apparently due to zooplankton migration. On the other hand, the vertical distribution of dissolved organic nitrogen shows two peaks, one at 20m depth and another at 100-170m depth. These are attributed to the death and decay of sinking phytoplankton cells and the metabolites excreted by zooplankton, resulting in the accumulation of soluble nitrogen in these depth ranges. The presence of a thin layer of nitrite-nitrogen, up to a concentration of 2μg atom/l in the thermocline, and a second nitrite maximum reaching a concentration of 5.0μg atom/l in the depth range 150-1500m have also been reported. The replenishment of nitrogen is correlated with the productivity of the waters and more particularly with the blooms of *Trichodesmium* which are credited with the ability to fix nitrogen in the sea.

Silicate-silicon (Figure 4.6 c): The silicate concentrations observed in the northern Indian Ocean are much higher than those observed in the Atlantic, but lower than those in the Pacific (McGill, 1973); the maximum values are 150-160μg atom/l for the Arabian Sea, 140μmol/l for the Bay of Bengal and 130-140μmol/l for the northern equatorial Indian Ocean. Silicates generally show a steady increase with increasing depth.

Eastern Indian Ocean and Bay of Bengal nutrients

Phosphate-phosphorus: Varying concentrations of nutrients have been reported from this vast area. While the surface waters of the Bay of Bengal show very low phosphate values (0.1μg atom/l), the Andaman Sea exhibits near-zero values for phosphates, as a result of high primary production and utilization.

In the southeastern Indian Ocean, however, upwelling has been reported along the continental slope off northwestern Australia and south of Timor, the surface waters showing relatively high concentrations of phosphate (0.2 to 0.3μg atom/l); and, in the Arafura Sea, 0.66μg atom/l during the period of southeast trade winds, when upwelling is reported.

Along the coast of Myanmar (Burma) and in the Andaman Sea, sometimes abnormally high phosphate values (>12.00μg atom/l) have been recorded and this is attributed to the proximity to the coral reefs and their high productivity. In the northern parts of the Bay of Bengal, including coastal areas off Waltair, upwelling is reported during the SW monsoon, resulting in enrichment of the surface waters and an increase in phosphorus values. They range from 0.99μg atom/l in April to a high 3.4μg atom/l in July. Curiously, the upwelling does not seem to contribute much in organic phosphorus to the surface waters and the high total phosphorus value is attributed mainly to the presence of dissolved organic phosphorus making up 80-90% of the total production.

Nitrate-nitrogen: Several workers have reported not-detectable to very low concentrations of nitrates in the surface waters above the permanent thermocline in the Bay of Bengal. However, off Visakhapatnam and Madras (east coast), during the SW monsoon, localized upwelling resulting in the enrichment of surface waters has been observed (De Sousa *et al.*, 1981). Similarly, the dissolved nitrate and nitrite values are also very low in the surface waters, but increase substantially with depth, to 15-20μg atom/l in the thermocline and 22-26μg atom/l at depth all over the eastern Indian Ocean.

Silicate-silicon: South of the equator, towards 100°E, the silicate values range from 4 to 148μg atom/l between the surface and depth. The silicate maximum is very much greater in the Indian Ocean than in the Atlantic, but less than in the Pacific. No intermediate maximum for silicate exists and silicate concentration below 2000m is nearly constant. It has been pointed out that the silicate values in the Bay of Bengal are much lower than those in the Arabian Sea (Sen Gupta and Naqvi, 1984). Apparently, the enormous river run-off into the Bay does not contribute much dissolved silicate, and much less of phosphate and nitrate (De Sousa *et al.*, 1981). The distribution of dissolved silicate is characterized by a continuous increase with depth starting from the thermocline. Its concentrations are from 150 to 160μg atom/l in the Bay of Bengal and from 130 to 140μg atom/l in the northern equatorial part of the Indian Ocean.

The northern Indian Ocean dissolved-oxygen distribution

In the sea, as on land, the presence of oxygen is indispensable for the life processes of all organisms. Although the concentration of dissolved oxygen in the sea water is about 9ml/l, as a maximum value, compared to about 200ml/l in air, it is nonetheless available in a free state all over the oceans and may not be a limiting factor in the distribution of animals and plants. However, variations in its distribution are brought about by differential rates of consumption by animals, liberation by plants during photosynthesis and utilization by bacteria for oxidation of organic matter. Where there is a low supply of oxygen, a reducing environment may develop causing hydrogen sulphide production and, in such cases, all aerobic forms may either be excluded or mass mortality may occur.

Against this background and the fact that the northern Indian Ocean is characterized by a low-oxygen layer across its mid-depths, a study of the oxygen distribution is very interesting. In the sea, the main source of oxygen is photosynthesis. The rate of photosynthesis is highest in well illuminated areas and confined to the euphotic zone whose depth varies depending on geographical location and water transparency. In the tropical parts of the Indian Ocean, where sunlight is intense, vigorous photosynthetic activity takes place just below the surface and therefore maximum concentration of oxygen also occurs there. Since the phytoplankton utilizes the nutrient salts also rapidly, leading to their depletion, the profiles of oxygen and nutrient salts, mainly the nitrates and phosphates, would be 'mirror images' of each other. Immediately following a phytoplankton bloom, aging, sinking and bacterial degradation of the cells ensue and, perhaps with some time lag, a similar fate overtakes the herbivorous zooplankton. The net result of all these activities would be the depletion of oxygen in the layer below the photosynthetic zone. At this point, the consumption of oxygen will be higher than its production and, as a result, an oxygen minimum layer develops, its extent and thickness depending on physical processes such as currents and convection in the area.

In their review of the chemical oceanography of the Indian Ocean, Sen Gupta and Naqvi (1984) have summarized what is known about the distribution of dissolved oxygen, which in some respects is unique. As the surface layer is well mixed down to the thermocline, dissolved oxygen values reach saturation levels and maximum concentration, particularly during February-May, when the intensity of solar radiation is very high, causing the maximum primary production to occur a few metres below the surface (Qasim, 1977). However, where upwelling occurs, or at divergences, dissolved-oxygen values could be lower, depending on the age of the upwelled water.

In general, there is a sharp fall in dissolved oxygen levels through the thermocline and the strong density gradients below prevent mixing of oxygen-rich surface waters with subsurface waters thereby causing an oxygen-minimum layer in the thermocline. In view of the semi-enclosed nature of the northern Indian Ocean, horizontal circulation is also limited in the north. These features, coupled with very high organic production reported in the area, result in a severe depletion of dissolved oxygen all over the northern Indian Ocean below the thermocline (Figure 4.6d). Dissolved oxygen values well below 0.2ml/l are recorded frequently at intermediate depths. Another interesting feature is the presence of two oxygen minima, one between 100 and 500m depth and the other between 1000 and 1500m depth in the Arabian Sea only north of 18°N. The deep oxygen-minimum layer (1.5ml/l) observed between 600 and 800m depth throughout the Indian Ocean (Figure 4.6d) shoals northwards and its dissolved-oxygen content becomes further depleted. Thus, there are two oxygen-minimum layers in the northern Arabian Sea and these are separated by an oxygen-maximum layer originating at or around 42°S which spreads northwards and was called the Subtropical Oxygen-Maximum Water by Rochford (1966). Equally interesting is the oxygen distribution in deep waters, where low concentrations are found in the northwestern part of the Arabian Sea, whereas, in the Pacific Ocean, such deficit layers are found towards the northeastern side. Warren (1981) attributes this to the pattern of flow of deep-water layers which are different in the two oceans. There is only one source of deep water in the Pacific, a deep western boundary current which could make the deep waters of the North Pacific the oldest, thereby accounting for the lowest dissolved-oxygen concentration. However, in the Indian Ocean, the complex ridge system divides the deep flow into three branches originating in the western boundary currents. The western basins are penetrated by water that flows into the Madagascar Basin from the Crozet Basin through the fractures in the Southwest Indian Ocean Ridge. The upper deep water (1000-3800m) in the Central Indian Ocean Basin is supplied by a mid-depth western boundary current flowing northwards along the eastern flank of the Central Indian Ocean Ridge, while the lower deep water (>3800m) is derived by overflow through deep saddles in the Ninety-East Ridge from the deep western boundary current flowing along the eastern flank of this ridge. As a result, the deep waters of the Bay of Bengal are probably younger than the waters of the Arabian Sea.

In a comprehensive paper on the oceanography of the northern Arabian Sea, Qasim (1982) also reported the occurrence of two oxygen minima, the first between 100 and 400m depth and the second between 800 and 1500m depth. The upper oxygen minimum is a result of biodegradation of organic matter in oxygen-poor water from the Persian Gulf, whereas the deeper minimum layer originated in the deep waters of the Gulf of Aden and mixes with similar water masses spreading from the southern regions of the Arabian Sea. Sen Gupta *et al.*, (1976) and Qasim (1982) have commented on the nutrient-oxygen relationships and derived the following ratio, by atoms, of Apparent Oxygen Utilization (AOU) to carbon, silicon, nitrate and phosphate: 280:108:40:16:1.

Using these relationships, the above-mentioned authors calculated the nitrate-reduction rates at the intermediate depth ranges and derived a residence time of the water in the oxygen-deficient layer of 43 to 51 years.

GEOSECS data from the Arabian Sea and the Bay of Bengal

GEOSECS sampling stations in the Arabian Sea and the Bay of Bengal are mostly located quite a distance from the influence of the land and, therefore, the data obtained can be considered as benchmark values for these areas. The hydrographic data from Station 417,

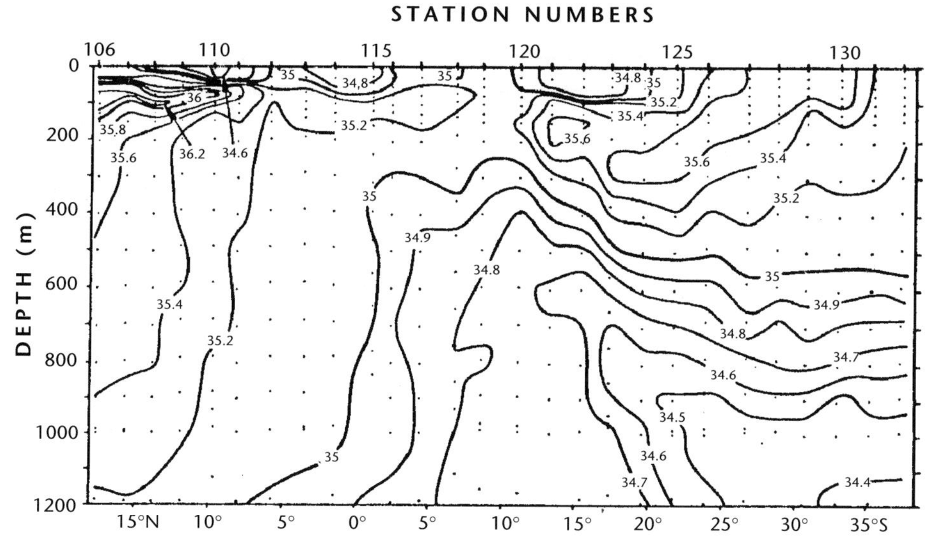

Figure 4.8a
Vertical distribution of salinity in a section along 70°E between about 10°S and 18°S during the cruise of RV *Anton Bruun* in May-June 1963. (After Wyrtki, 1973)

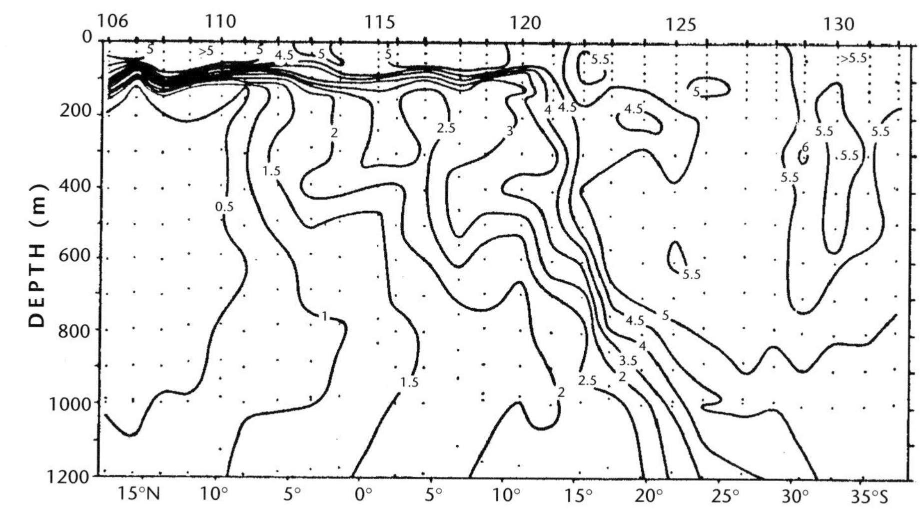

Figure 4.8b
Vertical distribution of dissolved oxygen (ml/l) in a section along 70°E between about 10°S and 18°S during the cruise of RV *Anton Bruun* in May-June 1963. (After Wyrtki, 1973)

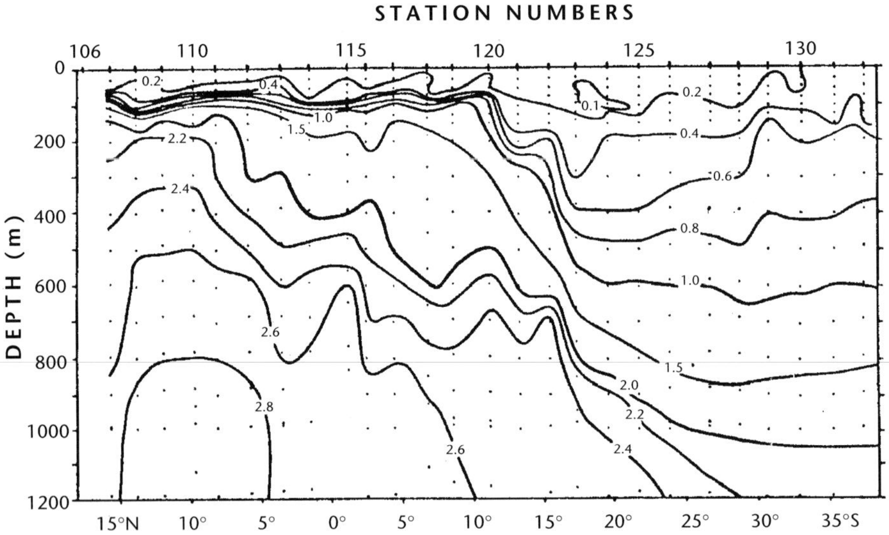

Figure 4.8c
Vertical distribution of phosphate-phosphorus (μmol/l) in a section along 70°E between about 10°S and 18°S during the cruise of RV *Anton Bruun* in May-June 1963. (After Wyrtki, 1973)

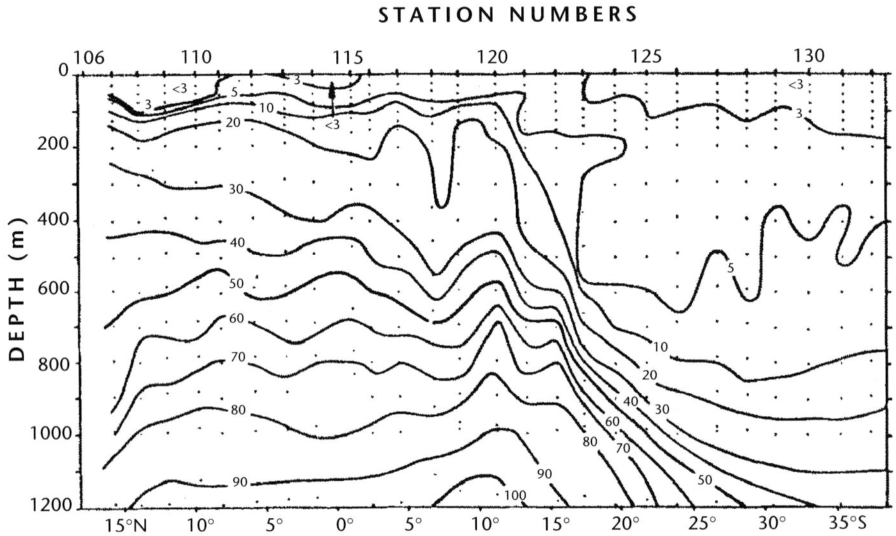

Figure 4.8d
Vertical distribution of silicate-silicon (μmol/l) in a section along 70°E between about 10°S and 18°S during the cruise of RV *Anton Bruun* in May-June 1963. (After Wyrtki, 1973)

in the Arabian Sea (Table 4.3), and Station 446, in the Bay of Bengal (Table 4.4), are presented for comparison, since they reflect the unique nature of the two seas.

In the Arabian Sea stations, surface values of dissolved oxygen are high in the upper 50m (highest value 200μM/kg ≈4.6ml/l), while the lowest values are found between the depths 100-1600m. In fact, at Station 416, no oxygen was detected at depths of 132 and 187m. In the Bay of Bengal, surface waters are equally rich in oxygen (209μM/kg ≈4.8ml/l) down to a depth of about 70m and the low dissolved-oxygen values are found between the depths of 120 and 1000m.

As far as the nitrates and phosphates are concerned, while the Arabian Sea has a certain amount of these nutrients in the surface waters, the Bay of Bengal nitrate values are nil in the surface waters down to a depth of about 70m, below which there is a gradual increase of both nutrients. Silicates show a uniform increase below the surface, down to great depths. The development of such a situation in the nutrient and oxygen distributions may be due, in part, to the lack of mixing, hence the stagnation of deep waters in these two seas.

The hydrochemical front at 10°S

The vertical profiles of salinity, oxygen, phosphate and silicate between 10° and 20°S right across the equatorial Indian Ocean are one of the most interesting features in their distributions. According to Wyrtki (1973), the monsoon gyre is separated from the subtropical gyre of the southern Indian Ocean by a strong and vertical front in its hydrographical and chemical structure; these are better developed in the subsurface layers than at the surface. This is attributed to the fact that, during the NE monsoon, the South Equatorial Current does not belong to the monsoon gyre, whereas, during the SW monsoon, its northern part forms part of the monsoon gyre. At the surface, the two gyres are separated by a horizontal salinity minimum extending from Sumatra to Africa, and the minimum separates the high-salinity waters of the northern Indian Ocean from the high-salinity waters of the subtropical gyre.

This front is even more pronounced in its chemical features. It separates the low-nutrient and high-oxygen waters of the subtropical gyre from the high-nutrient and low-oxygen waters of the monsoon gyre. Figures 4.8a, b, c, d make it clear regarding the sloping of the salinity, oxygen, phosphate and nitrate profiles from about 100m depth at 10-12°S to 800m depth at 16°-18°S. Near Australia and near Madagascar, the front is less pronounced because of the prevailing strong meridional flows. According to Wyrtki, this front is maintained by continuous formation of essentially different water masses in the northern and southern Indian Ocean gyres.

In a recent paper on the thermocline circulation and ventilation in the Indian Ocean, You and Tomczak (1993) report that 'Strong meridional gradients of all properties at about 13°S indicate the presence of a hydrological front south of the region of the Australasian Mediterranean Water inflow. The front indicates that, throughout much of the Indian Ocean, there is little communication between water in the northern and in the southern hemisphere, and that the region of the western boundary current along the coast of Africa is the only path for water to cross the equatorial current system'. The front also appears to act as a barrier to the free migration of some of the marine organisms between the northern and southern Indian Ocean.

Conclusion

The northern Indian Ocean is a fertile area and, more particularly, the northwestern sector is characterized by high productivity. The two-layered circulation discussed by Ryther *et al.* (1966) provides for an active recycling of nutrients, which causes high organic productivity in the euphotic zone and extremely low dissolved-oxygen and high nutrient concentrations below the thermocline. Profiles of phosphate and nitrate show maximal values between 1000 and 1500m depth, somewhat below the deep oxygen minima. The concentrations of phosphate and silicate are found to be higher in the Arabian Sea than in the Bay of Bengal; nitrate concentrations, however, are lower. In fact, marked regional variations in the distribution of nutrients were noticed by Sankaranarayan and Reddy (1968) in the northern Bay of Bengal. The absence of an increase in dissolved-silicate concentrations in this area close to the massive river run-offs is somewhat surprising. Silicate generally shows a steady increase with depth, but its surface distribution is highly variable. The silicate concentrations observed in the northern Indian Ocean are much higher than those found in the North Atlantic, but lower than those in the Pacific. The important findings are:

- Higher nutrient and lower oxygen contents in the bottom layers may be attributed to oxidation.
- Very high dissolved-silicate concentrations, particularly in the Arabian Sea Bottom Water, steadily decrease southwards, mainly due to dissolution of diatomaceous shells from the sea floor.
- The northern Indian Ocean has a significant level of nitrate reduction to nitrite in sea water.
- As a result of very low oxygen concentrations below the thermocline in the entire northern Indian Ocean, a reducing environment prevails at intermediate depths.
- Upwelling or mixing of oxygen-poor layers with the surface waters may create lethal conditions for biological processes and life in general. Future studies must concentrate on this aspect of oxygen distribution.

Such a pattern of concentration and distribution of nutrients has greatly influenced the biology of the Indian Ocean. The following chapter deals with the biological aspects, beginning with a description of the fauna by Alcock (1902), nearly a century ago, as an interesting sidelight to show the types of observations made by biologists in those pioneering days of marine biology.

CHAPTER 5

BIOLOGY OF THE INDIAN OCEAN

Introduction

The biology of the Indian Ocean is a vast subject. The majority of papers published deal with such aspects as plankton and productivity, nekton (including fisheries), and benthos. For the first time, an attempt is made here to synthesize the available information as far as possible; this section on the biology begins with a description of the coast and deep-sea fauna made by Alcock (1902) nearly a hundred years ago, in his book entitled *A Naturalist in the Indian Seas*, which is now out of print. As the only book so far published on the Indian Ocean seas, it is an interesting document on the types of observations made by marine biologists at the beginning of this century. A list of marine biologists who worked with Alcock, as well as their subjects, is given as a matter of general information in Table 5.1.

A naturalist in the Indian Ocean seas

The book entitled *A Naturalist in the Indian Seas or Four Years with the Royal Indian Marine Survey Ship, Investigator* is the first and perhaps only book on the biology of the Indian Ocean seas. The contents of the book are divided into three parts: (1) an introduction, an explanation and a narrative; (2) the deep-sea fauna of the Indian region; (3) appendices providing the list of dredging stations and the bibliography of the *Investigator*. Alcock's writing style is delightful to read and his narrative of the native customs, rituals observed in temples and the social structure is very interesting.

A total of 276 stations were dredged in the Bay of Bengal, the Andaman Sea and the Laccadive Sea, based on which Alcock and his co-workers have described the deep-sea life in these areas. The bibliography contains a complete list of the *Investigator* publications. There are 92 papers, of which 83 deal with the fauna, 6 with the flora and, of the remaining 3, one reports on some nodules dredged off Colombo, from a depth of 675fa (about 1350m), the second on the bottom topography of the Arabian Sea, and the third on the mean temperature of the deep water of the Bay of Bengal. Of the 92 papers, Alcock himself worked on and published 49 papers, some of them with his colleagues. These pioneering publications are mainly found in the *Journal of the Asiatic Society of Bengal* beginning in 1885, and in the *Annals and Magazines of Natural History, London*, from 1889.

The papers listed in Table 5.1 were published during the period 1885-1900. Alcock expressed his regret that no papers were published on the collections of Medusa, Annelida, Polyzoa, Tunicata and Pycnogonida. There is also no work on plankton. In fact, the term *plankton* was first used in 1889 by Professor Victor Hensen, the leader of the German Plankton Expedition; prior to that time, the *Investigator* had not paid any attention to the collection of plankton, until the arrival of Seymour Sewell on the scene in 1913.

After working along the Orissa coast during 1888-89, Alcock was deeply impressed by the abundance of fauna and made the following comments: 'To sum up my impressions of the Orissa coast from the zoological point of view, I look upon it as an ideal place for anyone who wishes to study the complete life histories of the Indian shore fishes and Crustacea, and I believe that a biological station, established at Puri, would be in the highway of great discoveries. To speak, finally, of the economic possibilities of this coast, so rich in certain kinds of marine produce and firewood, I should say that if the regulations of the salt excise could be modified and if capital on a liberal scale was forthcoming, it would furnish inexhaustible supplies of dried and smoked fish, fish oil, isinglass and gelatin for the world in general, and the shark fins for the China market in particular.'

Alcock's suggestion for having a biological station on the Orissa coast was realized 70 years later, when a university was established at Brehampur with a Marine Biology Department.

Alcock's description of coastal and deep-sea fauna

Alcock imagines a journey across the Bay of Bengal; assuming that all the water were removed, it would present a very different picture compared to land (Figure 5.1). Starting from Madras towards the Andamans, leaving behind the sandy beaches, rocky out-crops and trees and shrubs, one proceeds for the next 10-15 miles on a level land covered with dark bluish mud which is of terrigenous origin. This mud is carried down by rains and rivers from the mainland and deposited all over the continental shelf and sometimes beyond. After another 10 or 15 miles, the sea floor begins to dip sharply, with steep slopes into an abyssal plain spread right across the Bay of Bengal. Here, the mud becomes chalky and gritty; a microscopic examination would reveal that it is all composed of microscopic shells made of calcium car-

Table 5.1 A list of authors and their subjects in the field of Indian Ocean biology

Subject	No. of papers	Authors
General Zoology	4	A. Alcock and J. Wood-Mason, A. R. S. Anderson and A. F. M'Ardle
Protozoa	4	H. B. Brady, J. Wood-Mason, John Murray, A. Alcock and F. Chapman
Porifera	2	F. E. Schulze
Coelenterata	8	J. Armstrong, J. Wood-Mason, A. Alcock
Asteroidea	3	J. Wood-Mason, A. Alcock
Crinoidea	1	J. Wood-Mason, A. Alcock
Ophiuroidea	6	J. Wood-Mason, A. Alcock, A. R. S. Anderson
Echinoidea	3	J. Wood-Mason, A. Alcock, A. R. S. Anderson
Holothuroidea	2	J. Wood-Mason, A. Alcock, J. H. Tull-Walsh
Crustacea	20	G. M. Giles, J. Wood-Mason, W. Weltner, A. Alcock, A. R. S. Anderson, A. F. M'Ardle
Mollusca	8	G. Nevill, J. Wood-Mason, A. Alcock, E. A. Smith, G. B. Soubry, E. S. Goodrich
Brachiopoda	1	A. Alcock
Fishes	21	A. Alcock, J. Wood-Mason
Botany (of Andamans, Diamond Island, Coco Group and Laccadives)	6	G. M. Giles, D. Prain
Geology (nodular material off Colombo)	1	E. J. Jones
Topography of the Arabian Sea off the Laccadives	1	C. F. Oldham
Temperature of deep-sea areas of the Bay of Bengal	1	A. Carpenter

bonate, belonging to a protozoan group known as the Foraminifera. What is more, most of it is contributed by a single genus, *Globigerina*; hence the mud is called *Globigerina* ooze. This covers most of the sea floor of the Bay right up to the coast of the Andaman Islands where it gets mixed up with coral and shell remains contributed by the coral reefs. Apart from *Globigerina* ooze, there are other oozes contributed by pteropods, radiolarians and diatoms. The shells of the latter two are siliceous and are generally found in very deep waters. However, in the open ocean, another type of mud covers vast areas of the sea bed and is called the Red Clay, which is of volcanic origin. The physical conditions across the entire cross section of the sea water overlying the sea floor are quite different when compared to those on land. Here, in the oceans, below 200-300m depth, there is complete darkness, since the sunlight cannot penetrate farther and the pressure increases at the rate of 1 atmosphere for every 10m increase in depth, and therefore the pressure at a depth of 3000m would be equal to 300 atmospheres (≈304,500hPa). As for the temperature, it gradually decreases and, finally, below a depth of 2000m, the water is uniformly cold, registering near-freezing temperatures of 2-3°C.

Against this general background, Alcock has drawn our attention to some very interesting examples of sea life. Among the fauna of the Orissa coast, he mentions a species of swimming crab, *Matuta miersii*, which makes a chirping noise by rubbing its nippers (chelipeds) against the carapace. Another, the buckler crab, *Cryptopodia angulata*, if disturbed, feigns death by pressing itself closely to the sand surface. Among the fishes, males were found to be more colourful and ornamental than females. Among flat fishes *Arnoglossus macrolophus* and *Brachypleura xanthosticta*, the anterior rays of the dorsal fin form a fine erectile crest only in the males. In the case of *Rhomboidichthys azureus*, the male alone has its forehead adorned on one side with azure blue spots, which gleam like jewels. Alcock also found interesting free-riders on the drift logs floating in the coastal water, thus facilitating their dispersal. Besides a crust of the ubiquitous barnacles, the logs harboured two species of grapsoid crabs, one a swimming species, *Varuna littorata*, the other, a rock crab, *Plagusia depressa*.

Describing the fauna of the Andamans, comments are made on the colour changes in fishes of the coral reefs. *Balistes maculatus*, while swimming, exhibits a blue-black colour with numerous blue spots all over the body, but when confronted with a black (dark) background, it turns completely black. A species of *Monacanthus* is equally adept at changing colour. This is normally greenish black and scribbled over with fine yellow lines, but when placed in a white bucket full of sea water, the fish turns pale yellow.

The rocks of the sea shore are inhabited by multitudes of crabs of the species *Grapsus grapsus* and *Grapsus stigosus*. These crabs are extremely agile and difficult to catch. Even more hopeless to pursue are the little gobiid fishes of the genera *Periophthalmus*

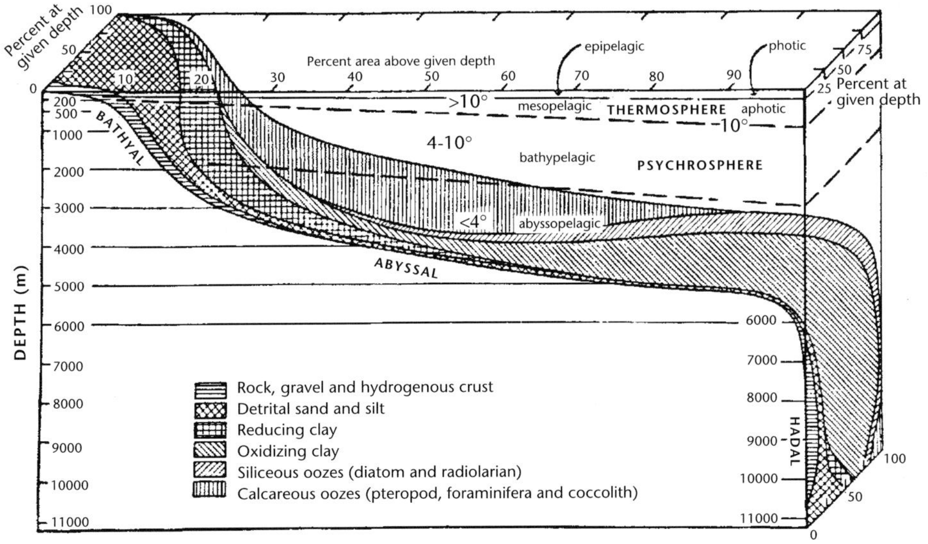

Figure 5.1
Ecological zonation of the deep sea. (After Bruun, 1956)

and *Boleophthalmus* (Figure 5.2a) which are commonly found in the mangrove swamps in the Andamans and all over the tropical Indo-Pacific area. Although they are typical fish with water-breathing gills, they generally bask in the sun and air.

In the South Sentinel Islands, Alcock was delighted to find the robber crab, *Birgus latro* (Figure 5.2b) in large numbers. This island is one of the few places where robber crabs are still found. It is a true hermit crab, but it is so big, no empty shell can hold it, in contrast to the smaller hermit crabs. As a result, its tail portion, instead of being soft, is adorned with hard plates; this crab is practically an air breather. Although it has the usual 14 pairs of gills, they are all small and the crab breathes through spongy, vascular and arborescent growths of the walls of the gill chamber. This crab is called a robber because of its alleged stealing of coconuts from the trees, but this may not be true. However, it may eat the flesh of fallen coconuts, using its large chelate legs to tear open the nut wall.

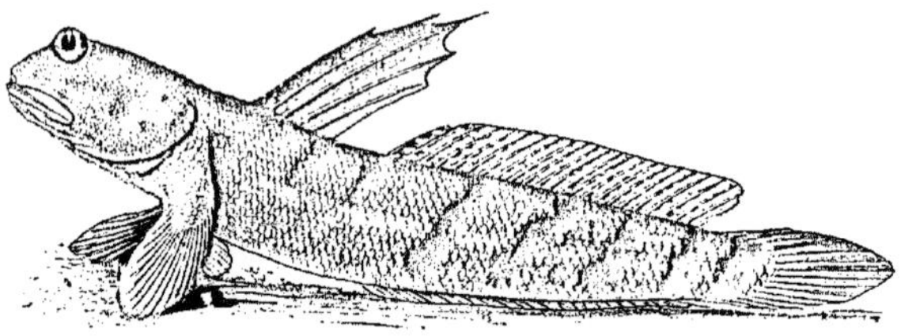

Figure 5.2a
***Boleophthalmus boddartii*, a fish common in estuarine mudflats in the Indo-Pacific region. (After Alcock, 1902)**

In his narration of Drake's voyage round the world, Francis Fletcher referred to this crab as one of the 'rare and admirable creatures' found on the uninhabited little island south of Celebes, where Drake's party took refuge for repairs and replenishment. To quote: 'They could not omit to speak of a certain kind of crayfish of such size that one was sufficient to satisfy four hungry men at a dinner, being a very good and restorative meat; the special means (as we conceived it) of our increase of health. They are as far as we could perceive, utter strangers to the sea, living always on land...'.

Coming back to the Ganjam (Orissa) coast, Alcock reports on the occurrence of solitary corals belonging to the genera *Heterocyathus* and *Heteropsammia*. These look like small discs. Also found commonly was a branching zoophyte, *Spongodes pustulosa*, which was found to harbour no less than four small crustaceans (an *Alpheus*, a *Galathea*, a *Porcellana* and a little spider crab, *Hoplophrys oatesi*), all of which in life are greyish-white with bright pink spots, so that they are practically invisible among the filaments of the zoophyte where they hide. Also found along this coast is a very venomous catfish, *Plotosus arab*, which is widely distributed from the Red Sea to the Polynesian islands. The young of the species have a rich purple-brown colour with two bright yellow bands running along each side of the body. The large barbed spine of the dorsal fin of this fish could inflict a venomous wound causing great pain and serious consequences. A wound inflicted on Alcock's forefinger made his forearm numb and useless for several days.

Relating the coloration and courtship among animals, Alcock describes the male of a fish, *Callionymus lineolatus,* to be found off the Madras coast, perhaps the most beautiful fish he had ever seen. Its head and body are coloured in many shades of brown, blue and green, set off with light blue spots and pearl-coloured stripes. The anterior dorsal fin, which can be erected as a high sail, is golden yellow studded with many white-edged blue ocelli; the tail fin is a blend of brown and yellow, set with deep blue spots; the belly fin is a dark blue velvet sewn with rows of turquoise spots; the pelvic fins are like golden green satin, fringed with dark blue and spangled with small blue spots, and the pectoral fins are of a delicate lavender-grey with serried dark-brown spots. The female is but a very poor imitation of the male. A related species in European waters, *C. lyra*, has been described by E. W. L. Holt, in the *Proceed-*

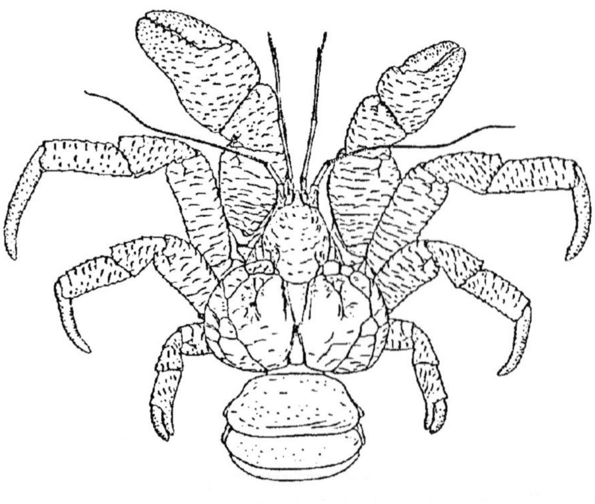

Figure 5.2b
***Birgus latro*, the robber crab or a hermit crab without a shell.**

ings of the Zoological Society of London, with a fine coloured plate of the male in an attitude of entreaty. It is a general rule that, among the lower groups of animals, it is the male that is flamboyantly coloured for the purposes of courtship; however, in one species of dragonet, *Callionymus carebares*, off the Madras coast, the female is more brightly coloured than the male and here it would appear that it is the female who makes the first advances.

Another fish belonging to the family Trachinidae (commonly called weavers), namely, *Bembrops caudimacula*, occurring below a depth of 200m, was first discovered in Japanese waters and was so named by Professor Steindachner in 1877. This fish was rediscovered three years later off the Atlantic coast of the United States and was not properly recognized, being described as a new species under the name *Hypsicometes gobioides*. Several years later, this fish was collected by the *Investigator* off the Madras coast and described as new, and named *Bathypercis platyrhynchus*. However, it was finally recognized that all three fishes were one and the same, but apparently distributed discontinuously at present; it can perhaps be assumed that, in the geological past, this fish was widely distributed.

Another very interesting fish that exhibits commensalism is the little rock-perch, *Minous inermis*. These rock-perches normally creep on the sea bottom and, because of their mottled coloration and profusion of wavy cutaneous filaments clothing their bodies, they are difficult to see among rock shingles, seaweed and zoophytes. Instead of frond-like filaments of its own, this fish is invested with living hydriform polyps belonging to the species *Stylactis minoi*. This, however, is an example not of parasitism, but of commensalism, since the polyps conceal the host from its predators, while the fish ensures the polyps' distribution.

Near the Laccadives, Alcock made four dredge hauls at depths of 738-1091fa (about 1500-2200m) and found an unusual mollusc and some deep-sea corals. The mollusc *Pleurotoma symbiotes* was always carrying small sea-anemones of the genus *Epizoanthus*. He was also surprised to find some species in common between the Laccadives and the Caribbean, such as the isopod *Bathynomus giganteus*, the blind spiny lobster *Phoberus coecus*, and the spiny hermit crab *Lithodes agasizi*. Another feature of geological interest is the finding of a bed of pumice on Cardamom Island, with no active volcanoes in sight. Alcock also made some observations on the finer qualities of the people in Minicoy, who are devoted to tuna fishing using live bait.

Alcock visited the uninhabited Pitti Island in 1889 and found it to be in a lamentable condition, with two species of tern crowding its sand banks, there being no nesting places for the birds. There were also two species of large crab mainly living on the terns, particularly the newly hatched birds. One of these crabs is a hermit crab belonging to the genus *Coenobita* and the other, a large ocypod crab, *Ocypoda ceratophthalmus*. These crabs attack young birds and literally eat them alive, which is very distasteful to witness. A similar situation was reported by Mosely in his narrative of the *Challenger* Expedition, where *Grapsus* crabs were found to eat terns on one of the islands of St. Paul. It is a sententious commonplace that nature is cruel and we had better not look too closely at the tooth and claw red with ravin. To quote Alcock in a philosophical strain:

> 'Are God and nature then at strife
> That nature lends such evil dreams
> So careful to the type she seems
> So careless of a single life.'

Before the voyage of the *Challenger* (1872-76), scarcely 30 species of deep-sea fishes were known. Now more than a thousand species are known, mainly from the findings of the later expeditions such as those of the British RRS *Challenger*, the US RV *Albatross* and *Blake*, the French research ships *Travailleur*, *Talisman* and *Candida*, and the Prince of Monaco's yachts *Princess Alice* and *Hirondella*. The most common deep-sea fishes described by Alcock from the Indian Ocean seas include *Macrurus investigatoris*, *Centrophorus rossi* and *Centroscyllium ornatum*, the latter two being spiny dogfishes with extremely fragile bodies. In some fishes, as *Bathopteris guenheri*, *Onirodes glomerulus*, *Dysomma* sp. and *Dysammopsis* sp. (the latter two being eels), the eyes were reduced and defective. In the electric ray, *Benthobatis moresbyi*, the eyes were rudimentary. However, in another fish caught at about 1500m off Kistna, namely, *Leptoderma affinis*, the eyes were of superlative size.

Many fishes and crustaceans possess light organs. These are generally modified slime glands and the reflectors are normally the modified scales of the fish or the chitinous skin. In *Diplophos corythoelum*, from depths of 390-810m in the Andaman Sea, two tiers of lanterns are found. In *Photostomias atrax*, in

Figure 5.3a
Head of *Malacosteus* sp. from the Andaman Sea (≈1300m depth); note the enormous mouth and formidable dentition.

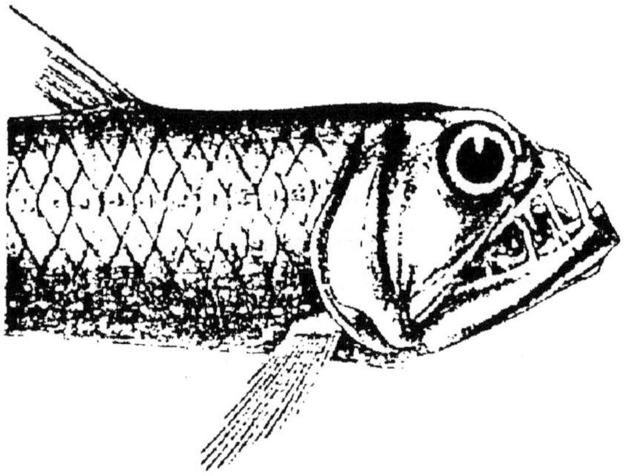

Figure 5.3b
Head of *Chauliodus pammelas* from the Laccadive Sea (≈260m depth); note the rapacious teeth.

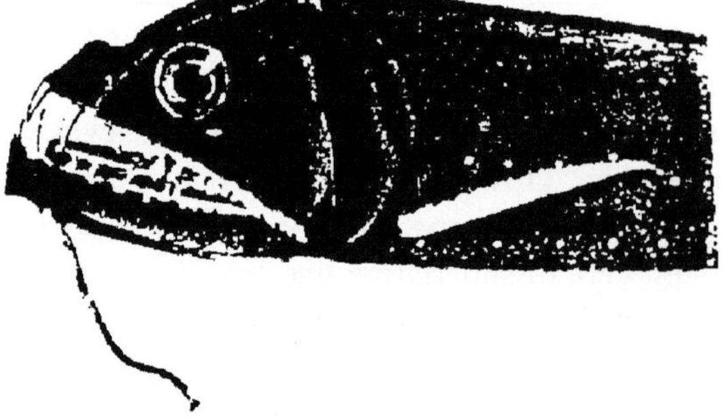

Figure 5.3c
Head of *Astronesthes* sp. from the Travancore coast; note the large teeth.

addition to two tiers, there is a large outstanding phosphorescent patch on either side of the cheek. In another fish, *Leptoderma* sp., the whole epidermis gets transformed into a luminescent coat. In *Neoscopetus*, the whole body is one dazzling sheen of purple and silver and burnished gold amidst which is a sparkling constellation of luminous organs. In some deep-sea fishes, the cleft of the mouth extends right up to the gills so that their mouth can be opened enormously so as to gulp down animals larger than themselves, as in *Odontostomius atratus*, *Malacosteus* (Figure 5.3a), *Chauliodus pammelas* (Figure 5.3b), *Astronesthes* sp. (Figure 5.3c) and *Photostomias atrax*. These are terrors of the deep sea.

Prior to the launching of the *Investigator* in 1881, nothing was known of the crustacea living in the depths of the Indian Ocean seas, with the exception of a blind lobster from the Andaman Sea described by the late Professor J. Wood-Mason, which he named *Nephropsis*. Since then, the *Investigator* has collected about 275 species of crustaceans between the meridians 60° and 90°E and from 200-2000m depth. Among them are the most brilliantly phosphorescent creatures: *Heterocarpus alphonsi*, from a depth range of about 950-1500m, and *Aristeus coruscan*, from about 1120-1650m depth. In some crustaceans from the depths, the eyes are large and black. For example, in the Andaman lobster, *Nephrops andamanica*, from 300-900m depth, the hermit crab, *Chionopagurus andersoni* (Figure 5.4a), from about 205m depth, the squeaker crab, *Psopheticus stridulans*, from 340-840m depth, and *Munida andamanica*, from 300-800m depth. However, there are also blind forms such as the blind shrimp, *Prionocrangon ommatosteres*, in which the eye stalks are reduced to a pair of useless scales; and in the isopod *Petalophthalmus armiger*, from about 1800m depth, there are no eyes at all.

Commensalism is also very common among deep-sea crustaceans. The hermit crab *Chionopagurus andersoni*, mentioned above, which is usually found blanketed with anemones, is commonly known as the blanket crab. Similarly, *Parapagurus pilosimus* (Figure 5.4b), a hermit crab with a worldwide distribution.

In Alcock's collections of Gastropoda, 22 percent belong to the genus *Pleurotoma* and one of its species, *P. simbiotes*, dredged off Cape Comorin at a depth of 2000m, always had a zoophyte, *Epizoanthus*, growing on it. More interestingly, *Turbo indicus*, dredged from 1200m depth, had two com-

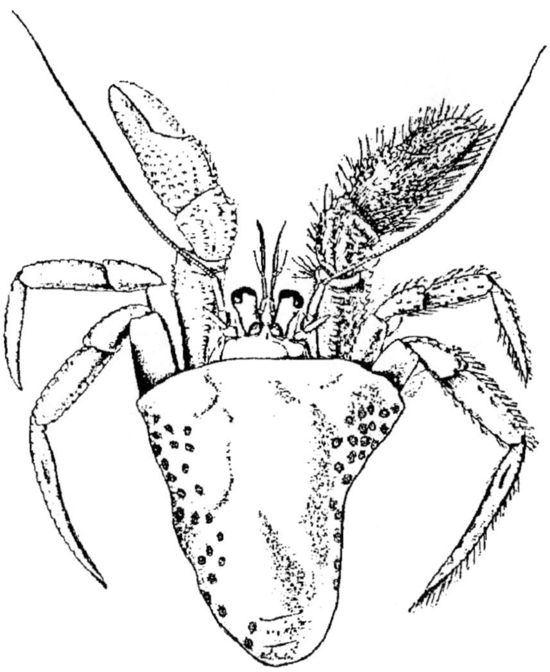

Figure 5.4a
***Chionopagurus andersoni,* a hermit crab.**

mensals, one a sponge and the other a barnacle, *Scalpellum*. It is most amazing that a barnacle larva is able to sink to such depths, and even more in some cases, and settle and metamorphose into an adult barnacle. Even telephone cables laid in deep water exceeding 4000m depth have been found encrusted with barnacles whose adults are normally found in coastal areas. This *Turbo* species is a variety of the Mediterranean and North Atlantic deep-water species, *Turbo paluritanus*. Another very common gastropod found in the Andaman Sea is the Japanese species, *Xenophora pallucida*. An example of discontinuous distribution is found in the keyhole limpet, *Puncturella astuarina*, which is found in the West Indies at a depth of some 780m, in the Bay of Biscay at a depth of 1200-2200m, and in the Bay of Bengal (southern end) at a depth of about 1218m.

Among the cuttlefishes, about 10 species were collected by Alcock and, of these, *Tonius abyssicola* was found to be the deepest-water form, generally occurring between 1200 and 2200m depth.

Among the echinoderms, about 200 species were collected and the most common of these was *Nymphaster florifer*.

Among other groups of animals, mention should be made of the corals, which, in the deep sea, are simple and single; some of them, as *Flabellum japonicum*, grow to a large size, as big as a coffee cup. Two hundred specimens of *Caryophyllia ambrosia* were taken from a depth of 2000m. In another place, at about 850m depth, nearly half a ton of *Caryophyllia*, *Desmophyllum*, *Lophohelia* and *Solenosmilia* was collected. Another noteworthy coral of the Andaman Seas is *Deltocyalthus andamanicus*, dredged from a depth of 345-605m; it has a most brilliantly sculptured disc-like corallum. More interestingly, a similar species, not really distinct from *D. andamanicus*, is *D. italicus*, a fossil species from the tertiary rocks of Italy and which still lives in the depths of the North Atlantic.

It is well known that the majority of the deep-sea sponges belong to the family Hexactinellida. These sponges are rooted in the ooze by a loose tuft of thread-like spicules resembling a wisp of spun glass, as in *Pheronema raphanus*; or by a stiff hollow stem,

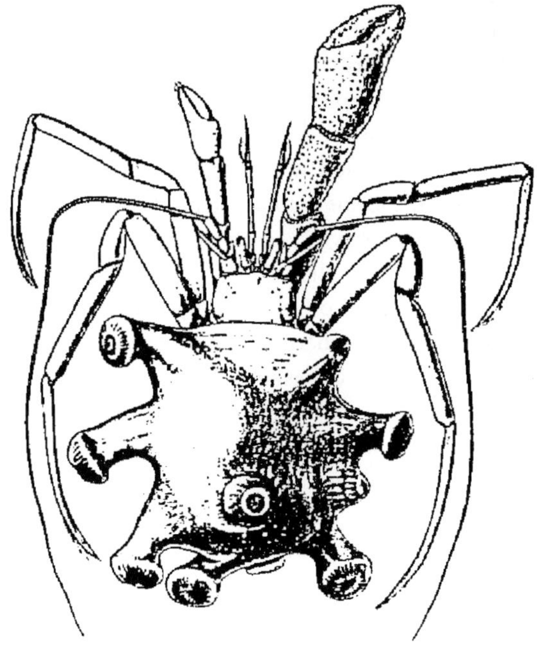

Figure 5.4b
***Parapagurus pilosimus,* bearing zoophytes.**

as in *Saccocalyx pedunculata* (Figure 5.4c), or by a long anchor rope made of strands of 'glass' tightly twisted together as in *Hyalonema* (Figure 5.4d). Sometimes we find these anchoring ropes are encrusted by zoophytes of the genus *Polythoa* and by barnacles of the genus *Scalpellum*.

To conclude, Alcock's narrative, as a naturalist on board the *Investigator* for four years (1889-92), is not only informative, but also a pioneering contribution on the deep-sea life of the Indian Ocean seas, perhaps the only publication of a century-old vintage.

Plankton studies

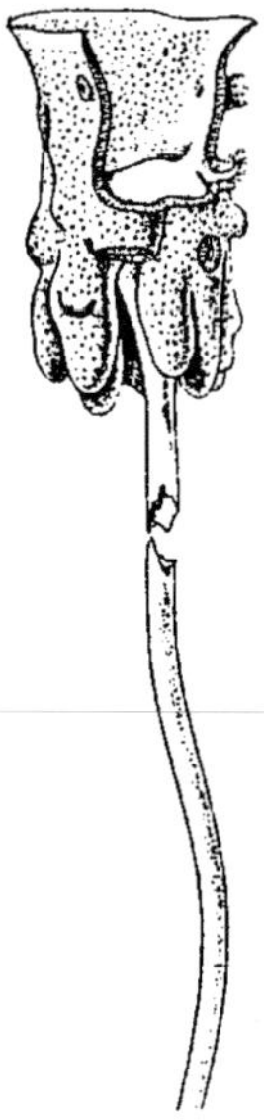

Figure 5.4c
***Saccocalyx pedunculata*, from the Bay of Bengal (≈1600m depth).**

Figure 5.4d
***Hyalonema masoni*, with encrusting barnacles, from the Bay of Bengal (≈3500m depth).**

Plankton includes two major groups. The phytoplankton comprises mainly the unicellular microscopic plants, such as the Diatomea, the Dinoflagellata, some blue-green algae (or bacteria), other photosynthetic flagellate groups, Coccolithophoridae, brown algae (floating seaweeds) and higher plants such as Sargassum weed. Phytoplankters populate the sunlit surface layers of the sea and produce carbohydrates by photosynthesis. The phytoplankters are termed *primary producers*; it is on their photosynthetic production that the entire food chain or food pyramid in the sea rests. The zooplankton constitutes the next group in the food chain and comprises a variety of organisms of weak swimming ability (though not always of small size), of which, the predominant groups are, generally speaking, the Copepoda, the Euphausiacea, the Chaetognatha, and several other, less important groups, such as the Foraminifera, Amphipoda, Ctenophora, Pteropoda, Radiolaria, and some Mollusca and Tunicata.

It was mentioned when dealing with the history of the Indian Ocean studies, that, almost up to 1947, there were very few research publications on plankton. Seymour Sewell (1929), before leaving India, published a paper on the *Copepods of the Indian Seas: Calanoida*, wherein he described nearly 400 species of copepods based on the collections of the Indian Museum. Running into 407 pages, plus 6 beautiful plates and 131 text figures, the monograph is a monumental piece of systematic work on the copepods of the Indian Ocean seas. Sewell continued this work into the John Murray Expedition (1933-34), reports of which were published in later years.

Commenting on the distribution, Sewell remarks that, as far as deep-water copepods are concerned, they are alike in all the three major oceans, but the surface-dwelling forms in the tropical zones have a common distribution between the Indian and Pacific Oceans, but a discontinuous one with the Atlantic. This may be due to the free exchange of surface and subsurface waters that takes place between the Pacific and the Indian Oceans through the various channels and straits connecting the two oceans in the Indo-Malayan and Indonesian archipelagos, and particularly during the northeast monsoon when there is a strong flow from the Pacific into the Indian Ocean through the Straits of Malacca and the Timor Sea. The absence of some of the Indo-Pacific surface-dwelling copepods in the Atlantic may be more apparent than real. Future investigations may discover many species common to these oceans. Coming to the coastal waters of India, no significant report on the plankton was published until 1950 when Chacko gave an account of the marine plankton from the waters around the Krusadai Islands. Prior to this, K. S. Menon (1931) described the plankton of the Madras coast, followed by M. A. S. Menon (1945) on the Travancore coast, and a systematic account of the diatoms of the Madras coast by Subrahmanyan in 1946. Some of the other papers are: *Contributions on the life-history of sardines with notes on the plankton of the Malabar coast*, by Hornell and Nayudu (1923); *On plankton records for the years 1929 and 1930, from Madras*, by Aiyar et al., (1936); *On the plankton of the Bombay Harbour (a preliminary note)*, by Bal and Pradhan (1945); and *On the swarming of the planktonic algae,* Trichodesmium erythraeum *in the Pamban area and its effect on the fauna*, by Chindambaram and Unni (1944). These contributions provide some basic information on the plankton, its systematics, general distribution and significance along the Indian coast. Between 1924 and 1939, Delsman

(1924-1939) published a series of papers on the fish eggs and larvae from the Java Sea, including a report on the plankton. Information on the plankton off the east African coast (western Indian Ocean) comes from Wickstead (1963) and De Decker (1973) and, in the Red Sea, from Kimor (1973) and others.

By the end of the 1950s, a large number of papers on all aspects of oceanography began to appear in some of the new journals, such as the *Proceedings of the Indian Academy of Sciences*, the *Indian Journal of Fisheries*, the *Journal of the Marine Biological Association of India*, the *Indian Journal of Marine Science*, and *Mahasagar*, the Bulletin of the National Institute of Oceanography, Goa. By the time the International Indian Ocean Expedition was over, in 1965, worldwide contributions to our understanding of the Indian Ocean were quite substantial. This is well reflected in the two International Indian Ocean Symposia, one held in Cochin from 12 to 18 January 1971 and the other held in Kiel from 31 March to 6 April 1971. These are all published and the Indian Ocean can no longer be considered as an unknown ocean (for details, see the *Biology of the Indian Ocean*, edited by Zeitzschel (1973) and the *Journal of the Marine Biological Association of India*, from Volume 14 onwards). Besides, UNESCO arranged for the issue of a reprint collection of all the papers published during the IIOE, in 8 volumes with an index, and this forms an important source for Indian Ocean studies.

In practically all the earlier papers dealing with phyto- and zooplankton, there is generally a description of the hydrography followed by a report on the occurrences and distribution of the various groups, and discussion. A good example of this approach is the paper by Chacko (1950), on the plankton of the Krusadai Islands (9°14'N, 79°12'E). Practically all research studies prior to IIOE, with the possible exception of the oceanographic work done by the Andhra University, were of inshore and coastal waters in the Indian Ocean, because of the lack of ocean-going facilities in this area. Since coastal areas are the most productive, these publications have considerable relevance to our present use and understanding.

Chacko's observations reveal that, as far as general hydrographic conditions are concerned, the monsoonal effect can be seen very clearly in the salinity values. During the NE monsoon, which is prevalent in the Krusadai area in the months of November-January, the salinity shows lowest values during the year, there being no effect of the SW monsoon here. The most common and important diatoms, which occur throughout the year, are the species of *Coscinodiscus*, *Rhizosolenia*, *Chaetoceros* and *Thalassiothrix*. The maximum bloom occurs from June to November in the Krusadai waters and, in the same period, off Madras on the east coast and off Calicut on the west coast. It is also important to note that the 'red tide' caused by the blue-green alga, *Trichodesmium erythraeum*, occurs at Krusadai in the months of April, May, July and October. When this alga occurs in large numbers, the sea becomes coloured brownish-yellow and emits a strong odour of chlorine; the bloom usually lasts 3-7 days, during which time heavy mortality of many organisms occurs. The 'red water' or 'red tide', as it is called, appeared to have caused an abrupt setback to the local fisheries along the Malabar and Kanara coasts, according to a paper written by Bhimachar and George (1950). They noticed that, in the middle of October 1948, within two weeks of the commencement of a bumper season for fish, there was a severe setback in the fish catches coinciding with the appearance of the 'red water' due to sudden prolific multiplication of the dinoflagellate, *Noctiluca miliaris*, their subsequent death and decomposition. Recurrence of this along the Indian coast and elsewhere has attracted the attention of marine biologists for a great number of years in Japan, South Africa, California, and in the Gulf of Mexico, where, apart from *Noctiluca*, other dinoflagellates, as *Gonyaulax polygramma*, *Prorocentrum micans* and *Gymnodinium brevis*, are involved. Periodically, there are continuing reports of fish mortality attributed to these causative organisms of 'red waters', but even after the IIOE and other national efforts, the problem remains fully open to investigation.

In a series of papers, Subrahmanyan (1946, 1959) and Subrahmanyan and Sen Gupta (1963, 1965) have presented a detailed picture of the phytoplankton and zooplankton off the Calicut and Madras coasts, covering a period of nearly 8 years (5 years at Calicut and 3 years at Madras). The results of these investigations are generally applicable to the Indian coast and other littorals in the northern Indian Ocean area, and the various papers published by other workers in these areas substantially corroborate Subrahmanyan's findings.

The summary of the nutrient parameters indicates that nutrient concentrations in the Indian coastal waters are fairly high and do not normally constitute limiting factors for the primary producers.

Table 5.2 Summary of Subrahmanyan's findings (1959)

Place of work:	Calicut (4 stations in the Laccadive Sea)
Years of study	1949-1955
Physical background	Two rivers, Kalair and Elathur, open into the sea near the place of study. The coast is interspersed with low rocky cliffs and sandy beaches.
Seasons at Calicut	Four seasons are recognized: (1) wet season of SW monsoon, May-September (2) fairly dry, October-November (3) dry NE monsoon season, December-March (4) hot season, dry, April-May
Wind force	Average value in mph: 5.6 to 8.7 (2.5 to 3.9m/s)
Rainfall	Over 50% of the rain normally falls in July-August. 1949: 138.4" (3.52m) 1952: 92.16" (2.34m) 1953: 95.7" (2.43m) 1954: 143.2" (3.64m)
Sea bottom	Soft gray mud (terrigenous)
Coastal currents	Northerly during October-January Southerly during February-September
Number of phytoplankton cells/l	Average for 5 years: Highest: 3,404,683,000 (July); Lowest: 79,649,000 (May)
Magnitude of standing crop	0.063-12.28g C/m^2/day
No. of species of phytoplankton occurring at a time	12-74
Most common diatoms (these form 90% of the sample)	*Fragillaria, Chaetoceros, Rhizosolenia, Coscinodiscus, Asterionella, Nitzschia, Bacteriastrum, Thalassiothrix, Thalassiosira, Actinoptychus*
Common Dinophyceae	(armoured): *Ceratium, Peridinium, Ornithocercus, Dinophysis* (unarmoured): *Gymnodinium, Gyrodinium, Noctiluca*
Seasonal fluctuation of plankton	Bimodal: a major peak during the SW monsoon season and a minor peak during the NE monsoon
Phytoplankton maxima	Bombay: January, February Calicut: June, July, August Krusadai: June, November Madras: April, May Waltair: February, April
Zooplankton peaks	Madras: November, February Mandapam: February, April, October (Krusadai) Trivandrum: December, February, May Calicut: August
Hydrographical parameters at Calicut	
Temperature: (surface water)	Lowest in July (25.3°C) Highest in April (30°C); presents a double oscillation - with high temperatures in April and October
Solar radiation:	Plentiful
Salinity at Calicut:	Maximum: 35.0 Minimum: 31.0 Presents a double oscillation

(Other salinity values in Indian coastal waters are given in Table 5.3)

Table 5.3 Salinity (in parts per mille) of Indian coastal waters

Region	Minimum	Maximum	Remarks	References*
Bombay harbour	23.56	38.40	Lower value due to SW monsoon	Bal *et al.* (1946)
Calicut	31.00	35.00	"	Subrahmanyan (1959)
Gulf of Mannar	28.85	36-47	"	Jayaraman (1954)
Palk Bay	25.52	36.39	"	Jayaraman (1954)
Madras	24.00	34.50	Lower value due to NE Monsoon	Jayaraman (1951)
Vishakapatnam	25.00	34.00	Lower value due to both monsoons but mainly SW monsoon	Ganapati and Murthy (1954)

* *In the present publication, the references cited in Tables are not cited in the list of References unless they are also cited in the body of the text.*

The dissolved oxygen shows great fluctuation, the values at Calicut ranging from 3.5 to 5ml/l; at Falk Bay, 2.5-4.5ml/l; at Madras, 2.5-5ml/l, but rarely reaching saturation levels as, for example, in the Arctic waters, where the oxygen content sometimes reaches 8.0ml/l.

Subrahmanyan's important findings are summarized in Tables 5.2 and 5.3.

The biological year along the west coast of India starts in April, co-inciding with the pre-monsoon conditions. The standing crop, expressed as numbers of phytoplankters per litre and Harvey units (the ^{14}C method was just then being developed by Steéman Nielsen and used on the *Galathea* Deep-Sea Expedition, 1950-52), was at its maximum during SW monsoon months (May to October) and minimum in November, during the NE monsoon. The bulk of the standing crop in the SW monsoon was mainly composed of phytoplankton, whereas it was zooplankton that constituted the bulk during the northeast monsoon. The ratio between the two crops was 1: 0.9 and, by this, it was surmised that a mass of water can support only a certain quantity of living matter regardless of what the biomass is made up of. A close relationship was found between the fish landings and the standing crop of the phytoplankton and zooplankton, and a more direct relationship was found between the oil sardine and the diatom *Fragillaria oceanica*. Subrahmanyan also calculated that the standing crop of phytoplankton stood at about 1.8 billion metric tons, while the fish catch landed in India was about 562,000 metric tons and the ratio between the two worked out to 0.003%. The conversion efficiency from the phytoplankton standing crop to that of fish is therefore extremely poor, and that, moreover, in a region that is claimed to be one of the most fertile regions of the world. Subrahmanyan has also commented upon the nutrient values (Table 5.4).

The study of the ratios of Si, N and P found in Indian waters indicates a ratio of 20:15:1 which is the same obtained by others elsewhere. Most of the work has been done in temperate waters, in Europe and USA; but even so, it would appear that life's processes are almost the same everywhere and offer extremely interesting visions of Nature in action. To quote Redfield (1934, pp. 189-91):

'It appears to mean that the relative quantities of nitrate and phosphate occurring in the oceans of the world are just those required for the composition of the plants and animals which live in the sea. That the two compounds of such great importance in the synthesis of living matter are so exactly balanced in the marine environment, is a unique fact and one which calls for some explanation, if it is not to be regarded as a mere coincidence. It is as though the seas had been created and populated with animals and plants and all of the nitrate and phosphate which the water contains had been derived from the decomposition of this original population... Whatever its explanation, the correspondence between the quantity of the biologically available nitrogen and phosphate in the sea and the proportions in which they are utilized by the plankton is a phenomenon of greatest interest and further the general agreement indicates that the amounts of nitrate and phosphate present in the sea are controlled in the main by the same agency – the consumption by the phytoplankton and regeneration from its remains'.

But this ratio may not be true in isolated observations. In fact, Subrahmanyan found the N:P ratio values highly erratic in his studies. For example, the values for nitrate-nitrogen at the sea surface fluctuated between (all in µg atom/l) 2.37 and 16.2, in 1951-52, 1.40 and 34, in 1952-53, 2.25 and 6.27, in 1953-54, and 0.47 and 10.7, in 1954-55, at the surface. The silicates similarly fluctuated on the west coast between 8 and 15µg atom/l, during the NE monsoon, and 17 and 20 µg atom/l, during the SW monsoon. In the case of phosphates, the values (in µg atom/l) at the sea surface varied between 0.46 and 1.26, 0.31 and 0.85, 0.20 and 1.19, 0.13 and 1.02, and 0.41 and 1.68 during the years 1950-59, the highest values occurring during the SW monsoon.

Subrahmanyan's (1959) paper on the phytoplankton off Calicut marks an important milestone in our understanding of the physical, chemical and biological factors influencing the primary production. Papers prior to this were all of a preliminary nature and mainly dealt with seasonal changes in the general plankton biomass, more particularly that of zooplankton. What was required at that time was an estimate of the fishery resources of India and a search for any new stocks and new fishing grounds. Marine scientific research had to be directed to this end; apart from launching regular exploratory fishing surveys, it was essential to study the biology of the Indian seas. A thorough study of the phytoplankton of the Indian coastal waters and offshore was launched by the Central Marine Fisheries Research Institute, both at the headquarters in Mandapam and at its regional stations in Calicut and Madras. Against this background, the observations of Subrahmanyan on the plankton off the Calicut coast are important and relevant.

Phytoplankton systematics and distribution

Subsequent to Subrahmanyan's paper, there has been a spate of contributions on phytoplankton and productivity in the Indian Ocean area between 1960 and 1990, thanks to the excellent opportunities provided by the research vessels participating in the International Indian Ocean Expedition (IIOE) and by the various oceanographic institutes and marine biological stations established around the Indian Ocean rim. (More importantly, practically all the littoral countries bordering the Indian Ocean had obtained their independence from colonial powers, and many scientists in these countries were anxious to contribute to knowledge of the Indian Ocean.) Apart from a deep desire to unravel the unknown, the scientists were particularly keen to estimate the biological productivity of the Indian Ocean and its fishery potential,

Table 5.4 Nutrient values (µg atom/l) along the Indian coast

Nutrient	Region	Annual range	Authors
Phosphate-P	Bombay	0.18-0.53	Bal *et al.* (1946)
	Calicut	0.13-1.68	Subrahmanyan (1959)
	Cochin	0.50-1.00	Qasim (1973)
	Palk Bay	0.12-0.25	Jayaraman (1954)
	Madras	0.33-1.98	Ramamoorthy (1953)
Nitrate-N	Calicut	2.14-9.67	Subrahmanyan (1959)
	Cochin	2.00-28.00	Qasim (1973)
	Gulf of Mannar	0.00-5.00	Jayaraman (1954)
	Madras	8.00-16.00	Jayaraman (1954)
	Arabian Sea	20.00-35.00	Gilson (1937)
Silicate-Si	Bombay	5.25-32.50	Bal *et al.* (1946)
	Calicut	5.17-40.47	Subrahmanyan (1959)
	Cochin	5.00-48.00	Qasim (1973)
	Gulf of Mannar	3.00-30.00	Jayaraman (1954)
	Madras	2.00-22.00	Jayaraman (1954)

the 'red tide' phenomenon and the taxonomy and distribution of the marine fauna and flora. All these were important to know.

Subrahmanyan and Sarma (1960) have described the occurrences of 291 species of phytoplankton in the Indian coastal waters, based on collections made between May 1949 and April 1954, and they have indicated monthly environmental factors and a list of dominant species found in each month. Of the phytoplankters, 29 species of Bacillariophyceae, 7 species of Dinophyceae and 1 of Cyanophyceae could be considered as 'mass forms', all constituting the bulk (over 107 cells per vertical haul) of the flora along the west coast of India, particularly off Calicut. (See Table 5.5 for the list of 'mass forms' given by the above-mentioned authors).

These authors recorded a large number of species every month, the maximum being 190 and the minimum, 140 species; however, the bulk would be contributed by a few species, depending on the seasonal and hydrographical factors.

Most of the papers on the phytoplankton published in India, during the last 30 years, with the exception of papers by Gopinathan (1984), Chandran (1985) and Thorrington-Smith (1971), deal mainly with the geographical distribution and standing crop of the phytoplankton, not by species, but mainly based on higher taxonomic groups, pigments and assimilation rates of carbon, in relation to light, nutrients and hydrographical conditions.

When compared to the land, oceans in general may be considered as being deserts, except for areas of divergences and upwelling, which all together may constitute about 1 percent of the ocean surface. To some extent, the difference is made up by the fact that oceans cover nearly three-fourths of the planet's surface. The difference between the land and the ocean, in terms of productivity, was vividly brought out by Ryther (1959), using nitrogen as an index of productivity. According to Ryther, rich fertile soil contains some 5% organic humus and 0.5% nitrogen. This concentration, together with other nutrients and carbon, hydrogen and oxygen in a cubic metre of soil, can support a crop of some 50kg of dry weight of organic matter, an amount equivalent to more than 200tons per acre ($\approx$4050m^2) of soil 3ft ($\approx$0.9m) deep. In contrast, the richest unpolluted ocean water contains about 60µg atom/l or 0.00005% of nitrogen, four orders of magnitude less than fertile land. A cubic metre of such water could perhaps support a phytoplankton crop of no more than 5g.

Table 5.5 List of 'mass forms'

Bacillariophyceae

Asterionella japonica
Bacteriastrum hyalinum var. *princeps*
Biddulphia heteroceros
B. mobiliensis
Chaetoceros affinis
C. brevis
C. compressus
C. contortum
C. curvisetus
C. lasciniosus
C. lauderii
C. orenzianus
C. pelagicus
C. socialis
Coscinodiscus asteromphalus
C. occulus-iridis
Fragilaria oceanica
Guinardia flaccida
Lauderia annulata
Leptocylindrus danicus
Nitzschia seriata
N. sigma var. *indica*
Rhizosolenia alata
R. robusta
R. stolterfothii
Sceletonema costatum
Schröderella delicatula
Thalassiothrix frauenfeldii
T. longissima

Dinophyceae

Ceratium fuscus
C. macroceros
C. tripos
Dinophysis caudata
Glenodinium lenticula f. *asymmetrica*
Noctiluca miliaris
Peridinium depressum

Myxophyceae

Trichodesmium erythraeum

With this general background, we may now review the situation in the Indian Ocean. For the first time, the IIOE (1959-65) provided an opportunity to study the distribution of phytoplankton and its productivity, mainly based on the estimation of chlorophyll concentration and ^{14}C assimilation over a wide area of the Indian Ocean. There has also been a large number of studies on phytoplankton standing crop and productivity in the inshore waters along the Indian coast, made by the scientists of research institutions and university departments too numerous to mention here. All those studies very clearly indicate the biological productivity and standing-crop distribution. Based on the IIOE data, Krey (1973) and Aruga (1973) have summarized the distribution of phytoplankton standing crop in the Indian Ocean. Earlier, Nielsen and Jensen (1957-59) made a pioneering study of the productivity in all of the three major oceans (Pacific, Atlantic and Indian), during the *Galathea* Expedition (1950-52).

They used the ^{14}C method for measuring the productivity. This method is considered to be more precise than other methods of measurement such as pigment extraction and dark- and light-bottle estimations.

Just to recall that, of the various factors that influence phytoplankton distribution, the following are important. First, the incident radiation. This becomes particularly critical in higher latitudes, but not in the Indian Ocean where the sun shines most of the time. Next comes transparency of the sea water. Where the waters are turbid, owing to the presence of particulate matter or dissolved substances and high concentrations of phytoplankton itself, the depth to which the light penetrates is an important factor in determining photosynthetic rates. In the northern Indian Ocean, particularly in the coastal areas of the Bay of Bengal, transparency is greatly affected by the tremendous river run-off and monsoon floods, which bring in loads of sediments, particularly during the SW monsoon.

The presence and relative concentration of nutrients, such as nitrates, phosphates, silicates, and of carbon dioxide and oxygen, are critical for the growth and production of the phytoplankton. As already stated, these chemicals are at very low concentrations over vast areas in the oceans, making the oceans biological deserts. Wherever enrichment of the surface waters takes place, mainly through upwelling, phytoplankton productivity will be at its maximum, given sufficient stability, radiation and transparency of the water.

Regarding light, Qasim *et al.* (1972a) tested 11 species of marine tropical algae and found that saturation intensities for most species ranged between 0.006 and 0.09 langleys/min, but *Dinophysis miles* and *Rhizosolenia styliformis* had exceptionally high saturation intensities (0.21 and 0.14 langleys/min, respectively). Qasim *et al.* (1972b) also found that, for each species, the relationship between light intensity and photosynthesis was approximately linear at lower light intensities. At supersaturation intensities, many species (e.g., *Rhizosolenia styliformis*, *Planktoniella sol*, *Ceratium furca*) showed inhibition; in others (e.g., *Triceratium favus* and *Coscinodiscus radiatus*) no inhibition was noticed. They further reported that tropical organisms in the coastal and estuarine areas are not only euryhaline, but also show maximum rates of photosynthesis at low salinities (Table 5.6).

Table 5.6 Salinities at which maximum photosynthesis occurred in different organisms and their rate of photosynthesis at lowest and highest salinities (from Quasim *et al.*, 1972*a*)

Organisms	Salinity range for maximum photosynthesis	Rate of photosynthesis at salinity of 5	Salinity at which organism was collected	Rate of photosynthesis (%) at salinity corresponding to natural environment
Diatoms				
Nitzschia closterium	10-15	61	34.5	57
Planktoniella sol	15-20	27	34.5	55
Triceratium favus	10-20	16	34.5	46
Asterionella japonica	10-20	46	34.5	45
Chaetoceros lorenzianus	10-20	22	34.5	46
Coscinodiscus radiatus	10-20	3	33.8	49
Biddulphia sinensis	10-20	68	33.2	50
Rhizosolenia styliformis	15-20	41	33.2	62
Thalassiosira subtilis	14-24	60	34.5	56
Dinoflagellates				
Ceratium furca	6-10	99	33.2	37
Dinophysis miles	10-15	10	34.5	37
Dichtyocha sp.	20-25	59	33.2	75

Along the east and west coasts of India, salinity of the coastal waters changes rapidly with the onset of the monsoons; on the west coast the low salinities are recorded during June-September, whereas, on the east coast, it would be some time later, in October-January. It is during these months that maximum photosynthesis occurs in India's coastal waters. Subrahmanyan (1959) reported maximum phytoplankton crop during the SW monsoon at Calicut when the salinities are low; other workers along the Indian coast have made similar observations. Raymont (1980), in his revised edition of the book titled *Plankton Productivity in the Oceans,* has summarized all available information of importance on phytoplankton distribution, standing crop and productivity in all the oceans. Earlier, Krey (1973) presented a paper on the primary production in the Indian Ocean at the International Symposium on the Biology of the Indian Ocean held in Kiel in 1969, and in 1976 he and Babenard published an atlas of the Indian Ocean showing primary productivity (Figure 5.5 a, b). Qasim (1977) contributed a comprehensive paper on the biological productivity of the Indian Ocean based on all available data at that time. Apart from the previously cited contributions, mention should be made of the pioneering studies on phytoplankton production made by Nielsen and Jensen (1957-59) in the tropical parts of the three major oceans, during the *Galathea* Deep Sea Expedition, and by Ryther *et al.* (1966) on the productivity of the Arabian Sea using the *Anton Bruun* data.

Ecological provinces of the Indian Ocean

Some general remarks on the hydrography and seasonal monsoons prevailing over the Indian Ocean have already been made. To stress again, the northern Indian Ocean is a monsoonal ocean, where the opposing monsoons bring about reversal of surface currents, which in turn influences the distribution of salinity and temperature, nutrients and dissolved gases. Different areas become subjected to divergence or convergence, upwelling and mixing of water masses, depending on the prevailing monsoon (Figure 5.6a, b). Where this happens, enrichment of the surface waters takes place leading to the growth and abundance of phytoplankton.

Thorrington-Smith (1971) made a detailed ecological study of the species distribution in the western Indian Ocean, based on phytoplankton samples collected by the RRS *Discovery* during the IIOE. He identified 237 different species living in this large area and grouped them seasonally, one group during the SW monsoon and the other during the NE monsoon; he also related their distribution to the water types and masses: (i) equatorial subsurface water; (ii) the

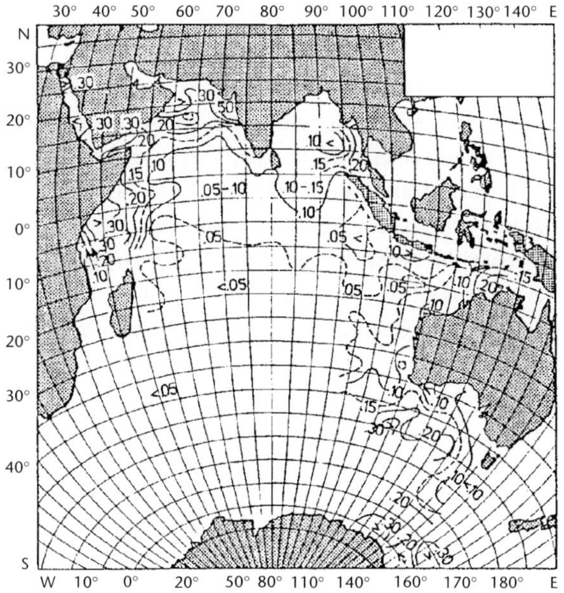

Figure 5.5a
Average concentration of chlorophyll *a* (mg/m³) in the surface layer (0-50m) from December to March. (After Krey, 1973)

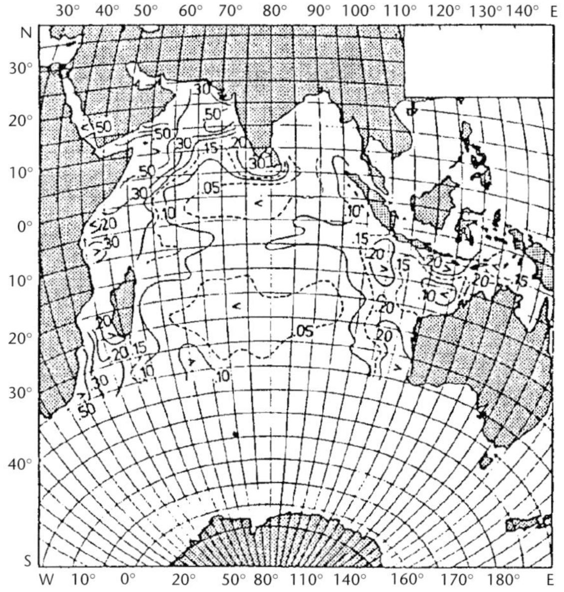

Figure 5.5b
Average concentration of chlorophyll *a* (mg/m³) in the surface layer (0-50m) from June to September. (After Krey, 1973)

equatorial under-current boundary; (iii) the South Equatorial Current; and (iv) the surface tropical water. Among the 11 floral elements identified, Smith found that the largest group comprised 50 species and was found in the equatorial subsurface water. Because of these group dominances in all samples, they were considered to be endemic Indian Ocean flora. Such studies are lacking from other areas of the Indian Ocean.

Reporting on the occurrence of more than 300 dinoflagellate species, belonging to 40 genera, collected by the RV *Anton Bruun*, particularly from the Bay of Bengal and the Andaman Sea, Taylor (1973) considers them to be very important and often indicative of upwelling zones. He supports the observations of Sukhanova (1962, 1964) that the following group of species, *Pyrocystis pseudonoctiluca*, *Ceratium cariense*, *C. massiliense* and *C. trichoceros*, forms the basic tropical complex of dinoflagellates. Of these, *P. pseudonoctiluca* is exceptionally widespread and *C. trichoceros* is extremely stenothermal in the area. The southern boundary of most of the dinoflagellates appears to be 30° S, with the exception of the Agulhas Current which carries dinoflagellates from tropical areas farther south.

Standing crop and productivity

As already mentioned, the main aim of the biological investigations in the sea is to estimate the living resources exploitable by Man, more particularly the annual yield of sustainable fish catch. Towards this goal, hundreds of papers have been published on the phytoplankton standing crop and productivity of the world oceans. In fact, one of the major aims of the International Indian Ocean Expedition (IIOE) was also to sort out this problem of productivity, particularly its estimation at different trophic levels, transfer efficiencies and the related fishery potential of the Indian Ocean.

The standing crop is the total biomass of the phytoplankton and this can be measured and expressed as wet or dry weight per unit volume or under unit surface (usually 1 m^2) covering the entire euphotic column. It is also measured in units of chlorophyll *a*, the green pigment universally present in all phytoplankton and the so-called 'green' bacteria. Direct counting of all the phytoplankton cells present in a unit volume (litre) of seawater is also followed. In the IIOE, pigment analysis was widely followed and these data form the basis of the phytoplankton atlases produced

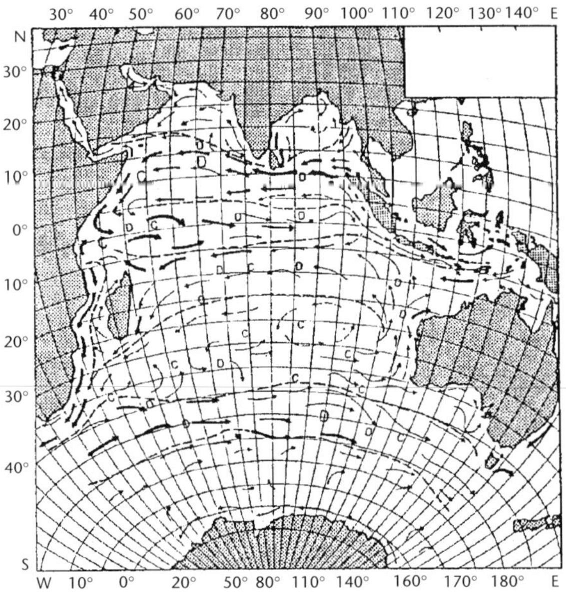

Figure 5.6a
Zones of divergence (D) and convergence (C) in the Indian Ocean, based on long-term hydrographic observations for the NE monsoon; arrows represent current direction. (After Krey, 1973)

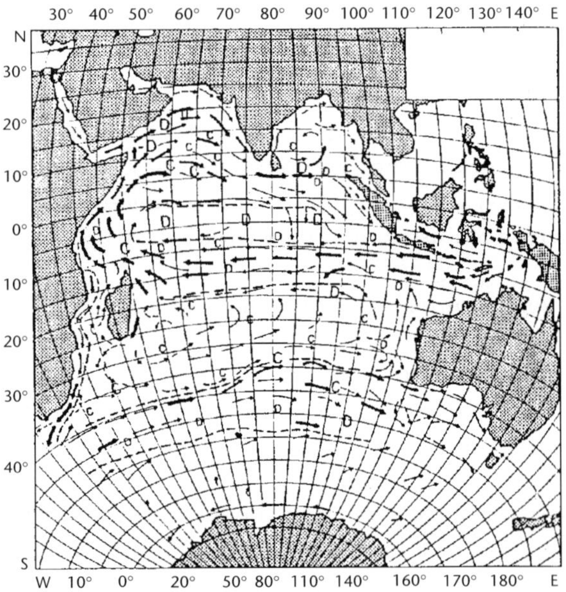

Figure 5.6b
Zones of divergence (D) and convergence (C) in the Indian Ocean, based on long-term hydrographic observations for the SW monsoon; arrows represent current direction. (After Krey, 1973)

by Krey and Babenard (1976) (see Figure 5.5a, b). It is clear from these Figures that phytoplankton abundance is restricted mostly to coastal areas of the Indian Ocean. It can also be seen that the extent of phytoplankton abundance is greater during the SW monsoon than during the NE monsoon season. This is because of the extensive upwelling taking place along the west coast of India, parts of its east coast, the Somali and Arabian coasts, in the Agulhas Current region, and on the leeward side of Sri Lanka. There are also high standing crops in the equatorial divergences (Figure 5.6a, b). In most of these areas, chlorophyll concentrations exceed 0.5mg/m^3 whereas, during the NE monsoon, the areas of upwelling shrink or almost disappear long the west coast of India, the Somali and Saudi Arabian coasts and in other areas, such as south of Java and the western Australian coast. As a result, chlorophyll concentrations here are less than 0.5mg/m^3. The rest of the Indian Ocean, both north and south of the equator, has very low standing crops and can be described as a biological desert.

While phytoplankton standing crop is a measure of the biomass, productivity gives the rate of photosynthesis of organic matter by plants in the sea. Nielsen and Jensen (1957-59) introduced a more precise method of estimating this productivity using a radio-isotope of carbon (^{14}C) as a tracer to record the rate of photosynthesis and productivity. As already stated, they had used this method successfully during the *Galathea* Expedition and, perhaps for the first time, obtained an estimate of productivity in the tropical and subtropical areas of the Indian, Atlantic and Pacific Oceans at 194 stations, with an average density of about one sample for every 2 million square kilometres.

Subsequently, many other expeditions used the same method, as a result of which Koblentz-Mishke *et al.* (1970) produced an atlas showing primary production of the world ocean. This was estimated to be 1.5-1.8 × 10^{10}tons. Apart from this estimate, Qasim, in India, and Ryther, in the USA, made specific estimates of productivity in the Indian Ocean area.

Kabanova (1968) and Aruga (1973) have summarized primary production rates in the Indian Ocean. Chlorophyll standing crop in the eastern part of the Indian Ocean was found to be generally low from December to January, being about 0.1mg/m^3 in the north and less than 0.05mg/m^3 in the south at the surface to the west of 100°E and about 0.05mg/m^3 at the surface to the east of 100°E.

Primary production measurements in the Indian Ocean indicate that the SW monsoon is more productive than the NE monsoon (Nair and Pillai, 1983).

Maximova (1971) summarized the distribution of nutrients in relation to primary productivity. The distribution of primary production north of 40°S followed the same general pattern as that of nutrients. In the regions where the intensity of phytoplankton photosynthesis was not restricted by light, high levels of primary production were observed in the areas with phosphate content in the 0-100m surface layer being above 60µg atom/m^2, and primary production was characteristically low where phosphate was below 50µg atom/m^2.

Based on these studies, the following general conclusions can be made:

1. Annual primary production of the open sea varies, for the most part, between 25 and 75g of carbon fixed below a square metre of ocean surface, and the productivity averages 50g of carbon per square metre per year. This covers an area of roughly 90% of the oceans, about 326 × 10^{6}km^2.
2. Higher levels of production occur in shallow coastal waters (less than 180m depth). The mean values of productivity for this area may be 100g of carbon/m^2/year (about 9.9% of the total ocean surface, covering an area of about 36 × 10^{6}km^2).
3. In a few restricted areas of the ocean, where upwelling takes place, particularly on the western coasts of the continents, the productivity may be (on average) as much as 300g of carbon/m^2/year covering an area amounting to 3.6 × 10^{6}km^2, or roughly 0.1 percent of the world ocean surface.

It is clear from the foregoing that 90% of the world ocean has very low productivity and produces only a small fraction of the world fish catch, with little or no potential yield in the future. Nor do Antarctic areas, south of the Antarctic convergence, appear to be as rich as was once made out by some biologists. While the short spring and summer season may produce a sharp burst of productivity leading to enormous abundance of krill and whales, the average primary productivity values are not higher than those in most open-sea areas. The total annual production in the Antarctic seas could be about 50 × 10^5 metric tons.

Previous studies of importance for evaluating the productivity of the Indian Ocean were made by the *Dana* Expedition (1928-30), the John Murray Expedition (1933-34), the Swedish Deep Sea Expedition (1947-48) and the RRS *Discovery* in 1964. Gilson (1937) used nitrate determinations from the John Murray Expedition to estimate the Indian Ocean productivity at 14.9g C/m^2/day, which Nielsen called unrealistic.

Subsequently, Nielsen's estimations showed that, in the middle latitudes (between 2 and 35°S), lowest productivity rates were observed, the values being 0.097g C/m²/day. In the upwelling areas of the Agulhas Current, over the shelf off East Africa and at the equatorial divergences, where the nutrient-rich subsurface waters were close to the surface, productivity values were highest, ranging from 0.43 to 0.59g C/m²/day.

During the IIOE (1959-65) and after, a number of workers made productivity studies on board ships and in the coastal areas, particularly in India. Practically no data are available from other littoral countries in the area. These studies were of one or two years' duration and seasonal studies extending over a longer period are lacking. Because of the great interest in the monsoonal studies, as also in the Somali upwelling, the majority of contributions cover the western Arabian Sea. The rest of the Indian Ocean has large gaps in observational coverage. The productivity rates from different areas of the Indian Ocean are given in Table 5.7 for comparison.

Critical examination of the productivity data would reveal that sharp variations in the coastal areas exist and these are easily related to the rapid changes taking place in the various factors affecting productivity. Qasim (1969) has summarized seasonal changes in phytoplankton production in relation to various parameters in a typical estuarine environment (Figure 5.7). Ryther (1965), discussing the geographic variations in productivity, states that variability in the environmental factors over space and time regulate or limit phytoplankton crop and productivity. These factors are incident radiation, water transparency, nutrient levels, stability of surface water, water salinity and temperature, and geographic location. Since these factors vary so much in coastal waters, interpretation of the productivity data may not be very realistic or meaningful. However, in the open sea, two factors mainly control marine productivity: the nutrient concentrations and incident radiation. As far as the Indian Ocean is concerned, incident radiation is no problem, since it is highest in the tropics. Nutrients, therefore, become a very important factor in determining productivity rates and their variability. It is a well documented fact that hydrographic conditions control the distribution and enrichment of nutrients in the surface waters (euphotic zone). They are: areas of upwelling, the duration and intensity of this upwelling, the pattern of currents and the concentration of nutrients. Based on the results of the IIOE (1959-65) and subsequent cruises of many

Table 5.7 Primary productivity rates in the Indian Ocean

Coastal areas	Value	Authors
Bombay	15-495mg C/m³/day	Krishna Murthy and Viswanathan, 1968
Goa	135-530mg C/m³/day	Dehadrai and Bhargava, 1977
Cochin	0.35-1.50g C/m²/day	Qasim, 1973
Vembanad lake (estuarine)	10-300mg C/m²/day	Nair *et al.*, 1975
Mangroves (east coast)	7.56g C/m³/day	Krishna Murthy and Sundarraj, 1973
Laccadives (coral islands)	12.92g C/m²/day	Qasim *et al.*, 1972
Palk Bay	1.0-8g C/m²/day	Prasad *et al.*, 1963
Saudia Arabia	6.4g C/m²/day	Ryther, 1966
Northern Arabian Sea	0.18-65.11mg C/m²/day	Radhakrishna *et al.*, 1978
Bay of Bengal (nearshore)	100mg C/m³/day	Radhakrishna, 1975
Andaman Sea	5.3-12.4mg C/m²/day	Bhattathiri and Devassy, 1981
Western Indian coastal waters	0.33g C/m²/day	Qasim *et al.*, 1978
Oceanic areas (average values)	**Value**	**Authors**
Oligotrophic waters (central and subtropical)	70mg C/m²/day	Moiseev, 1969
Transition areas and peripheral regions of divergences	140mg C/m²/day	Moiseev, 1969
Equatorial divergences	200mg C/m²/day	Moiseev, 1969
Upwelled coastal waters	340mg C/m²/day	Moiseev, 1969
Neritic waters	1000mg C/m²/day	Moiseev, 1969

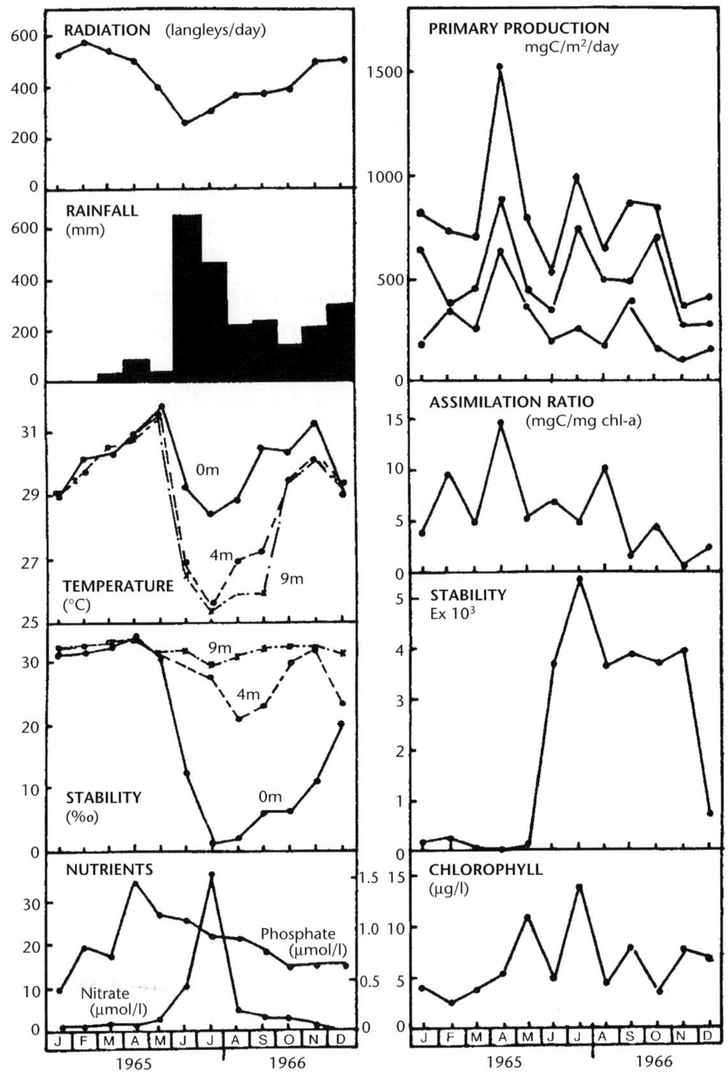

Figure 5.7
Seasonal changes in phytoplankton production relative to various parameters in a typical estuarine or inshore environment; note influence of the monsoons. (After Qasim, 1969)

research vessels belonging to different countries, Wyrtki (1962, 1971) has given an account of upwelling in the Indian Ocean, as well as depicting the distribution of nutrients and dissolved oxygen in his atlas. If the phytoplankton charts prepared by Krey and Babenard are superimposed on appropriate Wyrtki charts, various parameters match very well. For example, productivity charts during the summer monsoon match well with the phosphate or nitrate charts, indicating close correlation between high productivity and nitrates and phosphate-rich areas. Similarly, low-productivity areas match nutrient-poor zones. Qasim (1977) has also projected in six charts (his Figures 2-7) the close correlation between the distribution of nutrients and production values.

Upwelling has been reported from various locations in the Indian Ocean (Sverdrup *et al.*, 1942; LaFond, 1954, 1957; Banse, 1959; Jayaraman and Gogte, 1957; Rochford, 1966; Swallow, 1984) and, of these, the upwelling in the Somali coast has received much attention. In most of these cases, the alternating monsoon winds drive the surface waters away from the coast, thus aiding upwelling. We have very little information on the extent of upwelling areas, their intensities, duration and productivity estimates. In view of this, an exercise to forecast the biological productivity of the Indian Ocean may not be realistic. This is also apparent from the general discrepancy that exists between the productivity estimates, particularly at the tertiary level, and the actual fishery yield, at the level of the world ocean and of the Indian Ocean.

According to Ryther *et al.* (1966) and Holt (1969), the potential fishery yield of the world ocean is 200-240 million tons and what is annually harvested is about 80 million tons. As for the Indian Ocean, the annual fish catch is 3-3.5 million tons, according to

FAO statistics, whereas Qasim (1977) estimates the potential yield to be 15-17 million tons, based on primary production. Gulland (1971) also comes to similar estimates and gives a value of 14.25 million tons of potential yield. Against such high estimates and the actual catch, one is tempted to conclude that most of the areas of the Indian Ocean are underfished and that the total catch can be increased 3-4 times.

Perhaps an improvement in our estimates would be possible if attention were paid to the understanding of the role of the oxygen-minimum layer in the northern Indian Ocean in the biology of the major fisheries. Information on both these factors is inadequate. According to Sen Gupta and Naqvi (1984), the dissolved-oxygen concentrations decrease steadily northwards at all subsurface depths in the northern Indian Ocean; in many places the values are below 0.2ml/l. Besides, the dissolved oxygen exhibits two minima in its vertical distribution and the shallow minimum lies within the thermocline at about 150-200m depth; when upwelling or mixing takes place, it is possible that subsurface waters, poor in dissolved oxygen, dome up to the surface creating an almost anoxic environment, wherein most of the nektonic organisms fail to spawn, survive or grow. Although Ryther and Menzel (1965) have reported highest primary productivity rates in the western Arabian Sea and, similarly, Nielsen and Jensen (1957-59) have estimated higher rates of primary production in the tropical and subtropical areas of the Indian Ocean, it would appear that efficient translation of the primary productivity into secondary and tertiary production is perhaps impeded by the poor dissolved-oxygen levels over wide areas of the northern Indian Ocean.

Zooplankton systematics and distribution

Ranging in size from the microscopic ciliates to a medusa, which may measure a metre in diameter, zooplankters occur in enormous numbers in all parts of the world ocean, from the surface to the deepest parts, and comprise practically all classes of the animal kingdom. Interest in zooplankton studies stems from the fact that, in the sea, they occupy a pivotal position, between the phytoplankton (the primary producers) and the larger fish, whales and other higher animals which are sustained by zooplankton. The distribution and abundance of zooplankton follow closely the phytoplankton on which they graze, and thus often act as a limiting factor in the growth of their population.

Factors affecting zooplankton distribution are many: nutrients, salinity, temperature, density, diurnal rhythms in solar radiation, tides and currents, phases of the moon, geographical features etc. All are important. To discern a pattern in the interplay of all these factors among the innumerable zooplankters which also form discrete populations with specific compositions drawn from an estimated 200,000 species, would perhaps have a dreamlike quality.

Seymour Sewell's contributions are seminal to the oceanography, as to the biology, of the Indian Ocean. In the memoirs of the Royal Asiatic Society of Bengal, Sewell (1929) has given a complete bibliography of plankton, particularly copepod, studies made till then, as a part of an introduction to his monograph on the copepods of the Indian Ocean seas. He began to collect plankton from 1910 onwards, using a net at night time from HMIS *Investigator* while she was at anchor. The passive net, through which tidal currents passed, collected a fairly good quantity of plankton, which forms the basis of his copepod studies in the Indian Ocean. It is said that the plankton net was first used in the North Sea by Johannes Muller in 1846, and this method spread to other parts of the world, and Sewell appeared to be the first to use it in the Indian Ocean. However, commencing with the *Challenger* during 1872-76, and subsequent expeditions the world over, he made plankton-net collections and reported on them.

Some of the earliest publications on the Indian coastal plankton come, however, from Hornell and Nayudu (1923) who reported on the sardines and plankton of the Malabar coast. Aiyar *et al.* (1936) made a record of the plankton from Madras waters for the years 1929-30. Later papers on the plankton began to appear slowly from 1949 onwards. Chacko (1950) gave an account of the general distribution of phytoplankton and zooplankton and provided a list of 77 species of phytoplankton and 81 species of zooplankton, including fish eggs and larvae, occurring in the waters of the Krusadai Islands. He found the plankton to be rich and abundant, the period of maximal abundance for the phytoplankton being from June to November and, for zooplankton, from October to April, corresponding fairly well to the NE monsoon. In Bombay, the phytoplankton maximum was in January and February and zooplankton, from January to March. Along the Calicut coast, phytoplankton was abundant from May to October, and zooplankton, from November to April.

In a fairly comprehensive paper, Prasad (1954) described the marine plankton in an inshore area near Mandapam, based on regular collections made from January 1950 to December 1951. He reported on the distribution and abundance of phytoplankton (Diatomea, Dinophyceae and *Trichodesmium*) and zooplankton (comprising tintinnids, chaetognaths, cladocerans, copepods, decapods, pteropods, tunicates and some invertebrate larvae). The general trend appears to be that the zooplankton components are closely tied to the hydrographical conditions and show two maxima in the year, and the specific months in which the maxima occur vary from place to place along the coast. In most cases, the major zooplankton peaks occur in March or April along the west coast and in November-January along the east coast of India. Since none of the papers on zooplankton covers a longer period and collections come from different areas, the results are not comparable. The nets used are of different diameters and mesh sizes, and treatment of the data in most cases is group-wise and not species-wise. In spite of these drawbacks, the large number of contributions made on the various zooplankton groups provide us with some basic information as to the quality and quantity of zooplankton as a whole from the Indian coastal waters and the open ocean. (There are very few contributions from other littoral countries around the rim of the Indian Ocean, apart from South Africa, Zanzibar, Pakistan and Australia).

An International Symposium on the Biology of the Indian Ocean was held in Kiel, Germany, in April 1971 and its proceedings were edited by Zeitzschel (1973). The papers presented at the Symposium contain interesting information on zooplankton: Aravindakshan reported on the distribution of the pelagic gasteropod *Pterotrachea coronata*; Brinton and Gopalakrishnan, on the distribution of euphausiids; De Decker, on the Agulhas Current plankton; Fenaux, on appendicularians; Fleminger and Hulsemann, on the relationship of Indian Ocean epiplanktonic calanoids to the world ocean; Haq *et al.*, on the distribution and abundance of zooplankton along the Pakistan coast; Kasturirangan *et al.*, on copepods; Kimor, on plankton relations of the Red Sea, Persian Gulf and Arabian Sea; Kurian, on the crustacean order Cumacea; Lenz, on zooplankton biomass in relation to particulate matter in the Arabian Sea; Nair and Rao, on the chaetognath distribution; Nair *et al.*, on the planktonic amphipods (crustaceans); Nellen, on fish larvae of the Arabian Sea and the Persian Gulf; Rao, on the zooplankton of the Indian Ocean; Sakthivel, on the distribution of the gasteropod *Limacina inflata*, and the bi-subtropical pteropod *Styliola subula*; Saraswathy, on the distribution of *Gaussia*; Vannucci and Navas, on the distribution of Hydromedusae in the Indian Ocean.

Subsequently, in October 1976, an International Symposium on Warm Water Zooplankton was organized, with financial support from UNESCO, at the National Institute of Oceanography, Goa, where again a large number of papers on various aspects of zooplankton, mainly in the Indian Ocean, were presented. The proceedings were published in 1977 by the Institute. The papers form a valuable complement to the proceedings of the earlier Symposium held in Kiel in 1971.

A review of all the above-mentioned papers would reveal a gross pattern in the distribution and abundance of zooplanktonic groups in the Indian Ocean. It may also be stated that, apart from Sewell's (1929, 1947, 1948) monumental work on the systematics and the distribution of the free-swimming copepods of the Indian Ocean seas, in which he has described nearly 400 species, the detailed systematics of many groups of zooplankton are not available for the Indian Ocean, outside some of the expedition reports.

Zoogeographically, the Indian Ocean is a subregion of the Indo-West Pacific area, which is one of the main regions of the warm-water fauna. In this large area, the following zoogeographical zones can be recognized (Rao, 1979): the Red Sea and the Persian Gulf; the Arabian Sea, its southern limit being placed at 10°N; the Bay of Bengal, its southern limit also being placed at 10°N; the equatorial region between 10°N and 20-25°S; the Somali Current and Agulhas Current regions; the entire Indian Ocean down to the Subtropical Convergence (35°-40°S). These geographical areas, as well as the area between the Subtropical Convergence and the Antarctic Convergence, and, finally, the Antarctic (Southern) Ocean itself, are the main contiguous zoogeographical provinces and are characterized by specific faunal associations, although they also share taxa with adjacent areas. The boundaries between these regions are no barriers to a large number of species that are carried out of or into them, depending on the prevailing current systems. The local abundance of a species is, therefore, a biohydrographical feature, whereas the occurrence or absence of a species is a matter of the palaeo-oceanography of the area.

These general observations may be illustrated by a few examples. Planktonic Foraminifera are very sensitive to environmental changes and are good ecological indicators. Bé and Tolderlund (1971) have described the distributional pattern and relative abundance of 27 species of living planktonic Foraminifera from the Atlantic and Indian Oceans. The following is a summary of their distributions in the Indian Ocean based on Bé and Tolderlund (*op. cit.*) (Figure 5.8a): *Globigerina pachyderma* is the best indicator of polar waters (0°-9°C), *Globigerina qinqueloba* prefers sub-Antarctic and Antarctic waters (1°-21°C), with its greatest frequency in 1°-5°C water; *Globigerina bulloides* occurs between 40° and 55°S (0°-27°C), with peak abundance in water of 3°-19°C; *Globorotalia* is a good indicator of the Subtropical Convergence and, in the southern hemisphere, it occurs in a distinct latitudinal belt; *Globorotalia truncatuloides* occurs mainly in the sub-tropical region of the central Indian Ocean, between 20° and 50°S, with peak abundance in water of 17°-22°C; *Globigerinoides ruber* is the most successful warm-water species in terms of distribution and abundance, occupying the entire central water mass between 10°N and 30°S, with peak abundance in the southern central gyre; *Globigerina sacculifera* has a similar distribution and extends farther north; *Globigerina dutertrei* is particularly abundant in the major currents nearer the continents. On the whole, the distribution of the various species is latitudinal and rarely meridional.

Copepods are the next important indicator of zoogeographical areas. They respond quickly to dominate areas of rich phytoplankton production. Sewell (1947, 1948) gave a comprehensive account of the world-wide zoogeography of the planktonic copepods with particular reference to the Indian Ocean. This is divided into two geographical areas: (1) a tropical zone north of 10°S, a subregion of Indo-Pacific nature; and (2) a southern subtropical zone, encompassing the central and southern Indian Ocean, extending from about 10°S to the Subtropical Convergence (35°-40°S). South of the Subtropical Convergence, the Antarctic regime is quite different. To summarize Sewell's findings for copepoda, it may be stated that the largest number of species in the Indian Ocean is of an Indo-Pacific nature. There is a decrease in species diversity from east to west, finally reaching very low diversity in the Red Sea and the Persian Gulf.

There is a gradual decrease in the number of species from northern latitudes towards the Subtropical Convergence and with increasing depth. Exchange between the copepod fauna of the Indian and Pacific Oceans takes place mainly through the Strait of Malacca and through the Timor Sea south of Java.

Exchange of copepods between the Atlantic and Indian Oceans is attributed by Sewell (1948) to the

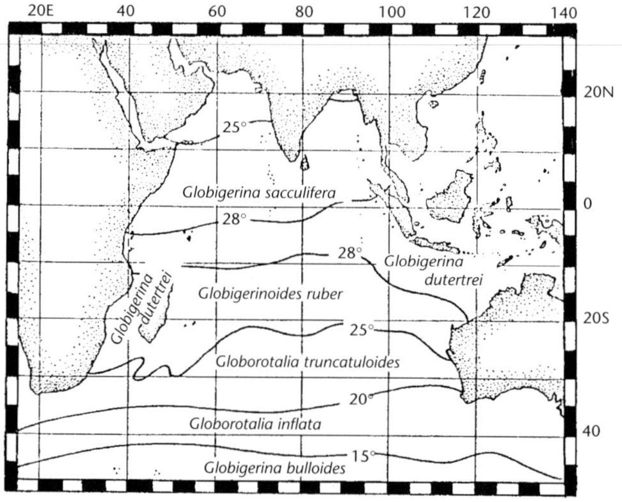

Figure 5.8a
Approximate distribution of some important foraminiferans in the Indian Ocean and the Indian Ocean sector of the Southern Ocean in relation to the mean isotherms in February. (After Bé and Tolderlund, 1971)

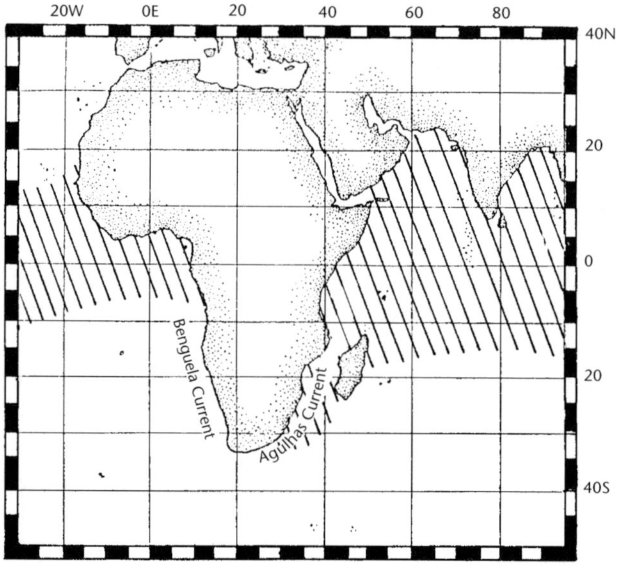

Figure 5.8b
Discontinuity in the distribution of *Candacia pachydactyla* (Copepoda) between the Indian and Atlantic Oceans. (Jones, 1966*b*)

flow of Antarctic Intermediate Water (AIW) at deeper levels; about 165 deep-water copepods are recorded from both oceans. At the surface level, however, according to Jones (1966a, b), in the western Indian Ocean and north to about 20°S, equatorial species, as *Candacia pachydactyla*, *C. catula* and *Paracandacia truncata*, predominate. These species also occur in the equatorial areas of the Atlantic, but the populations are separated by the cold waters of the Benguela Current off southwestern Africa. In fact, this current forms a barrier to the transport of Indo-Pacific forms into the South Atlantic. Lawson (1977) has made similar observation on the distribution of Candaciidae in the Indian Ocean (Figure 5.8b).

Sewell found in the Indian Ocean a gradual reduction of deep-sea and North Atlantic copepods in a northward direction, from a maximum of 70 (including 56 North Atlantic species) to a minimum of 20 (12 North Atlantic species). De Decker and Mombeck (1965) observed a similar reduction in the number of Atlantic species in the northward direction, from about 20 to 0. However, the observations of Grice and Hulsemann (1967) make it clear that the so-called North Atlantic species in the Indian Ocean are present at all latitudes and no significant decrease in the number of species in either ocean takes place either northwards or southwards as reported by Sewell (1947) and by De Decker and Mombeck (1965). Further, Grice and Hulsemann emphasize that the bathypelagic copepods living in depths below 1000m show extremely widespread distribution. They report 241 species common to the Atlantic and Indian Oceans which in itself is clear evidence for the effective dispersal of copepods by deep-water currents. Ninety-two percent of the deep-living species in the Arabian Sea are also found in the North Atlantic, some 8000 nautical miles away.

They further comment that there are few species in the Indian Ocean that are not found elsewhere. They found 310 species in their collection, of which 40 were not found in the Atlantic; their discovery of 78 species not reported till then in the Indian Ocean and their description of 17 new species would indicate the paucity of taxonomic studies in the Indian Ocean.

All other groups of planktonic organisms exhibit more or less similar distribution to that of the Foraminifera and copepods (see Van der Spoel and Pierrot-Bults, 1979; Van der Spoel and Heyman, 1983).

Seasonal changes

By virtue of its location, the Indian Ocean can be considered a tropical ocean, and, particularly north of the 10°S, there are no pronounced seasonal changes except for the conditions imposed by the alternating monsoons. Cut off from the northern polar areas, unlike the Atlantic and Pacific Oceans, the vast area of the northern Indian Ocean remains hot and humid and the daily temperature and salinity ranges may, at times, exceed annual changes (Sewell, 1925-1938; Rao and Rao, 1962). In fact, from the oceanographic point of view, the following four seasons may be recognized and tagged to the calendar months, more as a marker than anything else:

1. Southwest monsoon: June to August
2. Intermonsoon period: September to November
3. Northeast monsoon: December to February
4. Intermonsoon period: March to May

The actual effect of these seasons can be seen in the amount of rainfall, changes in the direction of current, salinity and temperature, and, where upwelling takes place, nutrient enrichment of the surface waters. These changes are immediately reflected in biological events such as plankton blooms and migration. Rao (1973) and Panikkar and Rao (1973) have reported on these changes at different locations along the Indian coastline (Figure 5.9). Coinciding with the two monsoon seasons, phytoplankton and zooplankton both show an annual bi-modal distribution, with two peaks and two lows; there is also an interesting movement of peaks of abundance from south to north along the west coast and north to south along the east coast, this again being a reflection of the monsoon effects. It would appear that upwelling along the west coast is initiated in the south and it strengthens gradually along the coast northwards, resulting in enrichment of the coastal waters and thus triggering plankton peaks. The reverse sequence appears to occur along the east coast of India.

In the open ocean, except for a section along the 100°E meridian (Tranter, 1973), seasonal coverage is not available. During the IIOE, all participating ships collected zooplankton samples using the Indian Ocean standard net (Currie, 1963) and hence it is possible to compare different locations in the Indian Ocean. It was also assumed that the day and night factors would not be crucial for reporting on the

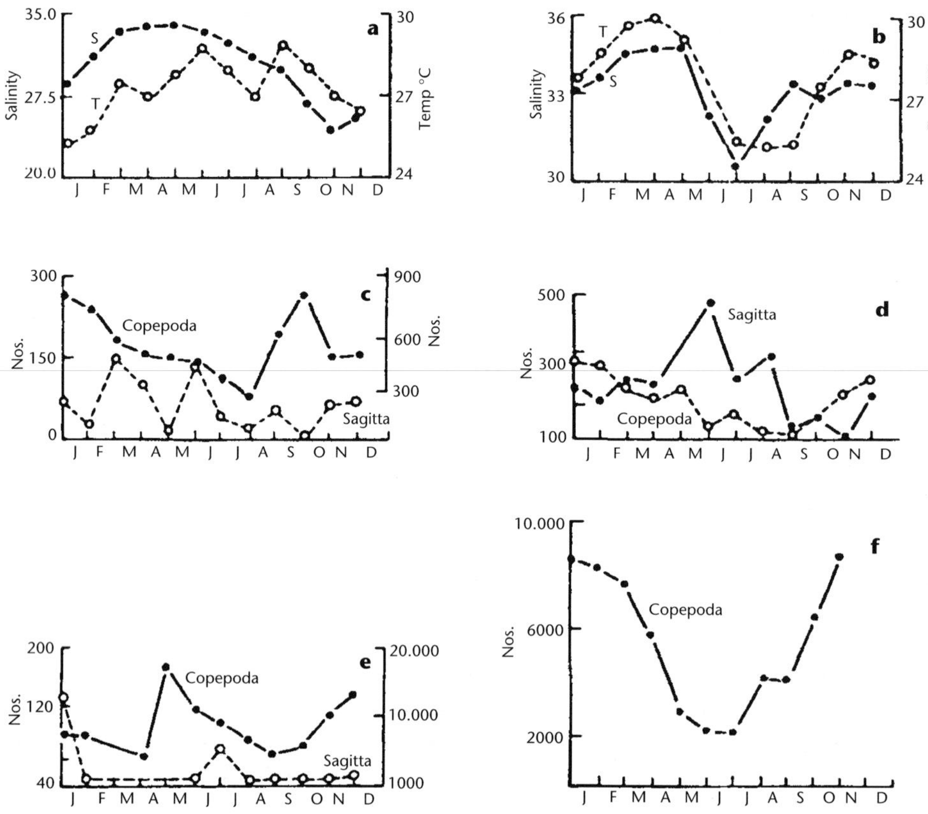

Figure 5.9
Annual variation in zooplankton numbers per haul, sea temperature and salinity in Indian coastal waters: (a) salinity and temperature off Waltair (east coast); (b) salinity and temperature off Calicut (west coast); (c) *Sagitta* and copepods off Mandapam; (d) *Sagitta* and copepods off Trivandrum; (e) *Sagitta* and copepods off Waltair; (f) Total copepods off Madras. (Pannikar and Rao, 1973)

density and vertical migration of the major groups of zooplankton, given that the Indian Ocean Standard Net (IOSN) was routinely hauled from a depth of 200m to the surface at all stations.

The seasonal distribution of the zooplankton biomass shows a clear response to the two monsoon seasons (Figure 5.10a, b). Also, reference should be made to the Indian Ocean Biological Centre (IOBC) plankton atlases (IOBC 1968, 1971, 1973) in which the distribution of various zooplankton groups is shown. Since the copepods make up 60-70% of the total numbers at any given station, the zooplankton biomass mainly reflects copepod distribution, the rest consisting of chaetognaths, ostracods, euphausiids, medusae and thecosomes, among others. The pattern of distribution of all groups, together with the total biomass, indicates that some of the Indian Ocean areas, such as the coasts of Somalia and Saudi Arabia, the west coast of India and the head of the Bay of Bengal, are rich in zooplankton. During the height of the NE monsoon (December-February), biomass values of 20-39.9ml/m^2 occur in the region of the Gulf of Aden and north of Mombasa. A few spots indicating the highest range (i.e., 80ml/m^2 and above) are found off Madras and Goa. The rest of the area shows low values, from 0.1 to 9.9ml/m^2, all over the Indian Ocean, except for the west coast of India and part of the Bay of Bengal, with values reaching 39.9ml/m^2. The beginning of the SW monsoon is in May-June and no data are available for the Somali and Arabian coasts. Scattered near the Gulf of Aden and off Sri Lanka are some values between 20 and 39.9ml/m^2, and the rest of the Indian Ocean has low values (10ml/m^2) during the height of the SW monsoon (July-August).

The biomass values are generally higher (40ml/m^2 and above) in most coastal areas, particularly off the Somali and Arabian coasts. In the Andaman Sea, off Cochin and Sri Lanka, along the 100°E longitude and around 5°N,90°E, the values tend to be between 20 and 39.9ml/m^2. During the post-monsoon months of October-November, most of the Indian Ocean has low values. In conclusion, it may be stated that, in the Indian Ocean, biomass values are consistently higher and cover a larger area during the SW monsoon, compared to the situation in the NE monsoon. This is attributed to upwelling reported off Somalia, Saudia Arabia, west and east coasts of India, and at the equatorial divergences, where enrichment of surface waters takes place over a wide area leading to

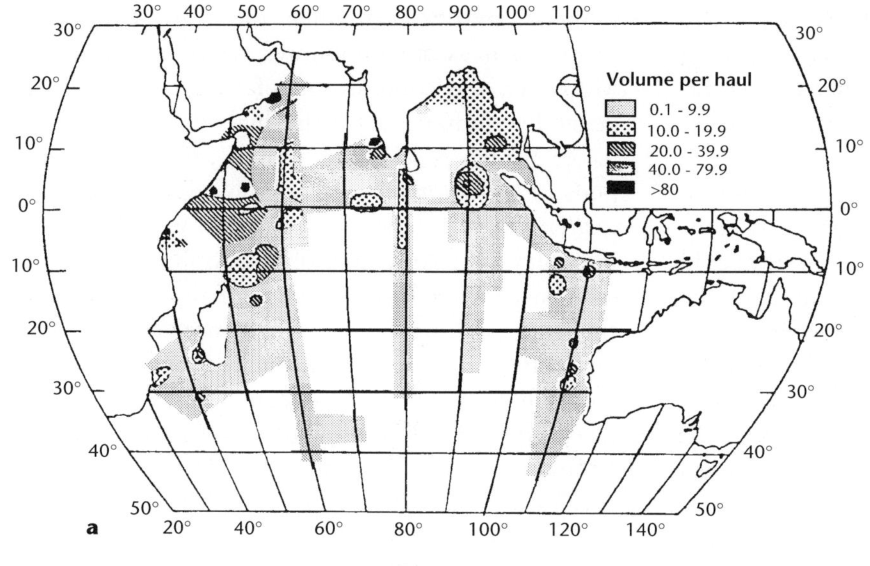

Figure 5.10a
Zooplankton biomass (ml/m²) distribution in the Indian Ocean for the depth range 0-200m in July-September (SW monsoon). (After Rao, 1973)

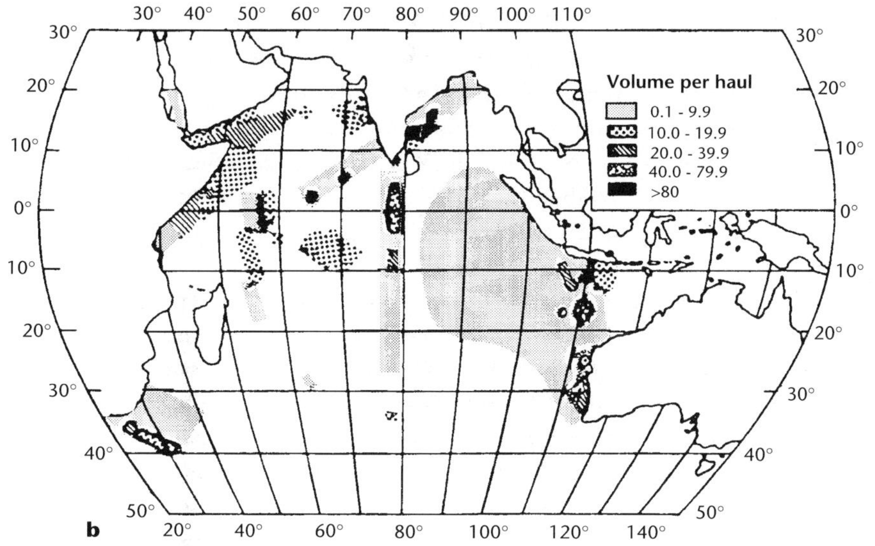

Figure 5.10b
Zooplankton biomass (ml/m²) distribution in the Indian Ocean for the depth range 0-200m in December-February (NE monsoon). (After Rao, 1973)

high levels of primary production followed by secondary production (Smith, 1989).

During the NE monsoon season (December-February), zooplankton biomass values are low at most locations in the Indian Ocean. This is attributed to the lack of upwelling or the cessation of upwelling which was active during the SW monsoon.

De Decker's (1973) account of the Agulhas Current plankton indicates that zooplankton biomass values are extremely high on the Agulhas Bank, and sometimes exceed 140cm³ settlement volume, as in 1966. Besides, the dominant group there appears to be the Thaliacea (salps), represented by two species, *Thalia democratica* and *Doliolum denticulatum*, which occur in incredible numbers, the swarms often extending over thousands of square miles; De Decker also reports that some species of chaetognaths, as *Sagitta minima*, *S. decipiens* and *Eukrohnia hamata*, could be considered as being indicators of mixing, by their occurrence in surface waters of the bank, given the fact that they are deep-water forms. Similar observations have been made by other workers in the Indian Ocean upwelling areas.

Smith, in a series of three papers (1982, 1988 and 1992) has made some innovative investigations of the distribution, abundance and feeding of zooplankton in the northwestern Indian Ocean, more particularly in the area of the Somali Current, which she calls a 'biological river'. It is now well known that, as the strong jet of the Somali Current veers off the coast of

Somalia towards the east at 5°-10°N, extensive upwelling is caused in the nearshore areas, resulting in lowered sea-surface temperatures, high concentrations of nutrients, such as nitrates and phosphates, and in phytoplankton blooms.

Smith found that, among the zooplankton, *Calanoides carinatus*, by virtue of its large size, ability to feed actively and reproduce in the cold upwelled waters during the SW monsoon, could be considered as an indicator of Somali upwelling. This apart, its ontogenetic adaptation to the dynamics of the Somali Current and the offshore eddy (the Great Whirl) is extremely interesting in the sense that the species manages to restrict itself to the Somali Current area. It is assumed that the offshore eddy of the Somali current is an ellipse whose circumference would be about 5000km (Duing *et al.* 1980) between 5° and 10°N and lasts for 3-4 months. *C. carinatus*, with an average generation time of 21-22 days, would, when entrained in the eddy at 10°N, go through 4-5 generations during a complete transit of the eddy by remaining at the surface. Smith (*op. cit.*) found that the male and female adults, and the 5th copepodid stage were adapted to living at a temperature of about 20°C, and sink to a depth of 500-1000m following the same isotherm during the post-monsoon period when upwelling ceases. At that depth, the copepod survives on its lipid storage till the onset of the next SW monsoon when, with renewed upwelling, the species returns to the surface to continue its life history. As a result of this adaptation, which is also seen among other groups of zooplankton, the species ensures its distribution in the area to which it belongs.

Smith also reports that the abundance of *Oithona* spp. shows an inverse relationship with that of *C. carinatus* inasmuch as its abundance decreases during the SW monsoon and increases during the NE monsoon. This would imply that zooplankton abundance, taken as a whole, may mask the real contributors responsible for such decreases in space and time.

It is estimated that the Somali area upwelling is one of the world's largest, exhibiting very high rates of productivity, particularly at the primary and secondary levels, but there is very little fishery development in the region. Smith feels that, in the absence of a wide shelf area off Somalia, the primary production is lost towards the central parts of the Arabian Sea.

Haq *et al.* (1973), reporting on the zooplankton off the Pakistan coast, found that total zooplankton, expressed in numbers, does not correspond to the biomass, particularly when there are large-sized forms such as siphonophores, salps, medusae etc. Moreover, the composition of the zooplankton is generally dominated by the high numerical abundance of a few species. For example, among the copepods, *Pleuromamma indica* far exceeds *Euchaeta wolfendeni* in numbers. Another factor which comes into play in a large area of the Arabian Sea is the oxygen-deficient layer, which normally occurs below the thermocline; however, during the extensive upwelling along the Somali, Saudi Arabian, Pakistan and Indian coasts, this layer comes nearer to the surface, affecting the distribution of some species. Vinogradov and Veronina (1961) described six species which include three species of the genus *Pleuromamma* (*P. gracilis*, *P. xiphias* and *P. indica*) and, of these, *P. indica* is associated with a layer of water having less than 0.1ml/l of dissolved oxygen. This is corroborated by the Pakistan workers, who add that those species which tolerate oxygen-deficient layers, generally form a significant proportion of the zooplankton at subsurface levels.

Nair and Rao (1973a, b) have described, for the Arabian Sea, the seasonal variation and distribution of 19 species of chaetognaths, next only to copepods in abundance. The chaetognaths are carnivorous and prey upon copepods and other microplankters. *Sagitta pacifica* and *S. enflata* are widely distributed and show higher densities near the upwelling zones off the Somali and Saudi Arabian coasts; such species as *Eukrohnia fowleri*, *S. bombayensis*, *S. decipiens*, *S. lyra*, and *S. zetsios* occur in negligible numbers in the Arabian Sea. Towards the Gulf of Aden and the Red Sea, there is a reduction in the number of chaetognath species. Except for *S. bombayensis*, all other epiplanktonic species present in the Arabian Sea are present in the Gulf of Aden. Except for *Krohnitta subtilis*, all other forms found in the Gulf of Aden are also present in the Bab-el-Mandeb area, but *Sagitta enflata*, *S. pacifica*, *S. bipunctata*, *S. regularis*, *S. neglecta*, *S. ferox*, *S. hexaptera* and *S. bedoti* are present only in the Red Sea. The high salinity (36.66-40.50) and temperature (21.9°-29.5°C) in the upper 200m of the Red Sea may not be suitable for most of the planktonic forms present in the Arabian Sea (Kimor, 1973; Halim, 1969). Halim notes that, out of the 270 species of calanoid copepods known from the Indian Ocean, only 158 species are recorded from the Red Sea. This reduction in the copepod fauna mostly concerns deep-sea species which perhaps cannot cross the shallow sill at Bab-el-Mandeb from the

Arabian Sea into the Red Sea. Besides, seasonal changes may also be responsible for the apparent low diversity as one proceeds from the Arabian Sea into the Red Sea and Persian Gulf. During the NE monsoon, the surface current from the Arabian Sea enters the Red Sea and the Persian Gulf bringing with it a large number of surface-dwelling species, which, with the approach of summer, may migrate into deeper layers of the Red Sea. Reduction in the number of species from the Arabian Sea through the Gulf of Aden into the Red Sea and the Gulfs of Eilat and Suez is also seen in the case of chaetognaths, appendicularians and pteropods, among others. Fenaux (1964, 1973) reported the number of appendicularians found in the Oman Sea to be 12, the Persian Gulf, 6, the Red Sea, 14, and the Gulf of Eilat, only 3.

The eastern segment of the Indian Ocean has been the subject of seasonal studies by Tranter (1973), who draws our attention to the following main interacting factors that determine the changes in the ecosystem, particularly the ecological succession from phytoplankton upwards in the southern latitudes. They are:

- Enrichment of tropical surface waters in winter (June-August) by upwelling associated with the South Equatorial Current.
- Impoverishment of subtropical surface water in late summer by thermal stratification.
- Meridional transport of tropical waters to the south in autumn and winter, and subtropical waters towards the north in spring and summer.
- Heavy algal grazing and rapid nutrient cycling, as evidenced by the close correlation between the concentration of inorganic phosphate and zooplankton biomass.

The above-mentioned changes constitute an annual cycle which can be considered to begin in late summer (March) at which time the phytoplankton is at its minimum. The zooplankton at this stage has indulged in heavy grazing and, in turn, is preyed upon by the micronekton. In winter (June-August), the subtropical thermocline breaks down and the regenerated nutrients, lying between the seasonal thermocline and the 'permanent thermal discontinuity' (Rochford, 1969), become available for plant production. Meanwhile in the tropical area, the South Equatorial Current strengthens, with associated upwelling (dynamic uplift), and this again provides for plant growth and a rapid response from the zooplankton. By September, they reach their maximum abundance. And further, the excretion from zooplankton undergoes rapid remineralization of nutrients; this cycle continues into its second phase through November and December. The continuation of plankton production associated with the South Equatorial Current throughout the winter provides forage for the zooplankton and micronekton and thus the cycle is completed.

Such seasonal studies are required for other areas of the Indian Ocean. Most of the contributions from the Indian Ocean deal with numerical and/or biomass distribution of various components of the zooplankton. All over the Indian Ocean, as already stated, major holoplanktonic groups follow the distribution pattern of copepods, with their chief areas of abundance near the Somali and Arabian coasts. After copepods, in order of importance, are ostracods, chaetognaths, euphausiids and thecosomes.

Tranter (1973), commenting on seasonal amplitudes in terms of the standard deviation of the monthly (or quarterly) mean (as a percentage of the annual mean), showed that different taxa exhibited different amplitudes (Figure 5.11). The greatest amplitude was for fish eggs and larvae (120%). Appendicularians and thaliaceans showed an amplitude of around 80%. On the other hand, copepods showed only 40% at the group level, but were more variable at the species level. For some copepods, as *Euchaeta marina*, seasonal differences were small. But for others, as *Rhincalanus nasutus*, there was a greater amplitude in the subtropics than in the tropics, whereas the reverse was the case for *Eucalanus subtenuis*. Although the seasonal variation in the eastern Indian Ocean is less than that in the Sargasso Sea, it is greater than that in the eastern Pacific. Tranter believes that large amplitudes shown in high latitudes may be more apparent than real if the whole water column is examined for seasonal vertical migration.

Seasonal changes shown by other taxa are not clearly discernible from the IIOE collections, since they are, in most cases, 'one shot' collections.

Della Croce and Venugopal (1973) found cladoceran distribution and abundance to be erratic. Particularly, *Penelia avirostris* shows continuous geographical distribution from Cape Agulhas to Zanzibar from September to March. In the eastern Indian Ocean, the species occurs from the Bay of Bengal to the Singapore Straits, usually in March or April. It is characteristic of *Penelia* to show peak abundances sporadically at various places in different

Figure 5.11
Mean seasonal amplitude (as percentage of annual mean) of species, higher taxa, selected biomasses and nutrients in the eastern Indian Ocean. (After Tranter, 1973)

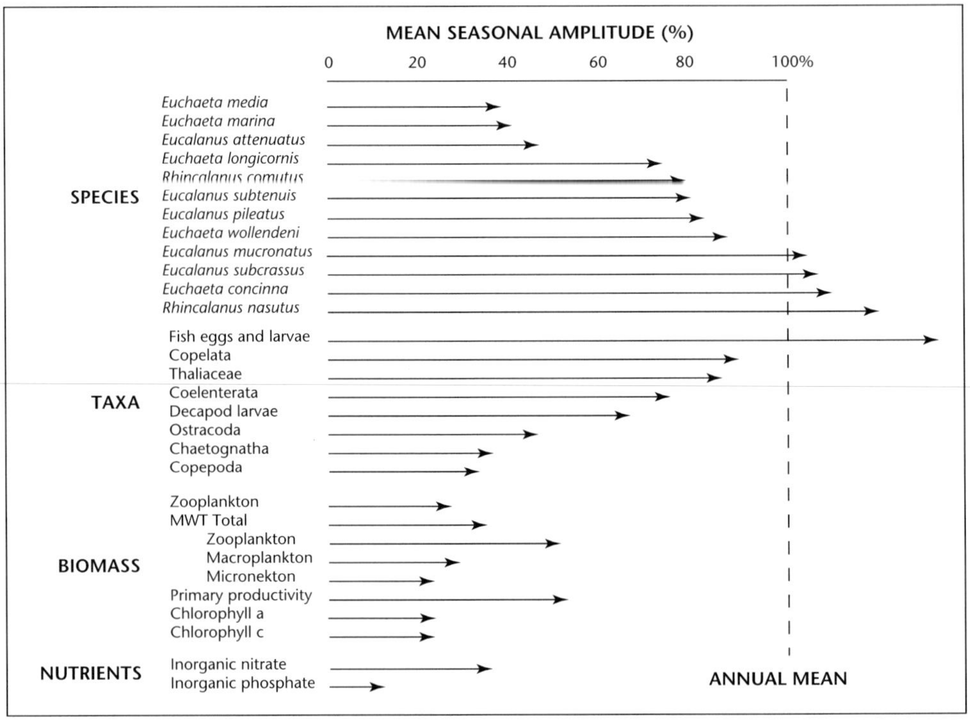

months, as off Madras in October, off Zanzibar in January, in Singapore waters, late September and early April, and on the Agulhas Bank throughout the year, and so on. It would appear that flowering of the ostracod populations has a close relationship with their life-history patterns, particularly with respect to the production of resting eggs, and with their ambient temperatures. They also stated that *Penelia* may produce 6 broods in 36-40 days and each brood may itself bear young in less than 7 days.

Rao and Kelly (1962) studied the breeding cycle of the chaetognaths over a three-year period in Lawson's Bay, Waltair, on the east coast of India, and found at least 9 broods from October 1956 to December 1957. Breeding was intense during June and July and maximum density of all stages occurred from June to September. Nair (1972) examined the chaetognaths of the Cochin backwaters, which is a tropical estuary, and found *Sagitta bedoti* and *S. enflata* to have restricted breeding because of monsoonal effects. During the SW monsoon, salinity was very low and most of the chaetognaths were absent from the estuary. Only during the premonsoon months, January to May, does *S. bedoti* breed continuously, with smaller peaks in breeding in September; i.e., after the SW monsoon. *S. enflata* showed similar behaviour.

The mesoplanktonic larvae contributed by a variety of benthic organisms, such as echinoderms, gastropods, polychaetes, crustaceans and others, are distributed over wide areas of the Indian Ocean and form an important component of the zooplankton biomass. Monsoonal effects on their distribution are equally pronounced, as also on the holoplankton. At the International Warm Water Zooplankton Symposium, held in Goa in 1977, there were several papers reporting on the life-histories of decapod crustaceans and other organisms with meroplanktonic larval stages, and their conclusion was that salinity, more than temperature, was the main causative factor influencing the breeding cycles in these waters. It was reported that some species of decapods, as *Uca annulipes*, *Portunus pelagicus* and *Metapenaeus affinis*, breed during most of the year except during the low-salinity period of the SW monsoon. Similar observations were made on the breeding cycles of Hydromedusae by Vannucci *et al.* (1970), who also point out that no medusae occurred in the estuary (Cochin) during the low-salinity period, but, with the return of the high salinity during the post-monsoon and pre-monsoon seasons, recolonization was achieved through the metamorphosis of resting stages. This phenomenon was also seen in copepods in the Cochin estuary.

Another ecosystem that has a rich meroplanktonic component is the coral reef, where a large variety of benthic organisms contribute significantly to the zooplankton biomass. Besides, it has been reported that the composition of zooplankton in the outside ocean is different from that in the lagoon. In the former, calanoids, chaetognaths and salps predominate, and in the latter, harpacticoids, mysids and other epibenthic animals predominate.

Since marine biologists are deeply interested in conversion efficiencies from the primary (phytoplanktonic) production level to the secondary (zooplanktonic) level, it is essential to know the selectivity and rate of feeding, excretion and respiration of zooplankton. The premise is that all zooplankton can be basically divided into two groups: the herbivorous zooplankters, which primarily feed on phytoplankton; and the carnivorous type, which feed on herbivorous zooplankton. Very little work has been done on these aspects in the Indian Ocean region. Mullin (1963) tested the food selection of several species of copepods regarded as herbivorous, omnivorous and carnivorous and taken from the Indian Ocean during a cruise on the RV *Anton Bruun*. His food mixture was composed of *Artemia* nauplii and three species of phytoplankton, *Coscinodiscus perforatus*, *Thalassiosira fluviatilis* and *Cyclotella nanna*. The difference in cell volume between the smallest and largest phytoplankters was nearly four orders of magnitude. When a mixture was offered containing all the four food species, *Neocalanus gracilis* consumed all of them, but selected *Coscinodiscus*; another copepod, *Eucalanus attenuatus*, appeared to feed only on the largest diatom, whereas *Nannocalanus minor* did not take *Cyclotella*. *Artemia* nauplii, however, appeared to be a favourite food item for many copepods. Mullin also examined many copepods from the Woods Hole (Massachusetts, USA) region and it would appear that copepods have the capacity to feed on variable diets. Mullin found grazing rates of 150 to 350ml/day in the case of *Calanus helgolandicus* feeding on dilute suspensions of *Dylilium* or *Asterionella* for 4-8 hours. He also demonstrated selective feeding by various species of copepods. Marked quantitative differences in the rate of food intake were found between the different species of copepods, and some qualitative differences in selective feeding were also indicated. Reporting on the feeding of zooplankton in the northwestern Indian Ocean during the SW monsoon of 1979, Smith (1982) observed that the dominant copepod, *Calanoides carinatus*, had the ability to ingest the species of diatoms that were common in the upwelling regions. Based on feeding experiments, Smith concluded that the feeding of copepods on natural assemblages of phytoplankton is a function of food concentration and of species composition. This implies that phytoplankton-zooplankton relationships are much more complex than is often supposed, particularly in respect of the size and age of phytoplankton species relative to those of zooplankton species and their life-history stages.

More investigations are required before we can attempt to calculate the secondary production with adequate accuracy, assuming that all herbivorous zooplankters necessarily feed on phytoplankton. They may not.

Nekton

General distribution

Animals grouped under the term nekton have effective swimming capabilities and are of great interest and value to mankind. They are at the apex of the food chain and occupy a central place in the food web of the oceans. Their bodies are streamlined and are often covered with slime so as to minimize friction when swimming at great speed and over long distances. Among them are the world's largest animals, the whales, other marine mammals, sharks, adult fishes, crustaceans and squid. Their life histories are most interesting. Some, like the salmon, spawn in rivers and migrate to the sea to grow, and others, as the eels, grow in rivers and migrate to the sea to spawn. Whales also migrate between breeding grounds in the middle latitudes and feeding grounds in the higher latitudes, involving distances of several thousand miles.

Many of the pelagic fishes, such as sardines, mackerels and tunas, living in the upper layers of the sea, where food is abundant in the form of phyto- and zooplankton, move in large schools or shoals and are harvested by man as an important source of food. According to FAO statistics, the present level of the harvest of marine fishes of all types has reached about 80 million tons annually. However, fishery experts and marine biologists have estimated that the world's oceans may support a sustainable yield of 200-240 million tons annually, and, of this, the Indian Ocean may contribute 11-14 million tons.

In the beginning of this chapter, some account of the observations made by Alcock on the deep-sea fishes in the Indian seas was given. Beyond that, hardly any observations are available on the deep-sea nekton of the Indian Ocean. In a recent edition of the book entitled *Marine Fisheries of India*, by Bal and Rao (1990), a detailed account has been given of all the major and minor commercial fisheries of India, based on the extensive work done by fishery scientists and institutes in the country. It may also be stated that there are very few papers describing the deep-sea fishes and marine mammals in the Indian Ocean north of the Subtropical Convergence, apart from an interesting paper on the dugongs of the Gulf of Mannar.

When the IIOE was launched, there were high hopes among many biologists and fishery scientists that new fishing areas would be located in the Indian Ocean, with possibilities of increased fish catches; but no new fishing grounds were discovered and, in that particular sense, the IIOE was a failure. The *Anton Bruun* did make an exploratory survey for fish and succeeded in hauling some spectacular demersal catches along the northern Arabian coast, but this area never became the site of a successful fishery after the IIOE was over.

There are two interesting aspects of Indian Ocean fishing. Firstly, there is the restricted geographical distribution of some of the major fisheries along the Indian coast (see Table 5.8).

For example, the oil sardine (*Sardinella longiceps*) and the Indian mackerel (*Rastrelliger kanagurta*) are together mainly confined to the west coast of India from Ratnagiri southwards to Quilon. Enormous shoals of these two fishes approach the nearshore areas with the cessation of the SW monsoon in September. Similarly, the Bombay Duck (*Harpadon nehereus*) has a major area of occurrence along the Gujarat and Maharashtra coasts. These three fishes sometimes contribute 50% of the annual marine fish landings of India. There are also some years when there is a setback in the sardine fishing, as in 1959 and 1963. In the case of the Indian mackerel too, in some years, there is a marked decline in the fishery and this appears to be cyclical. In the 22-year period (1958-1979) reported by Bal and Rao (1990), the highest catch was in 1971, amounting to 204,575 tons, and the lowest catch was in 1968, when it was 21,703 tons, forming only 2.3% of the total marine fish landings for that year.

Although much information has been collected on the growth and breeding of these fishes, we do not know fully their migration patterns and their response to seasonal changes. In a very interesting paper, Longhurst and Wooster (1990) have reported that the abundance of oil sardine (*Sardinella longiceps*) seems to be related to sea-level changes along the Kerala coast, just prior to the onset of the SW monsoon. At this time, sea-level change indicates remote forcing of upwelling and, if this happens early, the oxygen-deficient upwelled water may inhibit subsequent recruitment of the fish. There is also an interesting inverse relationship between the mackerel and the oil sardine (see Figure 5.12). There are years in which both these fishes increase in abundance and others in which the reverse is the case. Since these two fishes occur in enormous numbers, the factors causing such a distribution remain an enigma. Bal and Rao (*op. cit.*) further observe that the spawning grounds of these two fishes occur at a large number of places along the coasts of peninsular India and the Andaman Islands. In spite of this, it is difficult to explain how these fishes shoal up and become a major fishery mainly along the southwest coast of India.

There is a substantial fishery for *Sardinella longiceps* along the coast of Oman from September to December, very similar to that on the Kerala coast. As in other coastal upwelling systems, scombrids are prominent in the northern Arabian Sea, with the narrow-barred Spanish mackerel (*Scomberomorus commerson*) and the Indian mackerel (*Rastrelliger kanagurta*) being the dominant forms. Fishing in the entire northern Arabian Sea is seasonal because of the

Table 5.8 Geographical distribution of some commercial fisheries along the Indian coast

Group or species	Areas of abundance
Elasmobranchs	Tamil Nadu, Kerala, Gujarat and Maharashtra
Sardinella longiceps (oil sardine)	Kerala and Karnataka
Rastrelliger kanagurta (mackerel)	South Maharashtra, Karnataka and Kerala
Harpadon nehereus (Bombay duck)	Maharashtra and Gujarat
Otolithus argentius (croaker)	All along the Indian coast
Polynemids (threadfins)	Maharashtra and Gujarat
Tunas and billfishes	Kerala, Tamil Nadu and the Laccadives
Pampus argentius (pomfrets)	Orissa and Maharashtra
Prawns	All along the Indian coast, major centres off Orissa and Kerala
Molluscs (clams)	Tamil Nadu, Kerala and the Andamans
Pearl oysters	Tamil Nadu and Gujarat
Squids	Orissa and Kerala

vigorous SW monsoon during which (June to August), the rough seas are unfavourable for fishing. The peak landings occur from October to January and include a fair proportion of large pelagic fish, as tuna, barracuda, kingfish, large jacks and others.

Recently, attention has been directed towards assessing the abundance of mesopelagic planktivorous fishes in the northern Arabian Sea and the squid resources in the Bay of Bengal. Accordingly, extraordinary concentrations of mesopelagic fishes belonging to the myctophid family have been reported (Gjosaeter and Kawaguchi, 1980): about 60g/m^2 for the Somali Current and Oman coast; and about 30g/m^2 for the west coast of India. The central Arabian Sea has a very low biomass, about 0.5g/m^2. Abundances as high as 500g/m^2 were reported by FAO surveys conducted during 1977-78. In all the surveys, highest concentrations were found at or near the shelf-break. The dominant species were *Benthosoma pterotum* and *Diaphus arabicus*. The long-term prominence of myctophids in this region is shown by otoliths found in the deep-sea cores; the otoliths in these cores belong almost exclusively to myctophids (Martini, 1974).

Gjosaeter and Kawaguchi (1980) stated that the population of *B. pterotum* is the largest localized fish stock in the world, amounting to 100 million tons. This fish is associated with the continental slope all around the Indian Ocean and in the western Pacific as far north as Japan. It is essential to target future fishery and oceanographic surveys to assess the abundance of this fish.

Cephalopod potential in the Bay of Bengal appears to be attractive. At present, nearly 40,000 tons of squid and cuttlefishes are harvested in the Bay, according to a report in the March 1992 issue of *Fisheries News International* (quoting the *Bay of Bengal News* and K. Sivasubramanian, p. 16). The biomass of the most common type of oceanic squid (purpleneck flying squid) in the Bay of Bengal alone is considered to be around 2 million tons.

In contrast to the pelagic fisheries described for the nearshore waters, offshore pelagic fishes, such as tunas, also show special features of distribution closely related to hydrographical conditions (Suda, 1973). For example, the yellowfin tuna (*Thunnus albacares*) mostly occurs north of 10°S and expands southwards along the African coast, particularly during the first half of the year and towards the Australian coast during the second half of the year. The albacore (*Thunnus alalunga*) distribution occu-

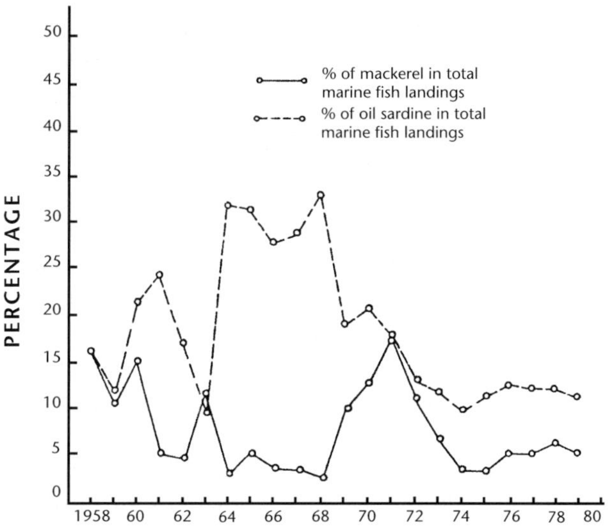

Figure 5.12
Percentage of oil sardine landings in total marine fish landings compared with the percentage of mackerel landings for each year from 1958 to 1979; note the inverse relationship between them from 1964 to 1968. (After Bal and Rao, 1990)

pies middle latitudes between 10° and 35°S, and does not overlap that of yellowfin.

The southern bluefin (*Thunnus thynnus maccoyii*) is generally found distributed in an east-west direction in the sub-Antarctic area, with a northward expansion along the west coast of Australia during the fourth quarter of one year and the first quarter of the next. This pattern of occurrence closely reflects the seasonal changes in the general hydrography of the areas concerned. Suda (1973), quoting other workers, indicates the following relationship between the distribution of catches of tunas and environmental conditions:

- In the South Equatorial Current area, albacore is abundant, but yellowfin is not.
- A remarkable concentration of yellowfin occurs during the northern summer in the Equatorial Current and in the zone with variable currents between the South Equatorial Current and the NE monsoon drift; albacore are absent in this area.
- Over the SW monsoon areas, which are other yellowfin habitats, albacore seldom occur.
- The distribution of bigeye tuna (*Thunnus obesus*) is complicated, since it occurs in two separate waters; the immature bigeye and albacore occur in the southern-half of the subtropical gyre and the mature fish, in the equatorial waters, along with yellowfin tuna.

Describing the zoogeography of the fishes of the Indian Ocean, Cohen (1973) has drawn our attention to the fact that water temperature, more than any other parameter, determines the fish distribution. The 20°C surface isotherm in winter generally demarcates the boundary between tropical and temperate regions and limits the distribution of reef-building corals and other shallow-water and coastal fauna. The geographical boundaries of the Indo-Pacific fishes, in general, are: from Natal to the head of the Red Sea and then to the east up to Taumotos-Hawaii in the Pacific; in the north, to Kiushin and Chekiang; and to the south, the limit is the Great Barrier Reef. This large area is by far the richest for the tropical shore fishes, containing practically all the families known to us. Ebeling (1962) divided the Indian Ocean into two almost equal halves, the eastern and the western Indian Ocean, the separating line being the longitude of Cape Comorin (Kanya Kumari), at 77°33'E. In the north-south direction, three areas are recognized, namely, equatorial, central and subantarctic; and each of these divisions is characterized by some endemic forms. Again there are further subdivisions exhibiting special hydrographic features which further favour endemism of the local fauna – for example, the Red Sea and the Persian Gulf.

Cohen (1973) also provides many examples of restricted distribution among fishes. In the genus *Blennius*, belonging to the family Blenniidae, out of the 20 species distributed around the world, in the coral-reef ecosystems, 10 are reported from the Indian Ocean, of which, 3 are restricted to the Red Sea, 2 to the western Indian Ocean, 2 to the eastern Indian Ocean and the Pacific; 3 species are found from the western Indian Ocean as well as in the Pacific. Similar distributions are shown by the members of the families, Gobiesocidae, Cirrhitidae and other families.

Among the continental slope and abyssal fishes, endemism becomes more pronounced. In the order Ophidioidei, of the 43 Indian Ocean species whose adults live close to the bottom (from about 250-4800m depth), 35% live in the western Indian Ocean, 16% in the eastern Indian Ocean, 14% are common to both, and 35% are wide-spread species.

A group of deep-water bathypelagic fishes surveyed by Ebling (1962) and others shows that, out of the 135 species, those living in the Indian Central and Indian Equatorial water masses have a 45% overlap with the fauna of the South Atlantic Central Water and a 75% overlap with the corresponding Indonesian fauna.

According to Cohen (*op. cit.*), melamphaids are the only deep-water Indian Ocean fishes for which salinity and temperature data have been published, and an attempt can be made to assign their distribution to water masses. The family Melamphaidae comprises about 35 species of bathypelagic fishes. Among them, 3 genera, *Melamphaes*, *Sio* and *Scopelogadus*, have 11 species (or sub-species) in the Indian Ocean, none of which is endemic. Six of them seem to have their centres of abundance in the Equatorial Water; 3 in the Indian Ocean Central Water; 1 may be considered to transcend the boundary between these two water masses; and 1 lives in the Sub-Antarctic Water.

Gibbs and Hurwitz (1967) have reported on the distribution of 2 species of *Chauliodus* found in the Indian Ocean (Figures 5.13 and 5.14); one of them, *C. pammelus*, is restricted to the north of 5°N and *C. sloani*, from 10°N southwards to the Subtropical Convergence. Compared to *C. sloani*, *C. pammelus* (Figure 5.14) can be associated with three oceanographic features: (i) well defined oxygen-poor areas; (ii) relatively shallow occurrence of high nutrient levels; and (iii) the Red Sea and Persian Gulf water mass-

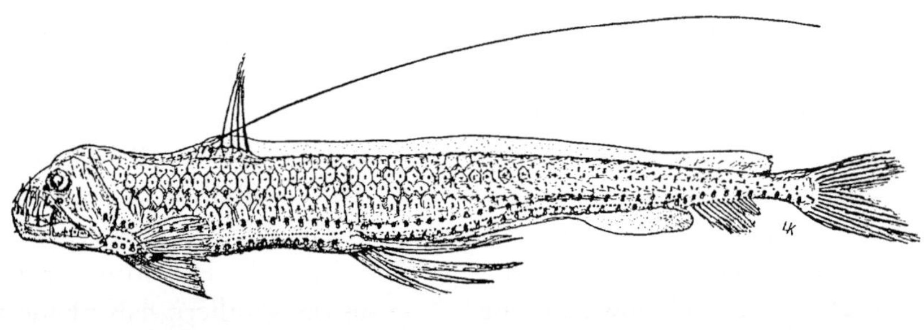

Figure 5.13
Drawing of a specimen of *Chauliodus* sp. (After Gibbs and Hurwitz, 1967)

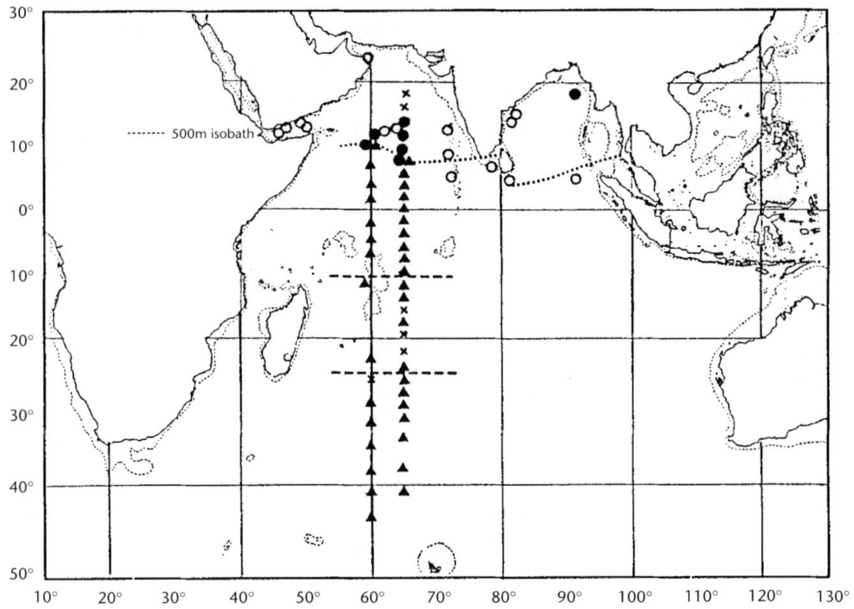

Figure 5.14
Mid-water trawl stations on the RV *Anton Bruun* cruises 3 and 6. x = unsuccessful trawl; ▲ = captures of *Chauliodus sloani*; ● = *Chauliodus pammelas*; o = literature records of *C. pammelas*; dashed line represents boundaries of three zones of *C. sloani* distribution; heavy dotted line is the northern limit of *C. sloani* as far as is known. (After Gibbs and Hurwitz, 1967)

es and their movements. The southern limit of the range of occurrence of *C. pammelus* is at about 5°N in the Arabian Sea and this corresponds closely to the southern extension of the oxygen-poor layer (1.0ml/l) which also marks the northward extension of *C. sloani*. The greater development of gills in *C. pammelus* is also indicative of its adaptation to living in the poorly oxygenated water layer. This layer also has high nutrient levels. For example, at 200m depth in the Arabian Sea, the phosphate level is 2.00µg atom/l, a concentration that is usually reached at 600-1000m depth south of 5°S. Rochford (1964, 1966) has dealt with the distribution of two high-salinity waters, those of the Red Sea and the Persian Gulf which enter the northern Arabian Sea and flow into the Bay of Bengal through channels between Chagos and the Maldive Islands. The Red Sea water sinks from the sill in the Bab-el-Mandeb and spreads in the Arabian Sea at 700-800m depth and, as it flows eastwards, it becomes shallower, reaching depths of 500-600m, then sinks deeper, to about 1000m, as it extends southwards. The Persian Gulf water also spreads all over the Arabian Sea and Bay of Bengal but at much shallower depths, of 200-300m. Either of these two water masses, which are high in salinity and nutrients and poor in oxygen, could transport or provide a pathway for the distribution of *C. pammelus* in the northern Indian Ocean.

Chauliodus sloani, on the other hand, has two distinct populations in the Indian Ocean. One is located north of 10°S and its occurrence is extremely sparse between 10 and 25°S, but it increases southwards to where the second population lives. It would appear that there is water movement down to a depth of 1000m towards 10°S both from the north and from the south, resulting in the development of a divergence called the Frontal Zone (Ovchinnikov, 1961; Rochford, 1966). Here, the isopleths of salinity and oxygen are essentially vertical down to a considerable depth. This zone is located on the north side of the South Equatorial Current and is at the boundary of the Indian Ocean Equatorial and Central Water masses. As a result of these interactions, a sterile region develops that extends from about 10°S to 25°S forming an effective, if not an absolute, barrier to the free migration of many organisms.

Upwelling and fisheries

Upwelling and dissolved-oxygen distribution, as factors controlling the biological processes in the Indian Ocean, have not received the attention they deserve. As early as 1959, just at the time of crystallizing plans for the International Indian Ocean Expedition, Carruthers *et al.* (1959) reported in *Nature* the occurrence of an oxygen-minimum layer and its effect on the marine biology and fisheries off Bombay. During the NE monsoon, with winds blowing away from the coast, upwelling is noticed off Bombay during October-November, resulting in the upswing of the oxygen-minimum layer and its incursion into the shallower parts of the shelf. The demersal fish, which are normally distributed over a wide area of the shelf in the earlier season, move inshore to escape the lethal low-oxygen conditions. In fact, the fishermen off Bombay are aware of these events and sometimes harvest a bumper crop of fish. If, on the other hand, the fish move offshore, they may come into the extensive low-oxygen areas beyond the shelf, resulting in mass mortalities; such mortalities are often reported from the Arabian Sea.

Since the upwelled waters are rich in nutrients, the resulting phytoplankton and zooplankton blooms deplete the surface water of its nutrients and further reduce the dissolved-oxygen content, which again will have repercussions on the fisheries of the area.

Banse (1959), working at Cochin at the same time as Carruthers in Bombay, had access to data on hydrography, plankton and fisheries for a longer period (from 1953-1964) and came to the conclusion that the upsurge of the oxygen-minimum layer (Figure 5.15a, b) all along the west coast of India, from Calicut to Cape Comorin, had a deleterious effect on demersal fisheries. Since the upwelling zone extended to a mean distance of 60 miles offshore, the damage done to the demersal fauna could be extensive. There are clear indications that many demersal fishes and prawns migrate to deeper waters with the onset of the SW monsoon in May and return to shallow waters after the vigour of the monsoon diminishes and upwelling of the low-oxygen layer subsides in September-October. Therefore, during the monsoon, trawling for demersal fish in the areas of depleted oxygen may not be profitable. The effect of the uplift of the oxygen-minimum layer on the migratory patterns of major pelagic fishes off the west coast is not known. Is it possible that the persistence of an oxygen-minimum layer close to the surface throughout the year may have a deleterious effect on biological activity at different depths, both in the Arabian Sea and the Bay of Bengal. This may be one of the reasons to explain the discrepancy between the estimated sustainable tertiary production of 10-17 million tons per year and the actual annual harvest of about 3-4 million tons in the Indian Ocean.

Benthos

Deep-sea fauna

Our knowledge of the benthos of the Indian Ocean is very limited. In fact, at the International Symposium on the Indian Ocean and Adjacent Seas, Their Origin, Science and Resources, held in Cochin

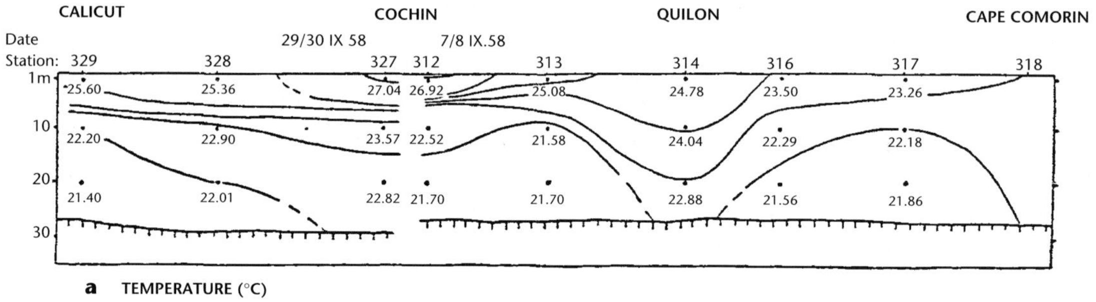

Figure 5.15a
Temperature section along the southwest coast of India. (After Banse, 1959)

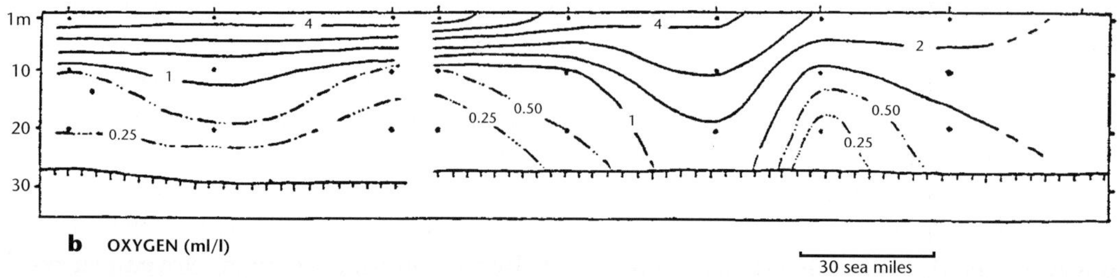

Figure 5.15b
Dissolved-oxygen section along the southwest coast of India. (After Banse, 1959)

in January 1971 and at the International Symposium on the Biology of the Indian Ocean, held in Kiel in March-April 1971, not many papers were read on benthos, except for four papers on marine algae, a few papers on benthic Foraminifera and one paper by Neyman *et al.* (1973) on some patterns of the distribution of bottom fauna in the Indian Ocean. However, in more recent years, this gap in our knowledge of the benthos in the Indian Ocean is slowly being bridged. While the work done in the Indian institutions mainly relates to the bottom fauna and flora from the littoral and nearshore areas, information on deep-water benthos is still very inadequate. It was only during the IIOE that some efforts were made by participating countries, notably USA, USSR and India, to study the benthic fauna of the Indian Ocean.

Benthic studies by the *Challenger* Expedition (1872-76) and the *Galathea* Deep Sea Expedition (1950-52) form, perhaps, the first and last effort to unravel the mysteries of the bottom fauna, their life patterns and distribution, in all three oceans. The *Challenger* did not cover the Indian Ocean except for taking some samples near Christmas Island before proceeding to the western Pacific. The published results of these two expeditions show that animals are found all over the ocean floor, even in the greatest depths of the Pacific Ocean trenches. Bruun (1956), describing the abyssal fauna, its ecology, distribution and origin, indicates that by far the largest known ecological unit in the world is the abyssal zone, occupying more than half the total area of the Earth's solid surface. The coastal fringe of the Indian Ocean (0-200m depth range) occupies about 4.2% of the total area of the Indian Ocean; the depth zone 2000-6000m encompasses 88.9% of its area. The most characteristic feature of the abyssal zone is the enormous hydrostatic pressure upon it, which increases from 200 atmospheres (≈203,000 hectoPascals) at 2000m to nearly 1000 atmospheres (≈1.01×10^6hPa) in the deepest trenches. Coupled with this are the low temperatures (<4°C), absence of seasons and light, all of which will have a great impact on the fauna, particularly in terms of its diversity and abundance. The littoral fringe has the greatest variety and abundance of benthic organisms, particularly among the coral reefs. As one progresses deeper, the diversity decreases, as do the numbers. In the deepest trenches, with a depth of more than 6000m, the fauna of which Bruun calls 'hadal fauna', the number of species may be 6-8.

In this largest ecological unit, the ecological parameters and subdivisions are quite distinct (Figure 5.1) and there are three major zones, apart from the littoral. They are the bathyal, abyssal and the hadal. The bottom is generally covered by soft siliceous or calcareous oozes and oxidizing clay. The boundary between the bathyal and abyssal zones could be the 4°C isotherm, there being no seasonal changes. The salinity of the waters would be about 34.8 and the variation, about ± 0.2. Oxygen is present at all depths, there being no anaerobic areas in the Indian Ocean. Food is a problem for deep-sea benthos. All organisms living below the photic zone have to depend on the amount of food existing or sinking into this zone. This is about all that is known. Unfortunately, what is not known is much greater, particularly regarding food and feeding, life histories, vertical distribution, metabolic rates, how each individual recognizes another, be it a mate, enemy, predator or prey – all in this primordial and perpetual darkness and high pressure under which these animals live and move.

The zoogeographical distribution of the deep-water bottom fauna in the abyssal zone of the ocean has been described by Vinogradova (1959). The analysis of 1031 species in depths greater than 2000m showed that 84% of the species are confined to a single ocean and only 4 percent occur in all the three major oceans. The Indian Ocean has the smallest proportion of endemic species. Another feature peculiar to the Indian Ocean is that only 2.4% of the deep-water species are found both in the northern and southern parts and a large number of species (47.7%) are common to the Indian Ocean and the western Pacific Ocean.

Dealing with the faunal diversity in the deep sea, Hessler and Sanders (1967) found, surprisingly, a very large number of individuals and species even in deep abyssal zones. They found a greater diversity of fauna in single environments than has ever been encountered. They observed that diversity in the deep sea is much greater than in equivalent shallow marine environments from temperate latitudes and is of the same magnitude as that of the shallow marine tropics. Marshall (1954) proposed a geographic rule that 'with an increasing distance from land, there is an increasing tendency for the deep-sea floor to be populated by fewer individuals belonging to fewer species'. Using an epibenthic sled, Hessler and Sanders obtained a record number of species (365) from a station at 14m depth on the continental slope of North

America. Even from 4700m depth at another station in the Sargasso Sea, they obtained 196 species. In contrast, one of the richest tropical shallow marine samples collected by them (6236 individuals belonging to 201 species) was from 15m depth off Madras, India, in the Bay of Bengal.

Summarizing the studies made by Russian research vessels, mainly the *Vityaz* in 1959-60 in the Indian Ocean, Neyman *et al.* (1973) describe quantitatively the patterns of the distribution of the bottom fauna. It would appear that the quantitative distribution of the abyssal fauna is a mirror image of the phytoplankton distribution in the sun-lit layers at the surface of the oceans. This is easily explained on the premise that, wherever there is phytoplankton production, the dead cells (often called 'marine snow') would rain down to the bottom, thus providing the food for the benthos. As a result, a rich benthic fauna biomass is found along the coastal and upwelling areas of the Indian Ocean, particularly in the Arabian Sea and the Bay of Bengal. According to Sokolova and Pasternak (1964), the Arabian Sea is characterized by a rich bottom fauna with biomass exceeding 500g/m^2, the maximum value recorded for the Indian Ocean. However, all along the East African coast and the Arabian peninsula, the biomass values range from 5 to 30g/m^2 and, similarly, off the west coast of India, the biomass values range from 5 to 30g/m^2. The abundance of bottom life on the shelf at 25-75m depth is attributed to the flow of low-salinity equatorial water causing a strongly expressed stratification. On the other hand, the fauna is extremely poor at 80-150m depth owing to an inflow, at this depth, of sub-surface water with a low dissolved-oxygen content. It may be recalled here that Banse (1959) also reported a similar adverse effect of low-oxygenated bottom waters on the demersal fisheries off the Kerala coast.

It was also found that the bottom fauna of the shelf of the Bay of Bengal and the Andaman Sea was much poorer than that of the Arabian Sea, its biomass varying from a fraction of a gramme to 10g/m^2; however, close to the mouth of the Ganges and Irrawady Rivers, a value of 42.8g/m^2 was recorded (Sokolova and Pasternak, 1962). In all other coastal regions in the tropical part of the Indian Ocean, the biomass of the bottom invertebrates is low and seldom reaches 10g/m^2. As one proceeds farther south towards the subtropical waters, there is an increase in the biomass values: 15g/m^2 on the west Australian Shelf and 16g/m^2 is on the shelf of the Great Australian Bight.

Neyman *et al.* (1973) further point out that the Bay of Bengal is extremely poor when compared to the Arabian Sea, where, even at a depth of 3000m and at considerable distance from the shore, biomass values may reach 1.92g/m^2. The corresponding values for the Bay of Bengal range from 0.11 to 0.45g/m^2. Another interesting feature observed in the Arabian Sea is the presence of a thick oxygen-minimum layer, extending from the layer of density discontinuity down to a depth of 1000-1250m, and, wherever this water washes the bottom, the fauna there would be extremely poor and impoverished; e.g., 0.01g/m^2 or less. Russian researchers have reported the presence of H_2S in some areas (0.1mg H_2S/l) in the Arabian Sea, but this has not been substantiated by other workers in the area.

In this connection, reference may be made to the findings of the John Murray Expedition (1933-34) as reported by Mohamed (1940) who found more or less azoic areas of the sea floor extending from about 100m to 1300m depth off the coast of Arabia and in the Gulf of Oman. The mud brought up in this area contained H_2S (in one case, 30mg/l in the interstitial water). Not much attention was paid to the further study of such an important discovery in subsequent investigations by other ships.

The western part of the Indian Ocean is comparatively rich in bottom fauna, the biomass measuring not less than 0.1g/m^2. However, in the open deep waters of the Indian Ocean, a very low biomass (less than 0.02g/m^2) was recorded in its tropical zone. Southwards, beyond the tropical zone, the biomass of the bottom fauna increases to a value of 0.2-0.4g/m^2 to the south of 30°S. According to Neyman *et al.* (1973), the composition of the bottom fauna at great depths is dominated by polychaetes, constituting nearly 50% of the biomass, followed by crustaceans, tanaidaceans and isopods. In shallower depths, crustaceans share equally with polychaetes. However, at all depths, sponges are well represented, whereas molluscs are poorly distributed.

Based on the productive nature of the waters, two major regions are recognized in the Indian Ocean (Figure 5.16a, b): the oligotrophic and the eutrophic; the former area is defined as the least productive and is generally located in the central parts of the Ocean, encompassing the southern parts of the Central Indian Ocean Basin, the Australian Basin and the northern half of the Crozet Basin, and coincides with the boundaries of the southern subtropical gyre, characterized by waters of extremely low productivity. This part of the Ocean is more than 3800-4000m deep. The sediments of the northern part of this area (down to 15-17°S) are represented by radiolarian oozes. The ferro-manganese nodules are widespread here and, in some places, the sea floor is literally paved with them. The content of organic carbon in the sediments of the oligotrophic area is 0.3% or even less.

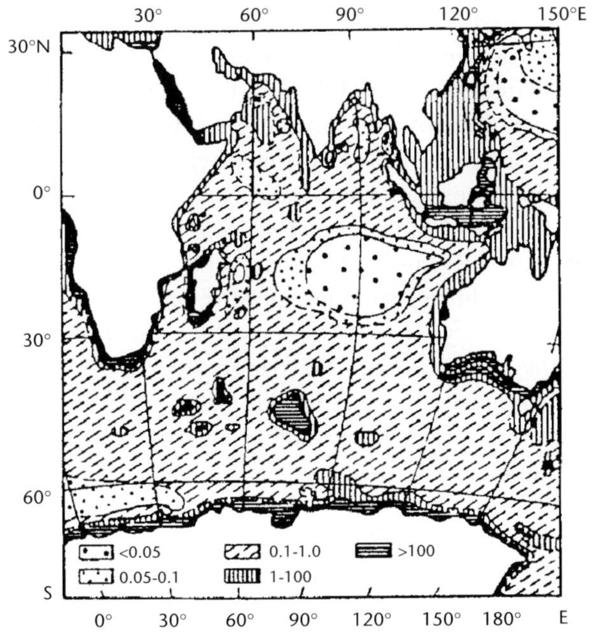

Figure 5.16a
Distribution of benthic biomass (g/m²) in the Indian Ocean.

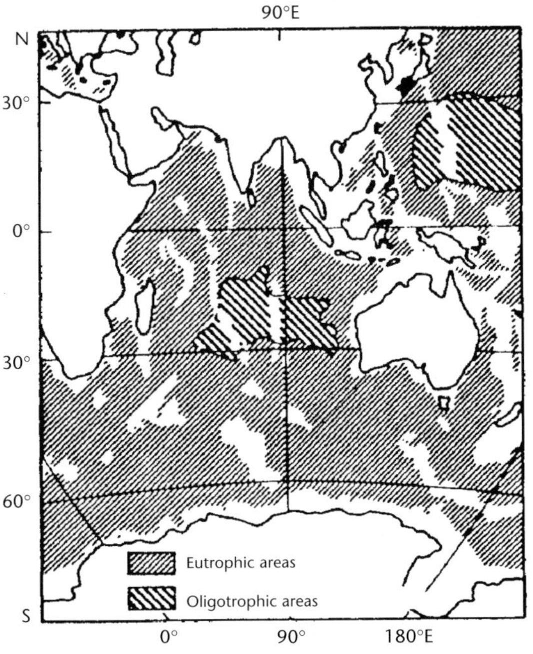

Figure 5.16b
Eutophic and oligotrophic areas of the Indian Ocean. (After Neyman *et al.*, 1973)

Eutrophic conditions exist all along the shores of the Indian Ocean, even down to depths of 3000m in some places. The eutrophic region encompasses: the Arabian Sea, with the adjacent Somali Basin, Mascarene and Mozambique Basins; the Bay of Bengal, with the adjacent parts of the Central Indian, Coco and Crozet Basins, the Java Trench, the greater part of the central rift zone, the southern part of the Crozet Basin and both the Antarctic Basins. The sediments of this vast area are composed of foraminiferan and radiolarian oozes to the north, and diatomaceous oozes to the south.

Seston and detritus-feeding invertebrates are widely distributed in the macrobenthos of the eutrophic region. The distribution is dependent on three factors: the food supply; the bottom topography; and the rate of sedimentation. Sokolova and Pasternak (1962) made a detailed study of the distribution of trophic groups in the Bay of Bengal and demonstrated a clear-cut succession of zones based on food and feeding habits. There is a predominance of seston feeders on the lower part of the shelf, followed by sorting detritophages on the slope, and indiscriminate detritophages in the deeper parts of the Bay.

Analysing the data collected on the RV *Gaveshani* (cruises 86 and 87) in the northern part of the central Indian Ocean, Parulekar *et al.* (1982) found a rich fauna and high standing crop associated with the occurrence of polymetallic nodules. Their studies corroborate the findings of Neyman *et al.* regarding the biomass estimates and faunal composition. The polychaetes (41.6%) made up the bulk of macrobenthos, followed by crustaceans (31.7%). Among the meiofaunal taxa, the nematodes (69.4%) formed the major component, followed by harpacticoid copepods (26.6%) and ostracods (4%).

The macrofaunal biomass varied from 0.47g/m² to 13.32g/m², with an overall mean value of 2.62g/m², whereas the meiofaunal contribution to the biomass was insignificant, the values ranging from 0.02 to 0.41g/m². The relatively high biomass values in the remote areas of the deep ocean is attributed to the transport of organic matter, by deep-water currents, from the shelf and shallow waters of the surrounding continents to the abyssal depths. In a comprehensive paper, Parulekar *et al.* (1982) have reported on the benthic biomass production for the entire Indian coast and the contiguous seas (Table 5.9).

Their values indicate that the Indian seas support a high level of benthic biomass and, based on this, they conclude that exploitation of the demersal fishery resources can be increased by an order of magnitude without adversely affecting their sustainable yields. Associated with the abyssal benthos are the ferro-manganese concretions (also called manganese nodules) which, in some places, literally carpet the floor of the Central Indian Ocean Basin. During the survey programme for these nodules by the National Institute of Oceanography, in Goa, thousands of seabed photographs were taken and analysed for benthic activity. Reporting on this, Sharma and Rao (1992) indicate extensive activities of benthic organisms based on their telltale marks (*lebensspurren*), particularly in sediment-covered areas (85 to 100%). The activity of benthic organisms may explain the occurrence of the manganese nodules and perhaps their growth.

Littoral benthos

The littoral benthos comprises a large number of organisms, plants (sea weeds and mangroves) and animals, of great nutritional value for mankind. Clams, oysters, shrimps, crabs, lobsters and seaweeds all form a renewable source of protein-rich food.

Fernando *et al.* (1983) made observations on the distribution of the benthic fauna in the Vellar estuary on the east coast of India. They found that polychaetes dominated the macrofauna (63.10%), whereas nematodes (≈52%) dominated the meiofauna. The macro and meiofauna showed great abundance during the months (February-June) before the onset of the SW monsoon, when salinity and temperatures were high. Ramana *et al.* (1987) surveyed the meiofauna of the Gauthami estuary and found it to be abundant during summer months, much as in the Vellar estuary. It was composed of the following groups: Nematoda, Harpacticoida, Ostracoda, Kinorhyncha, Turbellaria, Mollusca, Oligochaeta, Tardigrada, crustacean larvae, insect larvae and Amphipoda. The Nematoda were the predominant group.

From the west coast of India, Damodaran (1964) studied the benthic fauna of the mud banks of Kerala. Desai and Krishnan Kutty (1967) reported on the benthic fauna of Cochin backwaters. Ansari (1977) made a repeat study of the macrobenthos of Cochin

Table 5.9 Benthic (macro- + meio-) production (g/m^2) in the Indian Ocean seas

Depth zone (m)	Arabian Sea	Lakshadweep Sea	Andaman Sea	Bay of Bengal
0-20	0.01-147 (17.87)	-	2.46-43.7 (19.61)	0.01-37.06 (5.7)
20-40	0.01-601 (14.06)	-	4.5-74.3 (29.89)	1.09-176.7 (12.08)
40-60	0.01-190.5 (14.5)	1.80*	-	0.55-44.7 (3.75)
60-80	0.02-67.9 (5.13)	-	0.7-80.9 (3.85)	0.18-10.99 (1.40)
80-100	0.05-52.6 (9.13)	-	0.7-19.4 (4.42)	1.4-12.86 (2.40)
100-200	0.11-66.2 (13.66)	-	2.41-14.68 (8.54)	0.4-98.79 (4.20)
200-300	0.05-44 (20.41)	-	-	0.4-60.2 (6.02)
300-400	-	0.45*	0.15-12.47 (6.18)	0.72*
400-500	-	-	0.90-15.75 (4.73)	0.05-8.46 (3.10)
500-600	31.00*	-	0.50-10.41 (3.82)	0.76-4.52 (2.64)
600-700	-	-	0.40-28.25 (14.32)	-
700-800	-	-	0.7-23.9 (9.62)	1.11-10.87 (5.99)
800-900	1.92*	-	1.58-3.57 (2.57)	-
900-1000	9.20-44.13 (26.66)	-	1.72-3.11 (2.41)	3.28*
1000-2000	0.01-119.6 (20.02)	0.03-0.45 (0.37)	1.47-19.52 (2.32)	0.7-5.2 (2.32)
>2000	0.8-43.3 (14.29)	0.71*	1.04-6.80 (3.92)	6.56*

* = single observation

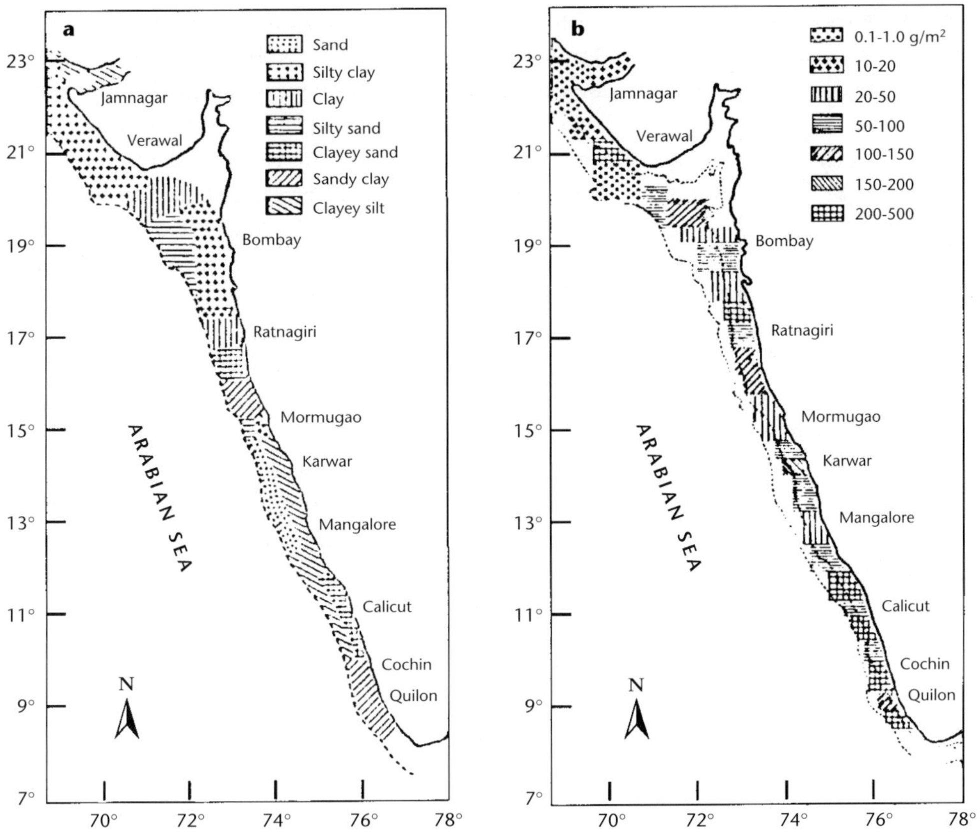

Figure 5.17
a. Sediment distribution along the west coast of India. (After Harkantra *et al.*, 1980)
b. Biomass distribution along the west coast of India. (After Harkantra *et al.*, 1980)

backwaters. His results, when compared with the studies made by Desai and Krishnan Kutty (1967) ten years earlier, revealed that, quantitatively, there were differences and some of the dominant groups, as gastropods and isopods, were either absent or occurred sparsely in 1977. Quantitative distribution of macrobenthos in five shallow bays (Vengurla, Goa, Karwar, Malpe and Mangalore), along the central west coast of India, was studied by Ansari *et al.* (1977) and they reported a wide range of densities among the populations. As usual, the polychaetes formed the most abundant group, constituting 45.3-79.9% of the total population, except at Mangalore, where molluscs (particularly *Gafrarium pectinatum* and *Pholas orientalis*) occurred in large numbers. In the sublittoral zone, the number of animals ranged from 251/m^2 at Goa to a maximum of 15800/m^2 at Mangalore. The annual average of the number of animals varied from 1279 to 5518/m^2. The minimum biomass (wet weight) was found to be 0.25g/m^2 and the maximum, 1109.72g/m^2, and the average biomass varied from 62.15 to 762.35g/m^2. Ansari (1978) also reported on the meiobenthos of the Karwar region. The sediments there were mostly composed of silt and clay, with sand contributing less than 20%. The annual variation in temperature was from 23.9 to 29.4°C and salinity ranged from 8.83 to 35.35. Dissolved oxygen varied from 2.06 to 3.76ml/l. The total number of animals ranged between 1022 and 1250/10cm^2, with an average of 1098 animals/10cm^2. The biomass varied between 8.42 and 11.60mg/10cm^2 (average 10.38mg/10cm^2). As reported elsewhere, nematodes constituted 68 to 85% of the total fauna, followed by foraminferans, bivalves, kinorynchs and turbellarians. Sediments with a low percentage of sand contained a high percentage of meiofauna.

Harkantra *et al.* (1980) contributed an informative paper on the benthos of the shelf region along the west coast of India. They found a definite relationship between benthic biomass and types of substrata, on the one hand (Figure 5.17a, b), and between this biomass and the demersal fishery, on the other hand (Figure 5.18). The area sampled stretched from 8° to 23°N and had seven types of sediments: sand, silty clay, clay, silty sand, clayey sand, sandy clay and clayey silt. Of these, sand contained the lowest percentage of organic carbon (0.47-3.26%) and clay had the highest percentage, amounting to 3.15 to 6.18%. The fauna comprised 16 groups, of which polychaetes formed the most dominant group (62.5%),

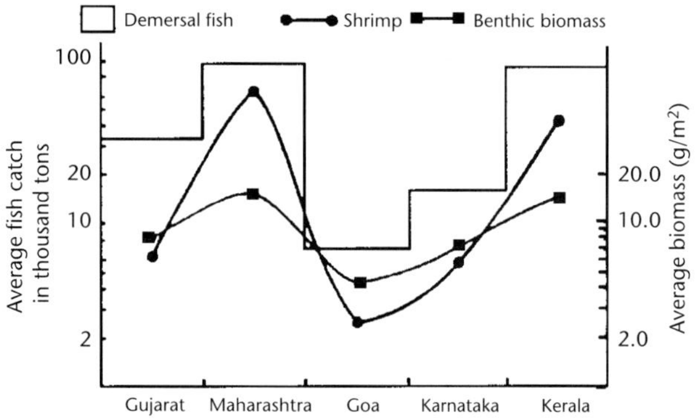

Figure 5.18
Demersal fish catch in relation to benthic biomass along the west coast of India, state by state. (After Harkantra *et al.*, 1980)

followed by crustaceans (17.62%), molluscs (12.28%), sipunculids (4.19%) and echinoderms (3.62%), and the rest, as nemertians, hydrozoans etc. The biomass varied from 0.12 to 190.5g/m^2 (average 11.59g/m^2). The population density ranged between 50 and 3175/m^2. There was also an abundance of fauna in sandy-clayey sediments, but not in clay.

In a paper on the marine fauna of Malvan, located on the central west coast of India, Parulekar (1981) has given a checklist of the animals from the rocky, sandy and mangrove areas. The list comprises 208 species belonging to 172 genera, 97 families, 16 classes and 9 phyla, thereby revealing the richness of the fauna at Malvan; it is also unique inasmuch as it includes 'red coral' and pearl oysters.

Coral reefs

As early as 1902, Alcock remarked the coral lagoons of the Laccadive Islands, and subsequently many scientific reports have been presented on the coral reefs and coral islands of the Indian Ocean. Sewell (1935) summarized all the information available on the corals and coral formations of Indian waters till then. Subsequently, in 1981, Gopinadha Pillai published a fairly comprehensive paper on the distribution of corals on a reef at Mandapam (Palk Bay), South India. The reefs are of a fringing type and are located around the various islands lying between Tuticorin and Rameswaram in the Gulf of Mannar. The environmental conditions are the following: annual rainfall varies from 762 to 1270mm, mostly falling during October-December; atmospheric temperature ranges from 25° to 31°C; monthly average surface water temperatures range from 24.6° to 29.1°C, the lowest and highest being in January and April, respectively. A total of 63 species belonging to 22 genera were found on the entire reef. Gopinadha Pillai (1983) has also updated the checklist of the corals (the recent Scleractina), their structure and genetic diversity, for the whole Indian coast, including the island groups of Lakshadweep, Andaman and Nicobar. This area has 199 species divided among 71 genera, none of them endemic, and all are widely distributed.

The Andaman and the Nicobar Islands are the richest and account for 59 genera with 135 species. Of these, 100 species belonging to 47 genera are hermatypes, the rest being ahermatypes. Among the hermatypes, *Acropora* alone formed 20% of the total and *Montipora*, 12%, the two being numerically rich genera, and the latter being mostly of an encrusting type.

Discussing the regional variation of the Indian coral fauna, Pillai reports that Lakshadweep, as far as is known, harbours 78 species belonging to 31 genera, which is much less than found in the neighbouring Maldive archipelago, where 241 species belonging to 75 genera are known. Since many of the islands of Lakshadweep remain to be studied, this difference between them may be more apparent than real.

In the Gulf of Cutch, the corals are in an early stage of development and the environmental conditions here are not benign enough for heavy and racemose growths. The coral fauna consists of 37 species in 22 genera; while *Acropora* and *Montipora* are present, *Pocillopora*, which is the most widely spread Indo-Pacific form, is not found. A single species of branching *Porites* was found on Pirotan Island in the Gulf of Cutch.

Qasim and Wafar (1979) have drawn attention to the occurrence of live corals at several places along the west coast of India, particularly in the rocky intertidal zones of Ratnagiri, Malvan and Redi. It is known that corals require fairly uniform temperatures and salinity, as well as clear water, for their colonization and growth. By their occurrence along the west coast, these mainly hermatypic corals belonging to the genera *Coscinaria*, *Turbinaria*, *Favites*, *Porites*, *Synerea*, *Pseudosiderastrea*, *Cyphastrea* and *Goniastrea*, have indicated their ability to adapt to highly varying environmental conditions. At Malvan, the colonies were found to be not more than 4-5 years old and appeared to tolerate and thrive in salinity between 20 and 25 and the high turbidity encountered during the monsoons.

In the absence of detailed work on the coral reefs of the Indian Ocean, with the exception of some of the expedition reports of the late 19th and early 20th centuries, there is very little to comment upon. Coral reefs can be considered as a biological indicator of environmental changes and it is absolutely essential to make an effort to assemble baseline information on coral systematics, growth and mortality; corals may also help in indicating any global changes manifested in changes in mean sea level, warming of the seas and their CO_2-carbonate chemistry. In this connection, we need sustained studies of coral reef fauna and flora, much on the lines of contributions made by Thomas (1976, 1983) on sponges, and by James (1976, 1982, 1983) on the echinoderms of the Indian Ocean seas.

Bakus *et al.* (1994) have published a book on coral reef ecosystems, which contains a wealth of information on the oceanography and coral reefs of India, Sri Lanka, the Laccadive, Maldive, Andaman and Nicobar islands, a bibliography and a check list of many groups of animals and plants associated with coral reefs.

Other benthic animals

Thomas has reported the occurrence of 481 species of sponges in the Indian Ocean, 44 species being assigned to the order Hexactinellida. All these species are widespread in deeper parts. Species of the order Hexasteophora are represented in the Australian region by 5 species (25%), in the Atlantic Ocean by 4 species (20%) and in the Pacific by 3 species (15%). The general distribution of the sponges reveals that 35.4% of the order Demospongiae from the Indian Ocean are also common to the Australian region, 21.1% to the Pacific, 20.4% to the Red Sea and 18.3% to the Atlantic Ocean. Among the sponges, 427 species (88.8%) belong to the Demospongiae, 44 species (9.1%) to the Hexactinellida and 10 species (2.1%) to the Calcarea.

The number of Red Sea species is now placed at 187. Of these, Row (1911) recorded only 73 species and, of these, 19 species were common to the Indian Ocean and the Red Sea. With the opening of the Suez Canal, there appears to be a migration of sponges from the Red Sea into the Suez Canal, but very few have migrated to the Red Sea from the Mediterranean. Of the 25 species occurring in the Suez Canal, 14 migrated from the Red Sea and 2 from the Mediterranean (Kimor, 1971).

James (1983) has reviewed the distribution of echinoderms in the Indian waters. This exclusively marine group has been the subject of study all over the world. He has listed 257 species from the Andaman and Nicobar Islands. More than 100 species are known from the Gulf of Mannar and Palk Bay, 24 species from Madras, 15 species from Bombay, and 6 from the Orissa coast. According to James, 59 species of the Asteroidea (starfishes), 55 of the Ophiuroidea (brittle stars), 54 of the Echinoidea (sea urchins) and 73 of the Holothuroidea (sea cucumbers) are known from the Indian Ocean seas. Whereas 178 species of echinoderms are known from the Sri Lankan shores, only 103 species are known from the adjacent coast of India. The reason for such a discrepancy requires an explanation, but it could be the result of current patterns and the nature of the coastline.

Animal associations, particularly commensalism, are quite common among many groups of animals. Enchinoderms present many examples, as the following from the Gulf of Mannar:

Pentoceros herdmanii (starfish)	*Podarke angustifrons* (polychaete worm)
Lamprometra sp. (feather-star)	*Horovia albolineata* (pea crab)
Actinopyga mauritani (sea cucumber)	*Lissocarcinus orbicularis* (pea crab)
Holothuria scabra (sea cucumber)	*Pinnotheres decanensis* (pea crab)
Holothuria arenicola (sea cucumber)	*Encheliophis gracilis* (carpet fish)

Marine algae (seaweeds)

Rao (1987, 1989) has reported on the marine algae of the Indian estuaries, and on the Indian seaweed resources and their management. He found the maximum occurrence of macro-algae in high-salinity areas of estuaries and open sea shores. The common forms were: *Enteromorpha chathrata*, *Ulva fasciata* and *Chaetomorpha linum*, belonging to the Chlorophyceae; *Gracilaria*, *Hypnea* and *Polysiphonia*, belonging to the Florideophyceae; and members of the Cyanophyceae, such as *Oscillatoria* sp. etc. There is very little information available on the seasonal variation and reproduction of the various species. However, Rao has summarized all available information based mainly on the reports of Rao (1969) and

Table 5.10 Seaweed resources estimated for various maritime states of India

Maritime State	Area/length of coast surveyed	Standing crop (tonnes wet wt.)[1]	Authors
Gujarat	548ha	446.2	Chauhan, 1978
Maharashtra	563km	278.3	Chauhan, 1978
Goa	-	2,000.0	Dhargalkar, 1981
Tamil Nadu	9,892ha	22,044.0	Anonymous, 1977
Andhra Pradesh	1,876ha	7,493.0	Subbaramiah *et al.*, 1987
Orissa (Chilka)	-	5.0	Mitra, 1946
Lakshadweep	1,334ha	7,524.0	Subbaramiah *et al.*, 1979

1. Total wet wt. = 39,790.5 tonnes; dry wt. = 11,937.2 tonnes.

Untawale *et al.* (1989). It would appear that Tamilnadu has a maximum abundance of seaweeds (Table 5.10).

It is estimated that the total seaweed resource of India is about 39,790 tons wet wt. Species of the genera *Gelidiella*, *Hypnea*, *Turbinaria*, *Ulva*, *Gracilaria* and *Sargassum* are extensively used for the extraction of polysaccharides in India, but there is still a need for importing these chemicals. Many seaweeds could be used either as food or fodder and as fertilizer, as in Japan, Korea, Norway and the former USSR. However, there appears to be no interest in this area for sustained research and development. The seaweed industry is practically non-existent in all the countries around the Indian Ocean.

There is very little information on the seagrasses in and around the Indian seas. Ansari *et al.*, (1991) have recently reported on the seagrasses of the Lakshadweep Atolls. The dominant forms include *Thalassia hemprichi* and *Cymodocea* sp. and the coralline alga *Halimeda* sp. The seagrass biomass showed the lowest value, of 405g/m^2, at Kadmat and the highest value, of 895g/m^2, at Agatti. This community had a close association with macroinvertebrates, their density showing wide variations, from 1041/m^2 at Kadmat to 8411/m^2 at Agatti. As in all other areas of benthos, polychaetes made up the bulk in numbers, 60.2-92.7% of the total macrofaunal numbers.

Untawale *et al.*, (1989) have reported on the benthic algae of the north Kanara coast. They recorded 65 species belonging to 42 genera. Maximum species were represented by Rhodophyta (25) followed by Chlorophyta (23) and Phaeophyta (17). They also studied the zonation and found local variations in the dominance of various species. For example, at Karwar, with a narrow inter-tidal expanse, the supralittoral fringe was occupied by species of *Ulva*, *Enteromorpha* and *Chaetomorpha*. The midlittoral fringe had *Dictyota*, *Stoechospermum*, *Hypnea* and *Gelidium* spp. The infralittoral fringe was dominated by *Gracilaria*, *Sargassum* and *Amphiroa* spp. Some algae showed marked seasonal changes; species of *Porphyra* and *Bangjia* completely disappeared as higher salinities and temperatures aided the appearance and increase of many species. Among the three dominant algae, *Sargassum* and *Gracilaria* were present from October to April, whereas the *Ulva* crop started deteriorating from January onwards.

Mangroves

Mangrove forests form a visible landmark along the estuarine shores of the Indian Ocean. They form an important fishery ground or serve as nurseries for the young of some species of fish; they also have a general economic importance. There is great ecological diversity among them.

The dominant genera are *Bruguiera*, *Ceriops*, *Lumnitzera*, *Sonneritia* and *Xylocarpus*; these are, however, absent from the Atlantic mangroves. James *et al.* (1983) have summarized the work done on mangroves by the Central Marine Fisheries Research Institute and the National Institute of Oceanography. More recently, Vannucci (1989) has summarized all useful information on mangroves in a delightful book entitled *Mangroves and us.*

UNESCO has carried out a research and development project on the ecology and distribution of the mangrove forests in the South Asian region.

The extent of mangrove areas on the mainland of India is estimated to be about 256,000ha and, in the Andaman and Nicobar Islands, another 100,000ha. The largest single area covered by mangroves is the Sunderbans at the mouths of the Ganges and Brahmaputra Rivers, at the head of the Bay of Bengal; it occupies an area of about 42,000km^2.

In the Indian subcontinent, there are 58 principal, halophilous non-parasitic species in the mangrove floral community. Moreover, the soils on the river banks are highly saline all over the peninsula, with the exception of the Cauvery delta where *Avicennia marina* grows in distinctly acidic soils with a pH between 3.5 and 4.5 (Blasco, 1977).

The dense mangrove forests of the Sunderbans are the richest in species. The pressure of human needs is also very intense on these forests. The important species are: *Excoecaria agallocha*, *Ceriops decandra*, *Sonneratia apetala*, *Avicennia* spp., *Bruguiera gymnorrhiza*, *Xylocarpus granatum*, *X. moluccensis* and *Rhizophora mucronata*. Two species, *Heritiera formes* and *Nypa fruticans*, have become practically extinct, victims of natural selection. However, *Nypa* is common in the Andamans. In the Krishna and Godavari deltas, *Sonneratia apetala* and several *Avicennia* spp. are dominant. In addition, there is a grass, *Myriostachia wightiana*, which is very common here, but unknown elsewhere.

Mangrove forests are of great value. Firstly, they reduce coastal erosion. The water beneath the forest is rich in organic debris and nutrients and so acts as a spawning and nursery ground for many organisms (particularly, finfish, shellfish and crustaceans). The varied fauna and flora of the mangrove community also provide valuable opportunities for education, scientific study and tourism (Saenger *et al.*, 1983).

At many places in the world, the mangroves are being destroyed to meet human needs. This could have serious consequences for the coastal areas, in the form of erosion and damage to the spawning and nursery grounds of many organisms (Teas, 1983).

Economic uses

Many benthic organisms are economically very valuable, both as a source of rich protein food and as products of commerce. Mussels, oysters and clams are an important source of food. The common mussels are *Perna viridis* and *Perna indica*, and the common edible oysters are *Crassostrea madrasensis*, *C. cucullata* and *C. gryphoides*. The edible clams are *Meretrix meretrix*, *M. casta*, *Katylysia opina*, *Paphia marmorata* and *P. malabarica*.

Pearls are produced by the oysters *Pinctada furcata* and *P. chemnitzi*. Rajendran *et al.* (1983) have written the history of the Indian pearl-oyster banks of the Gulf of Mannar. Pearls have been used in jewelry for many millennia in India, but no records are available as to when pearl fishing started in the Gulf of Mannar. The authors have provided an interesting table showing the number of pearl-bearing oysters fished from different oyster banks and the revenue to the government, from 1805 to 1961. For example, in 1805, the number of pearl-bearing oysters fished from the four pearl-oyster beds was 7, 16, 47 and 305, and the revenue was Rs39,109. In 1961, which was the last year there was a fishery, 1, 53, 96 and 928 pearl-bearing oysters were fished from the four beds, yielding a revenue of Rs288,860. With the Japanese flooding the market with artificially cultured pearls, the natural pearls no longer hold much attraction for people.

Lime-shell fisheries are an important, though minor, coastal industry. The Vembanad Lake on the Kerala coast and the Pulicat Lake near Madras on the east coast support many small-scale lime-making industries by harvesting shell deposits from the beds of these lakes. Rasalam and Sebastian (1976) have given a comprehensive account of the lime-shell fisheries of the Vembanad Lake, where this forms a chief source of calcium carbonate. Several surveys reveal that, in addition to surface clam beds, subfossil shell beds, which lie below a silt burden 20-60cm thick, are also excavated. The deposits are estimated to be 2 to 4.5 million tons. The main species of lime-shell are *Villorita cyprinoides* and *Meretrix meretrix* and their subspecies. There are many industries using these shells for the manufacture of cement, calcium carbide and lime bricks. Thangavelu and Sanjeevaraj (1985) have given an account of the shell deposits and their exploration in Pulicat Lake. Here also, there are extensive sub-fossil shell deposits found a few feet below the surface and over a wide area of about 2000ha, the main species being *Meretrix meretrix*. Other areas along the coastline of India, particularly in the deltas of many rivers, are also rich in clam beds, which are regularly mined for lime.

Ruben *et al.* (1984) state that the black sea urchin, *Stomopnuestes variolaris*, which occurs in large numbers along the rocky coasts of Waltair, could be a source of food. They estimate this resource to be 1224 tons with an average density of 8.5 individuals/m^2. It is believed that farming is a possibility, since the roe of these sea urchins is considered to be a delicacy.

Thomas and George (1986) have drawn our attention to the discovery of prostaglandins in gorgonids and to their clinical possibilities. Commercial exploitation of gorgonids appears to have started in India in 1975 and the material is being exported to many countries in Europe and America. Between 1974 and 1985, the total quantity exported was about 34.4 tons.

The Indian horse-shoe crab, *Tachypleus gigas*, is another potential source of a bioactive substance known as amoebocyte lysate (AL), which is being used to test bacterial toxins. These crabs can be bled and returned unharmed to the sea. A single crab can supply 20ml of blood which can yield 5ml of AL. The National Institute of Oceanography, in Goa, has been studying this crab and its populations along the Orissa coast. Debnath and Choudhury (1988) have estimated their population off the Orissa coast to be nearly 6000, at one full-moon event, these crabs being known to come ashore during specific phases of the moon for spawning.

Indo-US co-operation in benthic studies

In 1984, India and the United States of America launched a series of biological studies of the benthic fauna, their biology, bio-active substances and marine biodeterioration. The participating agencies were, on the American side, the Office of Naval Research, the American Institute of Biological Sciences, the Stevens Institute of Technology and the University of California, Los Angeles, and, on the Indian side, the Department of Ocean Development, the National Institute of Oceanography, the Central Drug Research Institute and the Marathwada University. Many scientists were involved in these studies and their results were presented at a series of three International Symposia held in India, as follows:

1. Biology of the Benthic Organisms; Aurangabad, 1984
2. Marine Biodeterioration; Goa, 1986
3. Bioactive Compounds from Marine Organisms; Goa, 1989

The proceedings of all three Symposia have been published (Thomson, 1988; Thomson *et al.*, 1986, 1991).

At the International Symposium on the Biology of the Benthic Marine Organisms, 60 papers were presented under 5 headings: physiology and endocrinology; productivity and larval development; ecology; mariculture and fisheries; and pollution and biofouling. In an interesting paper, Crisp compared the reproductive patterns of the high- and low-latitude barnacles, including *Balanus balanoides* and *Tetraclita (Tesseropora) pacifica*. The former is a circumpolar species, whereas the latter is tropical. It is most interesting to note that, as in the polar form, *Tetraclita* also produces very large eggs, contrary to the general pattern for tropical animals which normally produce small eggs in millions.

Another paper of interest was presented by Fingerman on the neurohormonal regulations in Crustacea. Recent advances in this area look very promising for research and development in mariculture and pharmacology. On the same lines, Morse drew the attention of the participants to the identification of the external molecular signals controlling reproduction, settlement and metamorphosis of benthic marine invertebrates. Further work in this area of molecular biology would be of great value, particularly from the view point of marine fouling and corrosion.

At the Second International Symposium, on Marine Biodeterioration, the emphasis was on the lack of proper and effective remedial measures against biodeterioration, even after several decades of research by various institutions, in India and abroad. In his welcome address, Zahuranec pointed out that, in the fouling cycle, an understanding was needed in the area of initial development of organic film and the subsequent biofouling processes at the molecular level. Qasim, in his inaugural address, emphasized the enormous economic losses sustained by the shipping and oil industries as well as by harbour authorities as a result of biofouling and related damage. It is estimated that the cost of removing biofouling organisms from an offshore oil platform is about $400,000; in India, the cleaning of merchant navy ships costs about $28 million annually and annual losses due to corrosion are placed at $117 million.

At this Symposium, 68 papers were presented under the following headings: introduction; fouling organisms; boring organisms; microbiology and biodeterioration; larval biology and settlement; ecology of macrofouling communities and biodeterioration; and protection of woods, cements and metals. Turner and Nagabhushanam reviewed, in general terms, the work done so far and pointed out that the solutions we are seeking to the problems of biodeterioration demand a multi-disciplinary approach. Wood borers continue their devastating activity, despite nearly three hundred years of human research thereon in many parts of the world, and there is a feeling that most of us are barking up the wrong tree as far as research and remedial measures are concerned. It is time to change course.

At the Third International Symposium, on the Bioactive Compounds from Marine Organisms, scientists from India and US met for the first time to discuss what is perhaps one of the most promising areas of research of great value to biotechnologists, genetic engineers, aquaculturists and medical professionals. There was also a naval interest in understanding the role of the bioactive compounds as anti-fouling and anti-corrosion agents. Fifty-seven papers were presented, all of a very high standard, under the headings: ecological interactions; toxins, their chemistry and pharmacology; non-toxic bioactive substances; chemical advances; physiological chemistry; anti-tumor, anti-viral and anti-microbial compounds; antifouling substances; and prospects and potentials.

As an example of ecological interactions, Hay reported that many seaweed secondary metabolites deter feeding by marine herbivores. Valiela and Buchsbaum pointed out the fact that phenolics produced by marine plants and macroalgae alter bacterial activities and reduce palatability to the consumers. Parulekar and Shirvoiker found many marine organisms to be rich sources of novel chemical compounds (see Table 5.11).

Gleason reported on another interesting compound, cyanobacterin, extracted from the Cyanobacterium, *Scytonema halfmanni*, which has herbicidal properties. This compound was found to be toxic to most Cyanobacteria at a concentration of approximately 5μmol/l, but had no effect on Eubacteria. Morse, speaking on morphogens and signal molecules, stressed the fact that they act as inducers of larval settlement, metamorphosis and differentiation leading to the structuring of the community, its distribution and settlement in the marine environment. Specifically quoting the work done on the settlement of *Haliotes*, *Phragmatoforma* and *Agaricia humilis* larvae, Morse concluded that in 'several groups of benthic marine invertebrates, larval metamorphosis and recruitment are not wholly stochastic (lottery-like) processes, as had been previously thought, but instead are determined by larval recognition of, and responses to, identifiable substratum-specific bio-

Table 5.11 Results of biological screening of extracts of marine molluscs, sponges and echinoderms

Species	Pharmacological activity	Habitat
Ischiochiton comptus	Toxic, hypoglycemic	High littoral rocks
Cellana radiata	Antifertility	High littoral rocks
Aplysia benedicti	Hypoglycemic	Sandy (seasonal)
Melbe rangi	CNS stimulant	Rock pools (rare)
Asteropecten indica	Toxic	Sandy
Holothuria leucospilata	Toxic, anti-implantation	Rock pools (exposed)
Holothuria atra	CNS stimulant; antiviral	Sandy (exposed)
Holothuria hilla	Diuretic	Sandy (exposed)
Tedania anhelans	Spasmogenic	Sandy
Suberites spp.	Antiviral	Sandy
Sigmadocia fibulata	Antiviral; toxic	Sandy, grows in association with seaweed *Ciratodictyon spongiosum*
Spongia sp.	NS stimulant	Sandy
Callyspongia sp.	Antiviral	Rock pools in association with *Teslesto* sp.
Ircinia ramosa	Antiviral; CNS stimulant	Rock pools
Dysidea herbacea	Diuretic	Rocks
Bioemma fortis	Spasmolytic	Sandy
Placospongia carinata	Antimicrobial (*Bacillus, Pseudomonas*)	Sandy, spreading

CNS = central nervous system; NS = nervous system

chemical signals that control the induction of settlement and metamorphosis in the oceans'. Summarizing the studies on bioactive substances, Bakus, D'Elia and Taylor drew the attention of the participants to the highlights of the research on this subject between 1981 and 1989, which included: the testing of a sponge (*Haliclona rubens*) toxin as a shark repellent; the discovery that 73% of all exposed common coral-reef invertebrates (42 species) at Lizard Island, Great Barrier Reef, were toxic to fishes; that many coral reef sponges were seasonal producers of antibiotics; and the successful application of computer image processing and measuring techniques to fouling studies. They also gave a list of important research and institutional questions to be pursued on bioactive compounds in marine organisms, the most important of them being: 'What are the compounds (and classes of compounds) organisms use in chemically mediated interactions in the marine realm? How are they unique and how do they have common effects in different taxa of organisms? How are the chemical signals perceived by the organisms they are meant to act upon? Where does the action occur in the target? Can we explain the evolution of bioactive substances? How are they important as factors influencing survivorship and reproductive success?'

CHAPTER 6

FUTURE DIRECTIONS IN THE OCEANOGRAPHY OF THE INDIAN OCEAN

Introduction

Social, political and economic history may repeat itself, or appear to, mainly because Man often makes the same mistakes in comparable circumstances. Natural history appears not to repeat itself, mainly because the Earth, and indeed *Gaia*, have a birth, a life and a death linked by a continuous evolution spread over a very long time (Lovelock, 1979). Forecasting natural history is therefore probably easier than forecasting human social, political and economic history. In which domain does forecasting oceanographic research and monitoring lie? In both, it seems, since it is a question of deciding what knowledge and understanding of natural history do we need to forecast better the future of the other kind of history?

We are not in a position to draw up a 'wish list' of research on all those phenomena and processes already partially known or suspected to exist, in each of the many disciplines of oceanography; there may be as many wishes as there are oceanographers – a few thousand – interested in the Indian Ocean partially or exclusively.

We therefore limit ourselves, initially, to what the oceanographic history of the Indian Ocean, briefly described herein, thrusts forcibly before our eyes. These 'indications' themselves are umbrellas for the many possible and feasible research projects that, taken together, will provide answers – improved, if not complete – to the basic questions and add to our knowledge of this unique ocean.

To these we must add a second level of requirements, not so much in response to the need to discover new processes (or phenomena) but to the need to understand better those already observed. Knowledge and understanding are intimately related, but once it is known that there is a process – the monsoons, to take the principal example – understanding will flow better from a knowledge of the process's variability over a long time; the more so since expectation and its *alter ego*, prediction, must always, or nearly always, other things being equal, remain within the limits of the process's known past variations. This is largely why there is still much hesitation in the scientific community regarding the so-called 'greenhouse effect', between those who project disaster from a variety of scenarios (if this..., if that...) partially rooted in current knowledge and understanding, but inevitably lacking the new knowledge and understanding that the process itself will generate as it is observed, and those who claim that even the most pessimistic predictions or projections fall well within the Earth's experience of the process in eons past. Perhaps both sides are right and wrong at the same time.

The solution to such problems as the 'greenhouse effect' and other manifestations of global climate change, as well as others more specifically oceanographic, appears to be ocean-observing systems, comprising the sea itself and the overlying lower atmosphere.

In the international sphere, the Intergovernmental Oceanographic Commission (of UNESCO), with the close collaboration of other international organizations, notably the WMO and UNEP, is charged by its Member States with the development of a Global Ocean Observing System (GOOS). If the name is relatively new, the idea is not, inasmuch as GOOS is being based, initially at least, on a number of existing systems that have been in operation, and more or less successful, for many years. GOOS and its components are briefly described later in this chapter.

Important discoveries during the last 30 years or so have revolutionized our view and understanding of the Earth and its oceans. Some of these are: the many details of the ocean's surface waters, (e.g., wave patterns and wavelets, sea-surface topography and temperature) by satellite remote sensing; the eddies and meanders of major currents and their impact on global heat transfer, sound propagation and biological productivity; confirmation of sea-floor spreading and the associated but unexpected dense faunal associations, living symbiotically by chemo-synthesis, in the vicinity of hydrothermal vents in the Earth's crust at considerable depths in the ocean.

There are also three major socio-economic developments ushering in a revolution – often referred to as the 'blue' revolution – in the use of the oceans with a long-term potential good for, and great impact on, human society, but requiring extensive research in basic sciences to develop each one of these aspects along the right lines at the national and international levels.

The first of these socio-economic developments is aquaculture. The present output of marine fisheries from the Indian Ocean amounts to about 4-5 million tons annually and this is not likely to increase in the future. The predicted level of 10-11 million tons of marine fish for the Indian Ocean may never be realized unless all the countries bordering the Indian Ocean take part in fishery survey and development and, more particularly, in coastal aquaculture and marine ranching.

Great strides have been made in successfully culturing many finfish and seaweeds, particularly in Asian countries. It is predicted that a major portion of our seafood will be produced by aquaculture, rather than by capture fishery, in the years to come. Very few Indian Ocean countries are involved in aquaculture production. None of the Indian Ocean countries cultured seaweeds, as of 1980. India, Indonesia, Malaysia, Nepal, Singapore, Sri Lanka and Thailand produce some quantities of finfish but are no match for China which produces nearly 3 million tons of finfish and 1 million tons of seaweeds. Japan and Taiwan also produce considerable quantities of finfish and seaweeds by aquaculture.

It is estimated that Asia accounts for about 85 percent of the world's aquaculture production, utilizing just 2 million hectares of wetlands, whereas the available wetland suitable for aquaculture is placed at 20 million hectares in that area alone. Many countries in the Indian Ocean region have extensive wetlands and brackish water areas that would be excellent for aquaculture.

In past millennia, Indians and Chinese have practised not only aquaculture, but have also combined fish culture with rice culture in the same field, and duck and fish production in the same pond culture. Simple innovations such as weed removal and control of predators have maximized production rates in these activities.

For the successful culture of a finfish, it is essential to know the biology of the fish, particularly its life history, environmental preferences, behaviour, physiology, pathology, biochemistry and genetics. We still know very little of the vast number of species that constitute the biodiversity of the Indian Ocean and some of which might be acceptable as culture species.

Seaweed cultivation is another area for intensive development because of its many uses. *Laminaria* and *Undaria*, among other species, are extensively cultured in China; *Porphyra* is cultivated in Japan, and *Euchemia*, in the Philippines. Seaweeds are not only eaten, but also provide biochemicals such as agar, alginin and carrageenin which are used as gels and stabilizers in a variety of industrial and food products.

Large masses of algae or seaweeds are potentially good renewable energy sources; for example, a strain of blue-green alga produces hydrogen as a by-product, and kelp, subjected to anaerobic fermentation, yields methane. Another use of the seaweeds may be in the biological control of metal pollution in aquatic systems. Seaweeds are known to extract from sea water and concentrate in their tissues mercury, cadmium, nickel, lead and other metals.

Innovative application of biotechnology (e.g., genetic engineering) may help in the development of seaweeds and animals with high nutrient and biochemical values, making them suitable for extensive culture. Thus, aquaculture has immense potential for human food and animal feed. This is an area totally neglected in most of the Indian Ocean countries.

The second major development relates to ocean mining. As aquaculture, ocean mining also has roots in the past. The kings of India had a Superintendent of Ocean Mines as early as the fourth century BC. This official was responsible for attending to the regulation of commerce in marine products such as conch shells, diamonds, pearls, corals, salt etc. The mining of tin, coal and oil from the beaches and sea floor is well known. More recently, there have been discoveries of vast deposits of polymetallic sulphides, manganese nodules and heavy placer deposits in practically all the oceans. The Japanese extract 10 million tons of coal from the seabed annually. The German research ship *Valdivia* appears to have discovered, off the coast of Mozambique, heavy sand deposits containing 50 million tons of recoverable ilmenite, 1.5 million tons of rutile and 4 million tons of zircon. India has fully surveyed and mapped an area in the Central Indian Ocean Basin, with rich deposits of manganese nodules which contain nickel, copper, cobalt and manganese in recoverable quantities.

Under the United Nations Convention on the Law of the Sea, India has acquired so-called 'pioneer' status to mine these nodules. Although no mining has taken place, there is an opportunity for other Indian Ocean countries to do likewise, not only for nodules but also for the phosphorites off South Africa, heavy sands off Mozambique, and so on. Thus, ocean mining may become a major human activity in the years to come.

Nevertheless, although it looks attractive, seabed mining requires a lot of highly sophisticated technology, such as satellite navigational systems and imagery, specialized ships and submersibles, underwater sensors and TV cameras, and a thorough knowledge of the environmental parameters, such as currents, water-mass density, chemistry and biology, and the geology and geophysics of the seabed etc.

The third major development is energy production from the sea. As early as 1969, France built a tidal power station at Saint Malo at the mouth of the Rance River. The production of about 500 gigawatts per year is based on utilizing the high tidal ranges there; seawater at high tide rushes into a man-made bay and goes out at low tide. During this inflow and outflow, turbines are turned to generate electricity. There are many such locations along the sea coasts ideally suited for tidal power plants. Some exploratory surveys have been made in the Gulf of Kutch in India with a view to constructing a tidal plant.

Another sustainable source of energy is Ocean Thermal Energy Conversion (OTEC). An OTEC plant utilizes a temperature differential between the sea surface and at depth to produce steam to run the generators. Similarly, salinity gradients near river mouths offer a potential source of energy. Some pilot plants have been tested with partial success, so there is hope that the exploitation of OTEC and of salinity gradients will have a high potential for energy generation.

The Indian Ocean has many locations suitable for aquaculture, mineral extraction and energy generation, and the countries bordering this Ocean have an extraordinary scope for research and development in these fields, particularly since the vast majority have declared Exclusive Economic Zones (EEZ) in which marine scientific research is primarily the coastal State's opportunity and responsibility.

Recognizing these aspects, the US National Research Council, through its Ocean Studies Board, prepared a report (NRC, 1992) on oceanography in the next decade; this report contains an excellent account of what is required in research and development in the oceans in the years to come. The report covers relevant national and international programmes. It would be extremely useful for Indian Ocean countries to study this report and develop programmes that are most relevant to their needs and understanding. Here is a great opportunity not only to study the local problems but also to participate in global programmes, since they are interrelated.

Oceanographic research in the Indian Ocean has been sporadic in the past. Major work was done during the IIOE (1959-65). Since then, very few countries on the Indian Ocean rim have established marine research stations, and their work is mostly confined to local areas. India is the only country that has extensive research programmes in the Indian Ocean. The next few years should witness the development of modern oceanographic institutions in all the Indian Ocean countries and their active participation in programmes at the regional and international levels.

Indian Ocean studies

The gross picture that emerges as a result of a general analysis of what is known about the oceanography of the Indian Ocean can be summarized as follows:

1. A major part of the Indian Ocean is monsoonal. As a result, the surface currents north of 10°-20°S are not permanently oriented as in the other major oceans. They reverse in their direction seasonally and abruptly.

2. The upwelling systems existing or developing in the eastern and western boundaries and along the equator are seasonal and do not persist for a sufficient time for an adequate nutrient base to be built up to ensure sustained productivity. This has resulted in relatively meagre fishery resources.
3. The development of a strong hydrochemical/hydrographical front near 10°S has a considerable impact on such features as the nutrient distribution and the nitrogen oxidation/reduction cycle.
4. The whole of the southern Indian Ocean has very low nutrient levels, lacks upwelling systems and is considered to be a biological 'desert'.
5. The development of an extensive oxygen-minimum layer is indicative of reduced deep-water circulation in the northern Indian Ocean.
6. Our knowledge of the geology and geophysics of the Mid-Indian Ocean Ridge is still fragmentary, even if fairly good information has been collected on sea-floor spreading and on polymetallic nodules in the Central Indian Ocean Basin.

Indian Ocean countries, as Australia, India, Kenya, Kuwait, Pakistan, Saudi Arabia, South Africa and Thailand, are now making considerable progress in their investigations of their respective coastal waters, but a vast majority of the region's countries are quiescent in this respect.

It is quite obvious that future studies in oceanography will be increasingly multidisciplinary. For example, palaeo-oceanographic studies involve geology, geophysics, chemistry, biology and physical oceanography, and deep-sea drilling cores are studied by geologists, geochemists and biologists to unearth the geological and faunistic successions in the Earth's history. Flora and fauna can no longer be successfully studied in isolation, of one species from another or from its marine environment, thus requiring the co-operation of biological, chemical and physical oceanographers.

Nor can most oceanographic problems be studied in isolation. In most cases, there are regional and global implications: the strength and direction of the Somali Current, for example, depend on SW-monsoon wind forcing which, in turn, is affected by low pressure areas in Tibet, the Sahara and Australia, which may also be related to the so-called *El Niño*-Southern Oscillation phenomenon.

Nevertheless, we start by outlining possible lines of development in the main disciplinary fields covered in this book, and then go on to describe briefly some inevitably interdisciplinary regional activities.

Marine geology and geophysics

The transformation that marine geology and geophysics have undergone during the last three decades is phenomenal. Until the 1960s, these scientific disciplines were not able to find a reasonable explanation for the geological evolution of the Earth and its oceans and the drifting of its continents first suggested by Wegener in 1910 and formalized in his classic book *The Origin of the Continents and Oceans* (Wegener, 1924). However, by linking marine magnetic anomalies to the geomagnetic reversals of the Earth's magnetic field, it has been possible to confirm accurately and quantitatively the origin of the sea-floor spreading and the rate of such spreading. Further confirmation of these processes came from the analysis of the sediment cores obtained by the circum-global Deep Sea Drilling Project (Simpson and Schlich, 1979) and its successor, the Ocean Drilling Programme (ODP), which extended the geomagnetic-reversal time-scale back to nearly 200 million years ago, and provided a framework for the reconstruction of past locations of present-day continents and the oceans. These studies have also revealed that the ocean floor is very much younger than the land and is destroyed in deep trenches at the site of the subduction of one crustal plate under another.

Also of great interest is the evidence of past climatic changes and the past prevailing ocean currents provided by fossils or organisms preserved in deep-sea cores.

Marine geologists and geophysicists are now undertaking a major global study of the mid-ocean ridge system: the Ridge Inter-Disciplinary Global Experiment (RIDGE). They plan to establish automated permanent observation stations on active ridge segments so as continuously to monitor the variability in tectonic, magnetic, hydrothermal and biological processes associated with sea-floor spreading.

Plate tectonics is the surface expression of the solid-earth geophysical cycle; it has had a major impact on, and has unified, the fields of marine geology and biogeochemistry. The four principal research elements are: (i) the oceanic ridge and lithosphere; (ii) off-ridge processes; (iii) ocean margins; and (iv) ocean-basin sediments.

In these fields of research, the Indian Ocean stands out as being an area that is the least studied. It is now known that the Indian Ocean has the most complex ridge systems, some active and others passive, with micro-continents making up a major portion of the southern Indian Ocean. The development of massive sedimentary fans is another unique feature of the Arabian Sea and the Bay of Bengal. These require much more detailed study.

Future research projects in this field are likely to include: improved measurement of mantle flow; magma transport beneath the ocean ridges; transformation of magma into oceanic crust; structure and composition of oceanic crust, fault dynamics and lithospheric deformation; distribution and intensity of mid-ocean hydrothermal venting; and studies of ocean-basin sediments, which carry records of global changes in climate and sea level. For these purposes, the Joint Oceanographic Institutions for Deep Earth Sampling (JOIDES) programme (Simpson and Schlich, 1979) and the Ocean Drilling Programme (ODP) will certainly continue for many years.

A most interesting study would be on the effects of a changing physical-chemical state on the evolution of marine life. The fossil records preserved in the sediments are the best source of information on evolutionary dynamics, as well as a powerful tool for forecasting the biological effects of global changes and vice versa. In the past, there have been slow as well as catastrophic changes in climatic conditions, sometimes resulting in mass extinction of organisms. We may be able to get biological evidence for or against the much talked-about impact of an asteroid on the Earth at the close of the Cretaceous.

In the long journey of India from the time it broke away from the Gondwanaland in the Mesozoic to its present location, it would be exciting to detail faunal and floral succession, the evolution of ridges and basins and their effect on ocean circulation and sedimentation. Another interesting question is whether monsoons were present when the Tethys Ocean existed?

Physical oceanography

There is an increasing social need for the accurate measurement and prediction of changes in global climate, by improving our understanding of the physical factors that maintain the overall physical, chemical and biological characteristics of the ocean.

In the Indian Ocean area, the origin, duration and strength of the summer (SW) monsoon have a tremendous impact on the land and the sea and on the life of billions of people.

The Indian Ocean is considered to be a strongly evaporative ocean, since the much heated tropical waters cannot be gradually cooled by flowing northwards to the Arctic as in the Atlantic and the Pacific. The strong insolation and evaporation create warm and highly saline water, particularly the Red Sea Water and the Persian Gulf Water, which, on entering the northern Indian Ocean, add heat and salt to it at intermediate depths.

Oceanographically, the ICSU-WMO-IOC World Climate Research Programme includes the World Ocean Circulation Experiment (WOCE) and the recently completed Tropical Ocean and Global Atmosphere (TOGA) study. TOGA is discussed briefly in a later section.

WOCE is a more classical type of study, though using all the latest available technology, to determine more thoroughly than ever before the ocean's currents – their velocities and their heat, salinity and mass (water) balances (ICSU-SCOR-IOC-WMO, 1991). Given that WOCE is designed to cover at least a decade of observations, it is hoped that useful information will be obtained on the variability of the various processes involved.

WOCE, as well as TOGA, like the great expeditions of yesteryear, are likely to generate as many questions as they answer. They will therefore have a significant future or follow-up, not only in the analysis of the data obtained but in the inevitable pursuit of answers to the ever-present outstanding questions.

Key areas of future research will certainly include: thermohaline circulation; water-mass mixing; deep currents (the so-called 'conveyor belt' circulation); upwelling; the evolution of the major dissolved-oxygen minimum and its relation to upwelling; fluxes of energy, momentum and mass; quasi-synoptic satellite remote sensing of sea-surface temperatures and the sea-surface topography; transport of chemical substances within the ocean and across its interfaces with the air, the land and the sea bed; and prediction of the onset of the monsoons. Modelling, which is expected to continue its phenomenal growth, will have to go well beyond basic physical oceanography. For example, biological communities in the upper layers of the Arabian Sea are closely adapted to the stability of the water column, hence to the relevant mixing rates and large-scale vertical and horizontal fluid movements; these should be linked to the oxygen-minimum layer and the age of the deep water, so relevant models will need to incorporate biological and perhaps biogeochemical elements in them. The Bay of Bengal also presents the modeller with an example of a tremendous influx of fresh water and the heaviest estuarine/coastal sedimentation in the world; it is also an area particularly subject to storm surges which, usually in synergy with high tides, may flood large areas of the coastal zone causing great damage. The many current ocean-atmosphere interaction models (see section on ocean-atmosphere interaction, below) may need to become more refined to explain adequately the fluxes of energy, mass and momentum across the air-sea interface and at the land-sea interface.

Marine geochemistry

The main aims of marine geochemistry are, *inter alia*: (i) to assess the contribution (to the oceans) of elements from the continents, the mantle (volcanism) and cosmic sources; (ii) to determine the distribution of the elements and their removal from the ocean to the sediments; (iii) to determine the chemical interactions between the atmosphere and the ocean; and (iv) to assess the role of marine organic compounds in the global carbon cycle.

Geochemical signatures in the sediments complement the sedimentological and palaeontological records. For example, oxygen and carbon isotopes in marine calcareous shells and in organic remains provide insight into the ocean-surface temperatures and ocean circulation in the geological past and the extent of ice storage during glaciation. An understanding of the geochemistry of elements such as calcium and barium adds to our understanding of deep-ocean circulation. These are fascinating studies which have not yet been taken up in the Indian Ocean.

The chemistry of the oceans is determined primarily by marine organisms and chemical kinetics, rather than by thermodynamics (Redfield, 1958; Redfield *et al.*, 1963). Essential nutrient elements cycle by way of metabolic processes in organisms. There is also evidence for an increase in the vertical flow of particles downwards in areas of high productivity, as in the Arabian Sea; the downward particle flux is directly related to the intensity of the monsoon (Nair *et al.*, 1989). These fields of research need to be developed much further in the Indian Ocean region.

During the Geochemical Ocean Sections Study (GEOSECS) in the Indian Ocean (1977-78), extremely precise baseline information on the geochemistry of the ocean water column at many stations was collected. It may be worthwhile to repeat the work in 1997-98 so as to measure what changes, if any, have taken place during the last 20 years. Here is an opportunity for Indian Ocean countries to organize and participate in such an important geochemical investigation.

Biological oceanography

Research trends in biological oceanography include studies on the productivity of organisms, food webs and dynamics of marine populations. The ocean is a biochemical system and the biotic and abiotic components of the sea water co-evolved (Redfield, 1958; Redfield *et al.*, 1963); the distribution of many key elements is largely dictated by biological processes in the sunlit surface waters. Measurement of the concentrations of key elements, as carbon, oxygen, nitrogen and phosphorus, and the rate of their transfer through biological systems, are very important. Too little is known in general of nutrient cycles (limiting factors in the NH_4-N → NO_2-N → NO_3-N cycle in the Indian Ocean is a case in point). The same is true of the role of bacteria and viruses in the marine food web. There is a need to know much more about primary productivity in the photic zone and the mechanisms by which phytoplankters become optimally or adaptively distributed. There remains a great deal of work to be done on the taxonomy and systematics of marine organisms in the Indian Ocean. In general, exploration of the biochemistry, physiology, behaviour, genetic diversity and abundance of individuals within populations and of species within ecosystems would assist in the development of accurate conceptual models of the nature and regulation of marine communities.

The role of the food web in global biogeochemical cycles, changes brought about in the coastal areas by pollution and eutrophication, as a result of human activities, and the role of dissolved organic material are some of the other areas for future studies. Recent reports of increased pollution of coastal areas call for increased support for microbiological studies.

One of the common phenomena signalling a temporary breakdown of phytoplankton community regulation is the so-called 'red tides' and similar outbreaks of certain phytoplankton species, which are sometimes harmful to fish directly and to human beings usually indirectly (via sea-food intake containing algal toxins). Much more work needs to be done on such phenomena in the Indian Ocean. The IOC's programme on harmful algal blooms (in the context of its programme of Ocean

Sciences in Relation to Living Resources – OSLR) may help to promote Indian Ocean studies of these phenomena.

There is a growing fear of an increase in the greenhouse gases in the atmosphere, such as carbon dioxide, nitrous oxide, methane and others (e.g., dimethylsulphide, carbonyl sulphide, possibly), and the role of the ocean and its biological systems in this increase would be a major area of study in the years to come. Scientists generally agree that the ocean forms a significant sink in the global carbon cycle and thus moderates the greenhouse effect due mainly to the build-up of atmospheric carbon dioxide. Its concentration, while in the sunlit layers of the ocean, is within 30 percent of its saturation level; it is supersaturated in the deep ocean by as much as 300 percent with respect to the present atmospheric carbon dioxide level. This is made possible by the 'biological pump' in the surface waters, which, through biological fixation, packaging and transfer, results in a net downward flow of carbon to the deep sea and the sediments. We are still uncertain about the quantitative role of marine organisms in the global carbon budget throughout the water column.

Another exciting area of biological research is the communities of organisms associated with the hydrothermal vents discovered in recent years. Here, in the deep sea and in the vicinity of hydrothermal vents and oil seeps, dense colonies of bacteria and animal communities thrive in situ by chemosynthesis and symbiosis. Yet the fact is that marine microbiological studies in the Indian Ocean are very scanty.

One of the major lacunas in oceanographic studies is the lack of time-series, and this is particularly true for the Indian Ocean. Most of the studies on hydrography, plankton and benthos are limited to one or two years and isolated in time and space and, as a result, long-term changes cannot be discovered. It would, therefore, be essential to establish and conduct time-series studies at one or two locations in the Indian Ocean or, even better, in each region thereof, to monitor key ecosystem variables, such as temperature, salinity, nutrients, dissolved gases, plankton and benthos etc. This might be possible in the framework of the Large Marine Ecosystem (LME) study being promoted by US marine biologists (Sherman *et al.*, 1990).

A related question, which is still far from being answered, is the relationship between, on the one hand, the abundance and distribution of fish and, on the other hand, the main hydrographic factors. In particular, why is the fishery potential of the northern Indian Ocean below what might be expected from the extent and degree of upwelling?

Multidisciplinary regional studies in the Indian Ocean

Monsoonal studies

The data indicate the profound influence of the reversing monsoons on the general hydrography, surface currents and biology. In the above-mentioned studies, the information generated is mostly descriptive, but not dynamic. It is therefore essential to collect quantitative data on the hydrography, chemistry and specific biomass and community structure in the Indian Ocean, to be able to predict changes in key parameters and to attempt to relate them to the monsoons.

Coastal-zone management

The coastal zone, in which the atmosphere, the sea and the land meet and interact, is a specialized environment in several respects. Human activity is maximum in this zone, in the form of urbanization, harbours, navigation, transportation, industry, domestic and industrial waste disposal, fishing, mariculture and recreation. Quite often, conflicts arise between some of these activities imposing significant constraints on the management of the problems. Natural forces, such as beach erosion, may pose a serious threat to human settlements, harbours and fishing. Over a longer term, sea-level changes may bring about catastrophic changes in the coastal zone. It is feared that such countries as Bangladesh, India, the Maldives and the Seychelles may suffer submergence of vast areas of their coastal lands if there is an increase in the mean sea level in the years to come. It is therefore extremely important to formulate coastal-zone management policies in all the countries in the area and to initiate critical studies on tidal regimes, storm surges and sea level, as major national and international programmes.

Marine pollution

Marine pollution in the Indian Ocean is a neglected area of study. There are periodic reports of metal and oil pollution near major cities along the Indian Ocean rim, more particularly in the coastal areas of the Persian Gulf, the Red Sea and along the oil tanker routes in the Arabian Sea and the Bay of Bengal. Reporting on the chlorinated hydrocarbons and DDT and its metabolites in the Indian Ocean, Sarkar and Sen Gupta (1992) observed that these toxic compounds are found throughout the Indian Ocean, with highest concentrations in the sediments of the Bay of Bengal, owing to large human inputs from the east coast of India. Very high concentrations of DDT are also reported for various tissues of Australian fur seals and South African fishes. A concerted region-wide study of marine pollution in the Indian Ocean is necessary. Continuous monitoring of the marine environment for pollution is essential to evaluate the state of health of this Ocean and to provide a basis for sound remedial measures against further pollution of the sea water and its organisms, particularly in coastal waters.

Ocean-atmosphere interaction

The relevance of TOGA and WOCE has already been mentioned. The SCOR-IOC Joint Global Ocean Flux Study (JGOFS), within the ICSU International Geosphere-Biosphere Programme (IGBP), is also relevant here.

The observational phase of TOGA was completed in 1994 and the results of the final scientific symposium have been published in two volumes (ICSU-WMO-IOC, 1995a). TOGA had three scientific objectives which have been largely achieved. They were, briefly, to: (i) describe the tropical oceans and global atmosphere system so as to determine the predictability of this system (on a time scale of months to years) and to understand mechanisms/processes underlying this predictability; (ii) study the feasibility of modelling this coupled system for predictive purposes; and (iii) provide a scientific background for designing the observation and data transmission system for operational prediction, if justified by the results of modelling. Thus, TOGA has laid the ground for future work in modelling the coupled ocean-atmosphere system and in developing an observation and data transmission system.

There is to be expected a considerable intensification of numerical modelling of ocean-atmosphere interaction, using TOGA and WOCE data in particular. The role of the Indian Ocean in the global climate system will be given particular attention in the framework of the World Climate Research Programme (WCRP) by the Indian Ocean Climate Studies Panel (ICSU-WMO-IOC, 1995b).

The exchange of gases between the sea water and the overlying air has important implications for the the biology of the oceans and for the assessment of marine and atmospheric pollution. The transfer – in both directions – of carbon dioxide, in particular, but also methane and possibly dimethylsulphide, for example, may depend on their generation by organisms in the sea and by industry and agriculture on land. The study of such transfers is still in its infancy, in general, and in the Indian Ocean, in particular. JGOFS is an attempt to address this sort of question; moreover, JGOFS, although concentrating on fluxes of specific substances, notably carbon dioxide, may also allow significant refinement of some of the above-mentioned models.

Ocean-observing systems

The IOC's Global Ocean-Observing System (GOOS), which was mentioned earlier in this chapter, is a scientifically based, comprehensive international system to collect and disseminate ocean data and data products with a view to providing economic, environmental, health, educational, research and other practical benefits to society (IOC, 1993, 1994, 1995a, 1995b). GOOS is essential to the understanding and forecasting of climate change, the health of the ocean, living marine resources, coastal-zone changes and marine weather. The System, as presently conceived, has five modules: (i) climate monitoring, assessment and prediction; (ii) monitoring and assessment of living marine resources; (iii) monitoring of the coastal-zone environment (to provide information for coastal-zone management); (iv) assessment and prediction of the state of health of the ocean – marine pollution monitoring; (v) marine meteorological and operational ocean services. It will be based, initially, on the following existing systems: IOC-WMO Integrated Global Ocean Services System (IGOSS); IOC Global Sea-Level Observing System (GLOSS); IOC International Oceanographic Data and Information Exchange (IODE); IOC-UNEP Global Investigation of Pollution in the Marine Environment (GIPME); WMO World Weather Watch (WWW); IOC-WMO Drifting Buoy Co-operation Panel; and IOC Training, Education and Mutual Assistance in the Marine Sciences (TEMA). All together, these systems have: set up mechanisms for data collection, quality control, archiving and dissemination; issued manuals and guides; developed standards for, and intercalibration of, relevant observational, sampling and analytical methods; and set up TEMA activities in support of the various systems. Moreover, the UK Continuous Plankton Recorder (CPR) surveys of the North Atlantic Ocean will provide an important input to GOOS and will, if feasible, be extended to other ocean areas. Also, pilot activities are already being undertaken jointly by UNEP, IOC and WMO to establish a long-term global system for monitoring coastal and near-shore phenomena related to climate change.

The names of the systems on which GOOS will be initially based are fairly self-explanatory; they have all undergone many years of trial and error which have exposed the current, but not the fixed and final, limits to international co-operation in observing the ocean and exchanging the related data freely, and which have shown what can effectively be done. Their inclusion in the newer but broader and better supported GOOS will promote their continuing improvement. They will add vast amounts of data and information to the existing body of knowledge of the processes they were designed to monitor, and from that knowledge, after careful evaluation and analysis of the data and information, they will add to our understanding of the processes.

While these systems have been developed, or will be developed, essentially on a global basis, they have advanced better in some areas than in others; for example, the North Atlantic and the North Pacific much more than the South Atlantic and South Pacific, for fairly obvious reasons; this fact may also bias knowledge and understanding, sometimes detrimentally.

Certain processes, because of their perceived importance to human economic activities and, nowadays, environmental protection, have received, or are intended to receive, preferential treatment; for example, the *El Niño*-Southern Oscillation phenomenon/process and sea-level rise (no one is concerned at present with sea-level drop because it is not expected). It may also be suggested, as well, that data-gathering by GOOS is also to some extent a function of current ocean-circulation models and coupled ocean-atmosphere models. Since such models are based largely on current understanding, there is some danger that the system (GOOS) and its subsystems could be substantially 'embarrassed' if understanding changes drastically or becomes conditioned by the physical limitations of the system itself. The world is not ready to place detectors (of mean sea level and tides, sea-surface and deep temperatures, salinities, currents, oceanic winds and so on) everywhere they would be desirable to improve the data on which clearer understanding would depend. There are still serious shortages of such data for the high seas, especially subsurface, and for the southern hemipshere. Remote-sensing satellites help, but their products (impressive images in some cases) depend more than is generally imagined on good 'ground truth' data; nor is the imagery as synoptic as one would wish where specific sea-surface features are concerned. The system may therefore generate its own bias as well as a possible bias in understanding. This may be less of a problem in the WMO's analogous Global Climate-Observing System (GCOS) for which the GOOS climate module or Ocean Climate Observing System (OCOS) will provide the ocean component, since the life of an environmental signal is much shorter in the atmosphere than in the ocean.

With respect to the Indian Ocean, the large-scale international research programmes already mentioned may help to remedy the possible shortcomings mentioned above. The World Climate Research Programme's Indian Ocean Climate Studies Panel (IOCSP) and the Ocean Observing System Development Panel (OOSDP) are working out the objectives and mechanisms for GOOS and OCOS in this Ocean.

In particular, the World Climate Research Programme's Indian Ocean Climate Studies Panel is currently developing a regional ocean-atmosphere interaction component of GOOS (Godfrey *et al.*, 1995). The Panel is considering at least two dozen key questions relating directly to the Indian Ocean, each of which could be a subject of future research. Many of the questions under consideration by the Panel are explicit or implicit in the information presented in this book. The Panel proposes to handle them from three different but related starting points: (i) analysis of available data (such as those generated by TOGA, satellite remote sensing etc.) and data to be generated by such international activities as WOCE, for the Indian Ocean; (ii) development of (more and better) numerical models (many of the questions posed have been generated by models already); and (iii) exploratory monitoring (of which the best example is, possibly, the ten or more years of IGOSS and TOGA Expendable Bathythermograph (XBT) data obtained from ships of opportunity, as well as research vessels, plying the Indian Ocean). In other words, GOOS and OCOS are going to take many years to be built up to an effective and operational standing.

The ICSU International Geosphere-Biosphere Programme (IGBP), which includes, *inter alia*, a core project on Land-Ocean Interaction in the Coastal Zone (LOICZ) and two others, the SCOR-IOC Joint Global Ocean Flux Study (JGOFS, in progress), and the Global Ocean Euphotic Zone Study (GOEZS, in preparation), the SCOR-IOC-ICES-PICES Global Ocean Ecosystem Dynamics study (GLOBEC), and the Joint NASA/French Centre National d'Etudes Spatiales (CNES) Venture to Measure the Surface Topography of the Ocean (TOPEX/Poseidon) can all be expected to contibute significantly to the development of ocean science and technology in the Indian Ocean.

Although all of the above-mentioned international programmes have an Indian Ocean component, it is still largely up to the scientists of the Indian Ocean coastal States to seek serious participation in them or co-operation with them.

REFERENCES

Aiyar, R. G., Menon, K. A. and Menon, M. G. K. (1936). Plankton records for the years 1929 and 1930. *J. Madras Univ.* 1936, 8: 1-43.

Alcock, A. W. (1902). *A Naturalist in Indian Seas or Four Years with the Royal Indian Marine Survey Ship Investigator*. John Murray, Albemarle Street, London. 318 pp.

Aleem, Anwar Abdel (1967). Concepts of currents, tides and winds among medieval Arab geographers in the Indian Ocean. *Deep-Sea Res.* 14: 459-463.

—— (1968a). Ahmed Ibn Magid, Arab navigator of the XVth century and his contributions to marine sciences. *Bull. Inst. Océanogr. Monaco*, Numero Special 2, pp. 565-580.

—— (1968b). Concept of marine biology among Arab writers in the Middle Ages. *Bull. Inst. Océanogr. Monaco*, Numero Special 2, pp. 359-367.

—— (1980). On the history of Arab navigation. In: Sears, M. and Merriman, D. (eds.), *Oceanography – The Past*. pp. 582-595. Springer-Verlag, Berlin, Heidelberg, New York.

Anderson, R. N. (1986). *Marine Geology*. John Wiley & Sons, New York. 328 pp.

Anon. (1st-2nd centuries AD). *The Periplus of the Erythrean Sea*. Translated and edited by G. W. B. Huntingford. The Hakluyt Society, c/o the British Library, London (1980). 225 pp.

Ansari, Z. A. (1977). Macrobenthos of Cochin backwaters. *Mahasagar, Bull. Nat. Inst. Oceanogr. India*, 10: 169-171.

Ansari, Z. A. (1978). Meiobenthos from the Karwar region (central west coast of India). *Mahasagar, Bull. Nat. Inst. Oceanogr. India*, 11(3&4): 163-169.

Ansari, Z. A., Parulekar, A. H., Harkantra, S. N. and Nair, A. (1977). Shallow water macrobenthos along the central west coast of India. *Mahasagar, Bull. Nat. Inst. Oceanogr. India*, 10(3&4): 123-127.

Ansari, Z. A., Rivonker, C. U., Ramani, P. and Parulekar, A. H. (1991). Seagrass habitat complexity and macroinvertebrates abundance in Lakshadweep coral reef lagoons, Arabian Sea. *Coral Reefs*, 10: 127-131.

Aruga, Y. (1973). Primary production in the Indian Ocean – II. In: Zeitzschel, B. (ed.), *The Biology of the Indian Ocean*. pp. 127-130. Springer-Verlag, Berlin, Heidelberg, New York.

Bal, D. V. and Pradhan, L. B. (1945). A preliminary note on the plankton of the Bombay harbour. *Curr. Sci.*, 14: 211-212.

Bal, D. V. and Rao, V. K. (1990). *Marine Fisheries of India.* Tata-McGraw Hill Publishing Co., New Delhi. xii + 470 pp.

Bakus, G. F. *et al.* (1994). *Coral Reef Ecosystems.* Oxford and IBH Publishing Co., New Delhi. 238 pp.

Ballegooyen, R. C. van *et al.*, (1991). Modelling of the Agulhas Current system. *S. Afr. J. Sci.*, 87(11/12): 569-571.

Banse, K. (1959). On upwelling and bottom trawling off the southwest coast of India. *J. Mar. Biol. Assoc. India,* 1: 33-49.

—— (1968). Hydrography of the Arabian Sea shelf off India and Pakistan and effects on demersal fishes. *Deep-Sea Res.*, 15: 45-79.

Bé, A. W. H. and Tolderlund, D. S. (1971). Distribution and ecology of living planktonic Foraminifera in surface waters of the Atlantic and Indian Ocean. In: Funnel, B. M. and Reidel, W. R. (eds.), *The Micropalaeontology of the Oceans.* pp. 105-149. Cambridge University Press.

Behrman, D. (1981). *Assault on the Largest Unknown.* UNESCO, Paris. 96 pp.

Bhimachar, B. S. and George, P. C. (1950). Abrupt set-back on the fisheries of the Malabar and Kanara coasts and 'red water' phenomena and their probable cause. *Proc. Ind. Acad. Sci.*, 31: 339-350.

Blasco, F. (1977). Outlines of the ecology, botany and forestry of the mangals of the Indian subcontinent. In: Chapman, V. J. (ed.), *Wet Coastal Ecosystems.* pp. 241-260. Elsevier, Amsterdam, The Netherlands.

Broecker, W. S. and Takahashi, T. (1985). Sources and flow patterns of deep ocean waters as deduced from potential temperature, salinity and initial phosphate concentration. *J. Geophys. Res.*, 90: 6925-6939.

Bruce, J. G., Fieux, M. and Gonella, J. (1981). A note on the continuance of the Somali eddy after the cessation of the southwest monsoon. *Oceanologica Acta*, 4(1): 7-9.

Bruun, A. F. (1956). The abyssal fauna: its ecology, distribution and origin. *Nature*, 177(4520): 1105-1108.

Bullard, E. C. (1969). The origin of the oceans. *Sci. American*, 221(3): 16-25.

Carruthers, J. N., Gogate, S. S., Naidu, J. R. and Laevatsu, T. (1959). Shorewards upslope of the layer of minimum oxygen off Bombay: its influence on marine biology, especially fisheries. *Nature*, 183: 1084-1087.

Chacko, P. I. (1950). Marine plankton from waters around the Krusadai Islands. *Proc. Ind. Acad. Sci.*, 31: 162-174.

Chandran, R. (1985). Seasonal and tidal variations of phytoplankton in the gradient zone of Vellar estuary. *Mahasagar, Bull. Nat. Inst. Oceanogr. India,* 18(1): 37-49.

Chindambaram, K. and Unni, M. M. (1944). Note on the swarming of the planktonic alga, *Trichodesmium erythraeum* in the Pamban area and its effect on the fauna. *Curr. Sci.*, 13: 263.

Clowes, A. J. (1938). Phosphates and silicates in the Southern Ocean. *Discovery Rept.* 19(1).

Cohen, D. M. (1973). Zoogeography of the fishes of the Indian Ocean. In: Zeitschel, B. (ed.), *The Biology of the Indian Ocean.* pp. 451-463. Springer-Verlag, Berlin, Heidelberg, New York.

Cooper, L. H. N. (1937). On the ratio of nitrogen and phosphorus in the sea. *J. Mar. Biol. Assoc. United Kingdom*, 22: 177.

Coumes, F. and Kolla, V. (1984). Indus fan: seismic structure, channel migration, and sediment thickness in the upper fan. In: Haq, B. U. and Milliman, J. D. (eds.), *Marine Geology and Oceanography of the Arabian Sea and Coastal Pakistan.* pp. 101-110. Van Nostrand Reynold, New York.

Cresswell, G. R. (1991). The Leeuwin Current – observations and recent models. In: Pearce, A. and Walker, D. L. (eds.), *The Leeuwin Current: an Influence on Coastal Climate and Marine Life of Western Australia.* pp. 1-13. *J. Roy. Soc. Western Australia,* F4, 140 pp.

Currie, R. G. (1963). The Indian Ocean standard net. *Deep-Sea Res.*, 10: 27-32.

Currie, R. G., Fisher, A. E. and Hargreaves, P. M. (1973). Arabian Sea upwelling. In: Zeitschel, B. (ed.), *The Biology of the Indian Ocean.* pp. 57-63. Springer-Verlag, New York.

Curry, J. R. and Moore, D. G. (1971). Growth of the Bengal Fan and denudation of the Himalayas. *Geol. Soc. Amer. Bull.*, 82: 563-572.

Damodaran, R. (1964). Studies on the mudbanks of the Kerala coast. *Bull. Dept. Mar. Sci.*, Cochin University, pp. 1-126.

De Decker, A. (1973). Agulhas Bank plankton. In: Zeitschel, B. (ed.), *The Biology of the Indian Ocean.* pp. 189-220. Springer-Verlag, Berlin, Heidelberg, New York.

De Decker, A. and Mombeck, F. J. (1965). South African contribution to the International Indian Ocean Expedition. A preliminary report on planktonic copepoda. *Invest. Rep. Div. Sea Fish. S. Africa*, 51: 10-67.

De Sousa, S. N., Naqvi, S. W. A. and Reddy, C. V. G. (1981). Distribution of nutrients in the western Bay of Bengal. *Ind. J. Mar. Sci.*, 10: 327-331.

Debnath, R. and Choudhury, A. (1988). Population estimation of horse-shoe crab *Tachypleus gigas* by capture/recapture method at Chandipur seashore (Orissa), India. *J. Mar. Biol. Assoc. India*, 30(1&2): 8-13.

Degens, E. T. and Ross, D. A. (eds.) (1969). *Hot brines and recent heavy metal deposits in the Red Sea.* Springer-Verlag, New York. 600 pp.

Della Croce, N. and Venugopal, P. (1973). Distribution of marine ostracods in the Indian Ocean. *Mar. Biol.*, 15: 132-138.

Delsman, H. C. (1924-39). Fish eggs and larvae from the Java Sea. *Treubia*, Vols. 5, 6, 8, 9, 11, 12 and 17.

Desai, B. N. and Krishnan Kutty, M. (1967). Studies on the benthic fauna of Cochin backwater. *Proc. Ind. Acad. Sci.*, 66(4): 123-142.

Dietz, R. S. (1961). Continent and ocean basin evolution by spreading of the sea floor. *Nature*, 190: 851.

Donguy, J.-R. and Piton, B. (1991). The Mozambique channel revisited. *Oceanologica Acta*, 14(6): 549-558.

Duing, W. (1970). The monsoon regime of the currents in the Indian Ocean. *Int. Indian Ocean Exped., Oceanogr. Monograph*, 1. East-West Centre, Honolulu. 68 pp.

Duing, W., Molinari, R. L. and Swallow, J. C. (1980). Somali current: evolution of surface flow. *Science*, 209: 588-590.

Ebeling, A. W. (1962). Melamphaidae. Systematics and zoogeography of the species of the bathypelagic fish genus *Melamphaeus* (Günther). *Dana Report*, 58: 1-104.

Emery, K. O., Hunt, J. M. and Hays, E. E. (1969). Summary of hot brines and heavy metal deposits in the Red Sea. In: Degens, E. T. and Ross, D. A. (eds.), *Hot Brines and Recent Heavy Metal Deposits in the Red Sea.* pp. 557-571. Springer-Verlag, New York.

Emery, W. J. and Meincke, J. (1986). Global water masses: summary and review. *Oceanologica Acta*, 9: 383-391.

Ewing, M., Eiltreim, S., Truchen, M. and Ewing, J. J. (1969). Sediment distribution in the Indian Ocean. *Deep-Sea Res.*, 16: 231-48.

Eyries, M. and Menaché, M. (1953). Contribution à la connaissance hydrologique de l'Océan indien entre Madagascar et la Réunion. *Bulletin du Comité d'Océanographie et d'Études des Côtes,* V: 433-438.

FAO(ACMMR)-SCOR-WMO(AGOR) (1969). *Global Ocean Research.* A report prepared by the Joint Working Party on the Scientific Aspects of International Ocean Research. Ponza and Rome, 29 April – 7 May 1969. Food and Agriculture Organization of the United Nations, Rome. 54 pp.

Fenaux, R. (1964). Appendicularians collected by the *Commandant Robert Giraud* in the Arabian Sea. *Bull. Inst. Océanogr. Monaco,* 62(1302): 1-14.

—— (1973). *Appendicularia* from the Indian Ocean, the Red Sea and the Persian Gulf. In: Zeitschel, B. (ed.), *The Biology of the Indian Ocean.* pp. 409-414. Springer-Verlag, Berlin, Heidelberg, New York.

Fernando, S. A., Ajmal Khan, S. and Kasinathan, R. (1983). Observations on the distribution of benthic fauna in Vellar estuary. *Mahasagar, Bull. Nat. Inst. Oceanogr. India,* 16(3): 341-349.

Fieux, M. (1975). Météorologie et océanographie – établissement de la mousson de sud-ouest en mer d'Arabie. *Comptes-rendus de l'Académie des Sciences de Paris,* 281(série B): 563-566.

—— (1988). Monsoon and currents in the Indian Ocean. In: Bruun Memorial Lectures, 1987. *Recent Advances in Selected Areas of Ocean Sciences in the Regions of the Caribbean, Indian Ocean and the Western Pacific.* pp. 17-32. Intergovernmental Oceanographic Commission, *Technical Series,* 34, 59 pp. UNESCO, Paris.

Fieux, M., Andrié, C., Charriaud, E., Ilahude, A. G., Metzl, N., Molcard, R. and Swallow, J. C. (1966a). Hydrological and chlorfluoromethane measurements of the Indonesian throughflow entering the Indian Ocean. *J. Geophys. Res.,* 101(C5): 12433-12454.

Fieux, M., Andrié, C., Delecluse, P., Ilahude, A. G., Kartavtseff, A., Mantisi, F., Molcard, R. and Swallow, J. C. (1994). Measurements within the Pacific-Indian oceans throughflow region. *Deep-Sea Res.,* 41(7): 1091-1130.

Fieux, M., Molcard, R. and Ilahude, A. G. (1996b). Geostrophic transport of the Pacific-Indian Oceans throughflow. *J. Geophys. Res.,* 101(C5): 12421-12432.

Fieux, M. and Schott, F. (1988). The boundary currents east and north of Madagascar. 1. Geostrophic currents and transports. *J. Geophys. Res.,* 93(C5): 4951-4962.

Fieux, M., Schott, F. and Swallow, J. C. (1986). Deep boundary currents in the western Indian Ocean revisited. *Deep-Sea Res.,* 33(4): 415-426.

Fieux, M. and Stommel, H. (1976). Historical sea-surface temperatures in the Arabian Sea. *Annales de l'Institut océanographique,* 52(1): 5-15.

Fieux, M. and Swallow, J. C. (1988). Flow of deep water into the Somali Basin. *Deep-Sea Res.,* 35(2): 303-309.

Fisher, R. L., Sclater, J. G. and McKenzie, D. P. (1971). Evolution of the Central Indian Ridge, western Indian Ocean. *Geol. Soc. Amer. Bull.,* 82: 553-562.

Gardiner, J. S. (1907). The Percy Sladen Trust Expedition to the Indian Ocean in 1905 under the leadership of J. S. Gardiner. Description of the expedition. *Trans. Limn. Soc. London,* Second Series, Vol. XII, 1907-1909, pp. 1-56.

Gibbs, R. H., Jr. and Hurwitz, B. A. (1967). Systematics and zoogeography of the stomiatoid fishes, *Chauliodus pammelus* and *C. sloani,* in the Indian Ocean. *Copeia,* pp. 798-805.

Gilson, H. C. (1937). The nitrogen cycle. *John Murray Expedition, Scientific Reports,* 11(2): 21.

Girdler, R. W. (1984). The evolution of the Gulf of Aden and Red Sea in space and time. In: Angel, M. V. (ed.), *Marine Science of the Northwest Indian Ocean and Adjacent Waters. Deep-Sea Res.,* 31(6-8A): 747-762.

Gjosaeter, J. and Kawaguchi, K. (1980). A Review of World Resources of Pelagic Fish. *FAO Fisheries Technical Papers,* 203729-E, 151 pp.

Godfrey, J. S. *et al.* (1995). The Role of the Indian Ocean in the Global Climate System: Recommendations Regarding the Global Ocean Observing System. Background Report of the Ocean Observing System Development Panel, 6. Texas A&M University, College Station, TX 77843-3146, USA. 89 pp.

Gonella, J. (1983). Les courants d'inertie au voisinage de l'équateur et leur stucture zonale. *Comptes-rendus de l'Académie des Sciences de Paris,* 297(série II): 721-724.

—— (1984). Inertial currents at the equator. *Tropical Ocean-Atmosphere Newsletter,* 26: 12.

Gonella, J., Fieux, M. and Philander, G. (1981). Océanographie dynamique – mise en évidence d'ondes de Rossby équatoriales dans l'Océan indien au moyen de bouées dérivantes. *Comptes-rendus de l'Académie des Sciences de Paris,* 292(série II): 1397-1399.

—— (1982). Evidence for equatorial Rossby waves in the Indian Ocean. *Tropical Ocean-Atmosphere Newsletter,* 10: 4-5.

Gopinathan, C. P. (1984). A systematic account of the littoral diatoms of the southwest coast of India. *J. Mar. Biol. Assoc. India,* 26(1&2): 1-31.

Gordon, A. L. (1985). Indian-Atlantic transfer of thermocline water at the Agulhas retroflection. *Science,* 227(4690): 1030-1033.

Gougenheim, A. (1961a). Campagne hydrologique de l'Aviso *Commandant Robert Giraud* (Marine nationale) dans le Golfe d'Aden, le Golfe d'Oman et le Golfe Persique. *Bulletin du Comité d'Études d'Océanographie et des Côtes,* XIII: 531-533.

—— (1961b). Observations hydrologiques effectuées à bord du Navire hydrographique *La Pérouse* dans l'Océan Indien au cours d'une traversée entre Djibouti et Diego Suárez. *Bulletin du Comité d'Études d'Océanographie et des Côtes,* XIII: 498-501.

Grice, G. D. and Hulsemann, K. (1967). Bathypelagic calanoid copepods of the western Indian Ocean. *Proc. US Nat. Mus. Smithsonian Inst.,* 122(3583): 1-67.

Gross, M. G. (1990). *Oceanography* (6th edition). MacMillan Publishing Company, New York.

Gründlingh, M. L. (1985). An intense cyclonic eddy east of the Mozambique Ridge. *J. Geophys. Res.,* 90: 7163-7167.

—— (1987). Cyclogenesis in the Mozambique ridge. *Deep-Sea Res.,* 34: 89-103.

Guibout, P. and Lizeray, J. C. (1959). Mesures effectuées dans l'Océan indien à l'aide du courantomètre à électrodes remorquées (GEK). *Bulletin du Comité d'Études d'Océanographie et des Côtes,* XI: 155-157 + 1 Figure.

Gulland, J. A. (1971). *The Fish Resources of the Oceans.* Fishing News Books Ltd. Surrey, England. 255 pp.

Halim, V. (1969). Plankton of the Red Sea. *Oceanogr. Mar. Biol. Ann. Rev.*, 7: 231-275.

Haq, B. U. (1988). Geological evolution of the Indian Ocean with special reference to the Arabian Sea. In: Thomson, M. F. and Tirmizi, N. M. (eds.), *Marine Science of the Arabian Sea.* Proceedings of the International Conference. pp. 9-37.

Haq, B. U. and Milliman, J. (1984). *Marine Geology and Oceanography of the Arabian Sea and Coastal Pakistan.* Van Nostrand Reynold, New York. 382 pp.

Haq, S. M., Alikharn, J. and Chugtai, S. (1973). The distribution and abundance of zooplankton along the coast of Pakistan during the monsoon and premonsoon periods. In: Zeitschel, B. (ed.), *The Biology of the Indian Ocean.* pp. 257-272. Springer-Verlag, Berlin, Heidelberg, New York.

Harkantra, S. N., Nair, A., Ansari, Z. A. and Parulekar, A. H. (1980). Benthos of the shelf region along the west coast of India. *Ind. J. Mar. Sci.*, 9: 106-110.

Heezen, B. C. and Tharp, M. (1965). *Physiographic Diagram of the Indian Ocean* (with description sheet). Geological Society of America, New York.

Heirtzler, J. R., Dickson, G. O., Herron, E. M., Pitman, W. C. and LePichon, X. (1968). Marine magnetic anomalies, geomagnetic field reversals and motions of the ocean floor and continents. *J. Geophys. Res.*, 73: 2119-2136.

Hessler, R. R. and Sanders, L. S. (1967). Faunal diversity in the deep sea. *Deep-Sea Res.*, 14: 65-78.

Hocult, C. H. (1987). Evolution of the Indian Ocean and the drift of India: a vicariant event. *Hydrobiologia*, 150(3): 203-229.

Holt, S. J. (1969). The food resources of the oceans. *Sci. Amer.* (Sept 1969), pp. 178-194.

Hornell, J. and Nayudu, R. M. (1923). A contribution on the life-history of the sardines with notes on the plankton of the Malabar coast. *Madras Fish Bull.*, 17: 129-157.

Huntingford, G. B. W. (1980). Translation of the *Periplus of the Erythrean Sea.* The Hakluyt Society, c/o British Library, London. 225 pp.

ICSU-SCOR-IOC-WMO (1991). *World Ocean Circulation Experiment.* International Council of Scientific Unions, Paris. 32 pp.

ICSU-WMO-IOC (1995a). *Proceedings of the International Scientific Conference on the Tropical Ocean Global Atmosphere (TOGA) Programme.* Melbourne, Australia, 2-7 April 1995. WMO/TD 717, Volumes I and II. World Meteorological Organization, Geneva. 911 pp.

—— (1995b). Report of the Eighth Session of the Indian Ocean Climate Studies Panel (Trieste, Italy, 14-18 May 1994). World Meteorological Organization, Geneva. 13 pp. + ten annexes.

IOBC (Indian Ocean Biological Centre, IIOE). (1968). *Zooplankton Atlases.* Panikkar, N. K. (ed.).

—— (1971). *Zooplankton Atlases.* Panikkar, N. K. (ed.).

—— (1973). *Zooplankton Atlases.* Panikkar, N. K. (ed.).

IOC (1984). *Ocean Science for the Year 2000.* Intergovernmental Oceanographic Commission (of UNESCO), Paris, France. 95 pp.

—— (1993). IOC Committee for the Global Ocean Observing System (GOOS). First Session, Paris, 16-19 February 1993. Intergovernmental Oceanographic Commision, *Reports of Governing and Major Subsidiary Bodies.* UNESCO, Paris. 8 pp. + seven annexes.

—— (1994). First Planning Session of the IOC-WMO-UNEP Committee for the Global Ocean Observing System (GOOS). Melbourne, Australia, 18-21 April 1994. Intergovernmental Oceanographic Commission (of UNESCO), Paris, France. 41 pp. + eight annexes.

—— (1995a). Joint Scientific and Technical Committee for the Global Ocean Observing System (J-GOOS). Second Session, Paris, 24-26 April 1995. Intergovernmental Oceanographic Commision, *Reports of Governing and Major Subsidiary Bodies.* UNESCO, Paris. 12 pp. + fifteen annexes.

—— (1995b). IOC Committee for the Global Ocean Observing System (GOOS). Second Session, Paris, 6-9 June 1995. Intergovernmental Oceanographic Commision, *Reports of Governing and Major Subsidiary Bodies.* UNESCO, Paris. 22 pp. + nine annexes.

James, D. B. (1976). The history of echinodermology of the Indian Ocean. *J. Mar. Biol. Assoc. India*, 18(2): 298-310.

—— (1982). Ecology of intertidal echinoderms of the Indian Seas. *J. Mar. Biol. Assoc. India*, 24(1&2): 124-129.

—— (1983). Research on Indian echinoderms – a review. *J. Mar. Biol. Assoc. India*, 25(1&2): 91-108.

James, P. S. B. R., Rao, D. S. and Selvaraj, G. S. D. (1983). A resume of marine biological and oceanographic research by the Central Marine Fisheries Research Institute, Cochin, India. *J. Mar. Biol. Assoc. India*, 25(1&2): 158-189. (Issued in 1987)

Jayaraman, R. and Gogte, S. S. (1957). Salinity and temperature variations in the surface waters of the Arabian Sea off Bombay and Sauvashtra coasts. *Proc. Indian Acad. Sci.*, B45: 151-164.

Jones, E. C. (1966a). The general distribution of the species of the calanoid copepod family *Candaciidae* in the Indian Ocean with new records. Proceedings of the Symposium on Crustacea. *J. Mar. Biol. Assoc. India*, pp. 390-405.

—— (1966b). Evidence of isolation between populations of *Candacia pachydactyla* (Dana), Copepoda: Calanoida. Proceedings of the Symposium on Crustacea. *J. Mar. Biol. Assoc. India*, pp. 406-410.

Kabanova, J. (1968). Primary production in the northern part of the Indian Ocean. *Oceanology*, 8: 216-225.

Kidd, R. B. and Davies, T. A. (1978). Indian Ocean sediment distribution since the late Jurassic. *Mar. Geol.*, 26: 49-70.

Kimor, B. (1971). The Suez Canal as a link and barrier in the migration of planktonic organisms. *Proc. Joint Oceanogr. Assembly* (Tokyo, 1970). 480 pp.

—— (1973). Plankton relations of the Red Sea, Persian Gulf and Arabian Sea. In: Zeitschel, B. (ed.), *The Biology of the Indian Ocean*. pp. 221-233. Springer-Verlag, Berlin.

King, L. C. (1983). *Wandering Continents and Spreading Sea Floors on an Expanding Earth*. John Wiley and Sons, Chichester, New York. 232 pp.

Koblentz-Mishke, O. K., Volkovinsky, V. V. and Kabanova, J. G. (1970). Plankton primary production of the world ocean. In: Wooster, W. S. (ed.), *Scientific Exploration of the South Pacific*. pp. 183-193. National Academy of Science, Washington, DC

Kolla, V. and Kidd, R. B. (1982). Sedimentation and sedimentary processes in the Indian Ocean. In: Nairn, A. E. M. and Stehli, F. G. (eds.), *The Ocean Basins and Margins*. Vol. 6, *The Indian Ocean*. pp. 1-45. Plenum Press, New York.

Krey, J. (1973). Primary production in the Indian Ocean – I. In: Zeitschel, B. (ed.), *The Biology of the Indian Ocean*. pp. 115-126. Springer-Verlag, Heidelberg, Berlin, New York.

Krey, J. and Babenard, B. (1976). *Phytoplankton Production Atlas of the International Indian Ocean Expedition*. Inst. für Meereskunde, Kiel.

Lacombe, H. (1951). Contributions à l'étude de l'Océan indien et du secteur adjacent de l'Océan antarctique, faite à l'occasion de l'exploitation des résultats des stations hydrographiques du *Commandant Charcot*. IV. Application de la méthode dynamique à la circulation dans l'Océan indien et le secteur antarctique adjacent. V. Conclusions. VI. Bibliographie. *Bulletin du Comité d'Océanographie et d'Études des Côtes,* III: 459-473.

LaFond, E. C. (1954). On upwelling and sinking off the east coast of India. *Andhra Univ. Ser.*, 49: 117-121.

—— (1957). Oceanographic studies in the Bay of Bengal. *Proc. Ind. Acad. Sci.* (B) 46: 1-46.

Laughton, A. S., Matthews, D. H. and Fischer, R. L. (1969). The structure of the Indian Ocean. In: Maxwell, A. E. (ed.), *The Sea*. 4: 543-586.

Lawson, T. J. (1977). Community interaction and zoogeography of the Indian Ocean *Candaciidae* (Copepoda-Calanoida). *Mar. Biol.*, 42: 71-92.

Le Floch, J. (1951). Contributions à l'étude de l'Océan indien et du secteur adjacent de l'Océan antarctique, faite à l'occasion de l'exploitation des résultats des stations hydrographiques du *Commandant Charcot*. III. Hydrologie et circulation de quelques masses d'eau de l'Océan indien. *Bulletin du Comité d'Océanographie et d'Études des Côtes,* III: 433-458.

Le Pichon, X. and Heirtzler, J. R. (1968). Magnetic anomalies in the Indian Ocean and sea floor spreading. *J. Geophys. Res.*, 73: 2101-2117.

Longhurst, A. R. and Wooster, W. S. (1990). Abundance of oil sardine (*Sardinella longiceps*) and upwelling on the south-west coast of India. *Can. J. Fish. Aquatic Sci.*, 47: 2407-2419.

Lovelock, J. E. (1979). *Gaia. A New Look at Life on Earth*. Oxford University Press, Oxford, vii+157 pp.

Lowrie, W. and Alvarez, W. (1981). One hundred million years of geomagnetic polarity history. *Geology*, 9: 392-397.

Lutjeharms, J. R. E. (1988). Examples of extreme circulation events of the Agulhas retroflection. *S. Afr. J. Sci.*, 84(7): 584-586.

—— (1994). Inter-basin leakage through Agulhas Current filaments. *Deep-Sea Res.* (submitted).

Lutjeharms, J. R. E. and Ansorge, I. J. (1997). The Agulhas Return Current. *J. Phys. Oceanogr.* (submitted).

Lutjeharms, J. R. E. and Ballegooyen, R. C. van (1984). Topographic control in the Agulhas Current system. *Deep-Sea Res.*, 31(11): 1321-1337.

—— (1988). The reflection of the Agulhas Current. *J. Phys. Oceanogr.*, 18(11): 1570-1583.

Lutjeharms, J. R. E. and Cooper, J. (1996). Interbasin leakage through Agulhas Current filaments. *Deep-Sea Res.*, 43(2): 213-238.

Lutjeharms, J. R. E., Bang, N. D. and Duncan, C. P. (1981). Characteristics of the currents east and south of Madagascar. *Deep-Sea Res.*, 28(9): 879-899.

Lutjeharms, J. R. E., Gründlingh, M. L. and Carter, R. A. (1989). Topographically induced upwelling in the Natal Bight. *S. Afr. J. Sci.*, 85(5): 310-322.

Luyten, J. R., Fieux, M. and Gonella, J. (1980). Equatorial currents in the western Indian Ocean. *Science,* 209: 600-603.

Markenday, S. and Srivastava, P. S. (1980). Physical oceanography in India. In: Sears, M. and Merriman, D. (eds.), *Oceanography – The Past*. pp. 551-561. Springer-Verlag, Berlin, Heidelberg, New York.

Markham, A. H. (1880). *The Voyages and Works of John Davis, the Navigator*. Works issued by the Hakluyt Society, No. 49.

Markham, C. R. (1878). A memoir on the Indian Surveys (2nd edition). Printed by the order of Her Majesty's Secretary of State for India in Council, London. 481 pp.

Marshall, N. B. (1954). *Aspects of Deep-Sea Biology*. Hutchinson, London, 380 pp.

Martin, J. (1956a). Utilisation du courantomètre à électrodes remorquées (GEK). 1ère partie. Ch. I. Généralités. Ch. II. Les essais effectués devant Cherbourg. *Bulletin du Comité d'Océanographie et d'Études des Côtes,* VIII: 355-398.

—— (1956b). Utilisation du courantomètre à électrodes remorquées (GEK). 2ème partie. Ch. III. Études faites dans la région de l'équateur magnétique dans le NW de l'Océan indien. Ch. IV. Études faites dans l'Océan indien et l'Océan austral. Détermination complète du courant superficiel par les boucles. *Bulletin du Comité d'Océanographie et d'Études des Côtes,* VIII: 456-505.

—— (1956c). Compte-rendu sommaire de travaux océanographiques français récents dans l'Océan indien et l'Océan austral. *Bulletin du Comité d'Océanographie et d'Études des Côtes,* VIII: 299-305.

Martini, E. (1974). Quaternary fish otoliths from sites 242 and 246 – Leg 25. Deep-Sea Drilling Project. In: Simpson, E. and Schlich, R. (eds.), *Initial Reports of the Deep-Sea Drilling Project*. pp. 647-650. Vol. 25. US Government Printing Office, Washington, DC.

Maximova, M. P. (1971). Nutrients in the troposphere and subtroposphere of the Indian Ocean and their relation to productivity. *Proc. Joint Oceanogr. Assembly* (Tokyo 1970), pp. 490-494.

McGill, D. A. (1973). Light and nutrients in the Indian Ocean. In: Zeitschel, B. (ed.), *The Biology of the Indian Ocean*. pp. 53-102. Springer-Verlag, Berlin, Heidelberg, New York.

McKenzie, D. and Sclater, J. G. (1971). The evolution of the Indian Ocean since the late Cretaceous. *Geophys. J. R. Astron. Soc.*, 25: 437-528.

Menaché, M. (1955). Étude hydrologique de la région d'Anjouan (Comores). *Bulletin du Comité d'Océanographie et d'Études des Côtes,* VII: 419-420 + 2 Figures.

—— (1958). Campagne du *Commandant Robert Giraud* dans le canal de Mozambique. *Bulletin du Comité d'Océanographie et d'Etudes de Côtes*, X: 114.

Menon, K. S. (1931). A preliminary account of the Madras plankton. *Rec. Ind. Mus.*, 33: 489-516.

Menon, M. A. S. (1945). Observations on the seasonal distribution of the plankton of the Travancoe coast. *Proc. Ind. Acad. Sci.*, (B) 22(a): 31-62.

Metzl, N., Moore, B. and Poisson, A. (1990). Resolving the intermediate and deep advective flows in the Indian Ocean by using temperature, salinity, oxygen and phosphate data: the interplay of biogeochemical and geophysical tracers. *Palaeogeography, Palaeoclimatology, Palaeoecology (Global and Planetary Change Section)*, 89: 81-111.

Mohamed, A. F. (1940). The distribution of hydrogen-ion concentration in the northwestern Indian Ocean and adjacent waters. *John Murray Expedition, Scientific Reports*, pp. 121-202.

Moiseev, P. A. (1971). The living resources of the world ocean. (Israel Programme for Scientific Translation, Jerusalem), 334 pp.

Molcard, R., Fieux, M. and Ilahude, A. G. (1996). The Indo-Pacific throughflow in the Timor Passage. *J. Geophys. Res.*, 101(C5): 12411-12420.

Montgomery, R. B. (1958). Water characteristics of the Atlantic and World Oceans. *Deep-Sea Res.*, 5: 134-148.

Mullin, M. M. (1963). Some factors affecting the feeding of marine copepods of the genus *Calanus*. *Limn. Oceanogr.*, 8: 239-250.

Munsch, M. and Schlich, A. (1989). The Rodriguez Triple Junction (Indian Ocean): structure and evolution for the past 1 million years. *Mar. Geophys. Res.*, 11(1): 1-14.

Murray, J. (1895). Summary of the scientific results. *Challenger Reports*, 1608 pp.

Murray, J. and Hjort, J. (1912). *Depths of the Ocean*. MacMillan and Co, Ltd., London. 821 pp.

Nair, P. V. R. and Pillai, V. K. (1983). Productivity of the Indian Seas. *J. Mar. Biol. Assoc. India*, 25(182): 41-50.

Nair, R. V. (1972). Variability in distribution of chaetognaths in the Arabian Sea. *Ind. J. Mar. Sci.*, 1: 85-88.

Nair, R. V. and Rao, T. S. S. (1973a). Chaetognatha of the Arabian Sea. In: Zeitschel, B. (ed.), *The Biology of the Indian Ocean*. pp. 168-196. Springer-Verlag, New York, Berlin, Heidelberg.

—— (1973b). Distribution of chaetognaths in the Arabian Sea. In: Zeitschel, B. (ed.), *The Biology of the Indian Ocean*. pp. 293-317. Springer-Verlag, Berlin, Heidelberg, New York.

—— (1973c). Chaetognaths from the Laccadives with a note on the record of *Spadella angulata*. In: Zeitschel, B. (ed.), *The Biology of the Indian Ocean*. pp. 319-328. Springer-Verlag, Berlin, Heidelberg, New York.

Nair, R. V. *et al.* (1989). Increased particle flux to the deep ocean related to monsoons. *Nature*, 338: 749-751.

Nairn, A. E. M. and Stehli, F. G. (eds.) (1982). *The Ocean Basins and Margins*. Vol. 6, *The Indian Ocean*. Plenum Press, New York.

Naqvi, S. W. A. and Qasim, S. Z. (1983). Inorganic nitrogen and nitrate reduction in the Arabian Sea. *Indian J. Mar. Sci.*, 12: 21-26.

NRC (1979). *The Continuing Quest: Large-Scale Ocean Science for the Future*. National Research Council, USA. National Academy of Sciences, Washington, DC. 91 pp.

—— (1992). *Oceanography in the Next Decade*. National Research Council, USA. National Academy Press, Washington, DC. 202 pp.

Neyman, A. A., Sokolova, M. N., Vinogradov, N. G. and Pasternak, F. A. (1973). Some patterns of the distribution of bottom fauna in the Indian Ocean. In: Zeitschel, B. (ed.), *The Biology of the Indian Ocean*. pp. 467-474. Springer Verlag, Berlin, Heidelberg, New York.

Nielsen, E. S. and Jensen, E. A. (1957-1959). Primary organic production of organic matter in the oceans. *Galathea Report*, Vol. 1, Scientific Results of the Danish Deep-Sea Expedition Round the World, 1950-52. pp. 49-136.

NIO (1977). *Proceedings of the Symposium on Warm Water Zooplankton*. Special publication, National Institute of Oceanography, Goa. vii + 721 pp.

Norton, J. O. and Sclater, J. G. (1979). A model for the evolution of the Indian Ocean and the break-up of the Gondwanaland. *J. Geophys. Res.*, 84(12): 6803-6830.

Open University (1988). *Ocean Basins, Their Structure and Evolution*. Prepared by Open University – Geology Course Team. Bearman, G. (ed.). 171 pp.

Ovchinnikov, I. M. (1961). Circulation of waters in the northern part of the Indian Ocean during the winter monsoon. *Okeanologia*, 4: 18-24.

Panikkar, N. K. and Jayaraman, R. (1956). Some aspects of productivity in relation to the fisheries of the Indian neritic waters. *Proceedings of the 8th Pacific Science Congress, Philippines, 1953*. 3A: 1112-1122.

Panikkar, N. K. and Rao, T. S. S. (1973). Zooplankton investigations in Indian waters and the role of the Indian Ocean Biological Centre. IIOE Handbooks, International Zooplankton Collections, IOBC, 5: 111-163.

Panikkar, N. K. and Srinivasan, T. M. (1972). Early concepts of oceanographic phenomena of the Indian Ocean. *Proc. Roy. Soc. Edinburgh*, Section B72: 263-272.

Parulekar, A. H. (1981). Marine fauna of Malvan, central west coast of India. *Mahasagar, Bull. Nat. Inst. of Oceanogr. India*, 14(1): 33-45.

Parulekar, A. H., Harkantra, S. N. and Ansari, Z. A. (1982). Benthic production and the assessment of demersal fishery resources of the Indian Seas. *Indian J. Mar. Sci.*, 11: 107-114.

Patterson, H. (ed.) (1957). Reports of the Swedish Deep Sea Expedition, 1947-48. Vol. 1. The Ship, Its Equipment and Voyage. Elanders Boktryckeri Aktiebolag, Gotesborg.

Pillai, C. S. G. (1969). The distribution of corals on a reef at Mandapan (Palk Bay, S. India). *J. Mar. Biol. Assoc. India*, 11(1&2): 62-72.

—— (1983). Structure and generic diversity of recent *Scleractinia* of India. *J. Mar. Biol. Assoc. India*, 25(1&2): 78-90.

Prasad, R. R. (1954). The characteristics of marine plankton at an inshore station in the Gulf of Mannar near Mandapam. *Indian J. Fish.*, 1(1): 1-36.

Premachand, K., Sastry, J. A. and Murthy, C. S. (1986a). Watermass structure in the western Indian Ocean. Part III. The spreading and transformation of Red Sea water masses. *Mausam*, 37(3): 317-325.

—— (1986b). Watermass structure in the western Indian Ocean. Part II. The spreading and transformation of Persian Gulf water. *Mausam*, 37(2): 179-187.

Qasim, S. Z. (1973). Productivity of backwaters and estuaries. In: Zeitzschel, B. (ed.), *The Biology of the Indian Ocean*. pp. 143-154. Springer-Verlag, Berlin, Heidelberg, New York.

—— (1977). Biological productivity of the Indian Ocean. *Indian J. Mar. Sci.*, 6(2): 122-137.

—— (1982). Oceanography of the northern Arabian Sea. *Deep-Sea Res.*, 29(9a): 1041-1068.

Qasim, S. Z., Bhattalhiri, P. M. A. and Devassy, V. P. (1972a). The effect of intensity and quality of illumination on the photosynthesis of some tropical marine phytoplankton. *Mar. Biol.*, 16: 22-27.

—— (1972b). The influence of salinity on the rate of photosynthesis and abundance of some tropical phytoplankton. *Mar. Biol.*, 12: 200-206.

Qasim, S. Z. and Wafar, M. V. M. (1979). Occurrence of living corals at several places along the west coast of India. *Mahasagar, Bull. Nat. Inst. Oceanogr. India*, 12(1): 53-58.

Rajendran, Isaac, A. D., Chandrasekharan, F. and Balakrishnan, N. (1983). History of the Indian pearl banks of the Gulf of Mannar. *J. Mar. Biol. Assoc. India*, 18(3): 549-576.

Ramage, C. S., Miller, F. R. and Jefferies, C. (1972). *Meteorological Atlas of the International Indian Ocean Expedition. Surface Climate of 1963 and 1964*. National Science Foundation, Washington, DC.

Ramage, C. S. and Raman, C. V. R. (1972). *Meteorological Atlas of the International Indian Ocean Expedition. The Upper Air*. National Science Foundation, Washington, DC.

Ramana, V., Murthy, K. and Rao, K. B. (1987). Survey of meiofauna in the Gautam Godavari estuary. *J. Mar. Biol. Assoc. India*, 29(1&2): 37-45.

Rao, S. R. (1970). Shipping in ancient India. In: *India's Contribution to World Thought and Culture*. (1st edition), Vivekananda Rock Memorial Committee. 705 pp.

Rao, T. C. S. (1976). Structural features of the ocean bottom off the west coast of India. *J. Mar. Biol. Assoc. India*, 18(2): 126-139.

Rao, T. S. S. (1967). An Indian views the International Indian Ocean Expedition (IIOE). *Oceanogr. Mar. Biol., Ann. Rev.*, 1967, 5: 111-118.

—— (1973). Zooplankton of the Indian Ocean. In: Zeitschel, B. (ed.), *The Biology of the Indian Ocean*. pp. 243-255. Springer-Verlag, Berlin, Heidelberg, New York.

—— (1979). Zoogeography of the Indian Ocean. In: Van der Spoel, S. and Pierrot-Bults, A. C. (eds.), *Zoogeography and Diversity of Plankton*. pp. 254-292. Halsted Press – A division of John Wiley & Sons Inc., New York.

Rao, T. S. S. and Kelly, S. (1962). Studies on the Chaetognatha of the Indian Seas. Part VI. On the biology of *Sagitta enflata* in the waters of Lawson's Bay. *J. Zool. Soc. India*, 14(2): 219-225.

Rao, T. S. S. and Rao, V. C. (1962). Short-term variations in the hydro-biological conditions of the Waltair coast. *J. Mar. Biol. Assoc. India*, 4(1): 23-43.

Rao, U. M. (1969). Coral reef flora of the Gulf of Mannar and Palk Bay. In: *The Proceedings of the Symposium on Corals and Coral Reefs. J. Mar. Biol. Assoc. India*, pp. 12-14.

—— (1987). Algae of Indian estuaries. *J. Mar. Biol. Assoc. India*, 29(1&2): 1-9.

—— (1989). Indian seaweed resources and their management. *J. Mar. Biol. Assoc. India*, 31(1&2): 234-238.

Rasalam, E. J. and Sebastian, M. J. (1976). The lime-shell fisheries of the Vembanad lake, Kerala. *J. Mar. Biol. Assoc. India*, 18(2): 323-355.

Raymont, J. E. G. (1980). *Plankton and Productivity in the Oceans*. Vol. I & II. Pergamon Press, New York. 625 pp. (2nd revised edition).

Redfield, A. C. (1934). On the proportions of organic derivatives in sea water and their relation to the composition of plankton. James Johnstone Memorial Volume. pp. 176-192. Liverpool.

—— (1958). The biological control of chemical factors in the environment. *Amer. Sci.*, 46: 205-221.

—— (1960). The distribution of phosphorus in the deep oceans of the world. Association d'Océanographie physique of the International Union of Geodesy and Geophysics. Procès-verbaux, 7(G16): 189-191.

Redfield, A. C., Ketchum, B. H. and Richards, F. A. (1963). The influence of the organisms on the composition of sea water. In: Hill, M. N. (ed.), *The Sea*. 2: 26-77. Wiley-Interscience, New York.

Reverdin, G. and Fieux, M. (1987). Sections in the western Indian Ocean – variability in the temperature structure. *Deep-Sea Res.*, 34(4): 601-626.

Reverdin, G., Fieux, M. and Gonella, J. (1982). Expérience SINODE (Surface Indian Ocean Dynamics Experiment). In: *Conférence des Utilisateurs ARGOS*. Paris, France, 20-22 Avril 1982. pp. 37-42.

Reverdin, G., Fieux, M., Gonella, J. and Luyten, J. (1983). Free drifting buoy measurements in the Indian Ocean equatorial jet. In: Nihoul, J. C. (ed.), *Hydrodynamics of the Equatorial Ocean*. Liège, Belgium, 1983.

Richards, F. A. (1965). Anoxic basins and fjords. In: Riley, J. P. and Skirrow, G. (eds.). *Chemical Oceanography*, 1: 611-645. Academic Press.

Rochford, D. J. (1964). Salinity maxima in the upper 1000m of the north Indian Ocean. *Austral. J. Mar. Freshw. Res.* 15(1): 1-24.

—— (1966). Source regions of oxygen maxima in intermediate depths of the Arabian Sea. *Austral. J. Mar. Freshw. Res.*, 17: 1-30.

—— (1969). Seasonal variation in the Indian Ocean along 110°E: hydrological structure of the upper 500m. *Austral. J. Mar. Freshw. Res.*, 20: 1-50.

Rodman, M. R. and Gordon, A. L. (1982). Southern Ocean bottom water of the Australian-New Zealand sector. *J. Geophys. Res.*, 87: 5771-5778.

Row, R. W. H. (1911). Report on the sponges collected by Cyril Crossland from the Red Sea in 1904-5. Part II. *J. Linn. Soc.*, 31(208): 287-400.

Ruben, S., Apparao, T. and Manickam, P. E. S. (1984). Sea-urchin resources of Waltair coast. *J. Mar. Biol. Assoc. India*, 26 (1&2): 37-41.

Ryther, J. H. (1959). Potential productivity of the sea. *Science*, 130: 602-608.

—— (1965). Geographic variations in productivity. In: Hill, M. N. (ed.), *The Sea*. 2(17): 347-380. Wiley-Interscience, New York.

Ryther, J. H., Hall, J. R., Pease, A. K., Bakun, A. and Jones, M. M. (1966). Primary production in relation to the chemistry and hydrography of the western Indian Ocean. *Limn. Oceanogr.*, 11: 371-389.

Ryther, J. H. and Menzel, D. W. (1965). On the production, composition and distribution of organic matter in the western Arabian Sea. *Deep-Sea Res.*, 12: 199-209.

Saenger, P., Hegerl, E. J. and Davie, J. D. S. (eds.) (1983). Global status of the mangrove ecosystems. Commission on Ecology, Paper No. 3, 88 pp. *The Environmentalist*. International Union for Conservation of Nature and Natural Resources.

Sankaranarayan, V. N. and Reddy, C. V. G. (1968). Nutrients of the northwestern Bay of Bengal. *Mahasagar, Bull. Nat. Inst. Oceanogr. India*, 38: 148-162.

Sarkar, A. and Sen Gupta, R. (1992). On chlorinated hydrocarbons in the Indian Ocean. In: Desai, B. N. (ed.), *Oceanography of the Indian Ocean*. pp. 385-397.

Scherzer, K. (1861). Narrative of the Circumnavigation of the Globe by the Austrian Frigate *Novara* in the Years 1857-58 and 59. Vol 1. Saunders, Otley and Co., London. 485 pp.

Schott, F., Fieux, M., Kindle, J., Swallow, J. C. and Zantopp, R. (1988). The boundary currents east and north of Madagascar. 2. Direct measurements and model comparisons. *J. Geophys. Res.*, 93(C5): 4963-4974.

Schott, F., Swallow, J. C. and Fieux, M. (1989). Deep currents underneath the equatorial Somali Current. *Deep-Sea Res.*, 26: 1191-1199.

—— (1990). The Somali Current at the equator: annual cycle of currents and transports in the upper 1000m, and connection to neighbouring latitudes. *Deep-Sea Res.*, 37: 1825-1848.

Sen Gupta, R. and Naqvi, S. W. A. (1984). Chemical oceanography of the Indian Ocean. *Deep-Sea Res.*, 32(6-8A): 671-706.

Sen Gupta, R., Rajgopal, M. D. and Qasim, S. Z. (1976). Relationship between dissolved oxygen and nutrients in the northwestern Indian Ocean. *Indian J. Mar. Sci.*, 5: 201-211.

Sewell, R. B. S. (1925-1938). Geographic and oceanographic research in Indian waters. *Mem. Roy. Asiatic Soc. Bengal*. 550 pp.

—— (1929). The copepods of the Indian seas. Calanoida. *Mem. Indian Mus.*, pp. 1-407.

—— (1935). Studies on corals and coral formations of Indian waters. *Mem. Roy. Asiatic Soc. Bengal*, 9: 461-539.

—— (1947). The free-swimming planktonic copepoda: systematic account. *John Murray Expedition, Scientific Reports*, 8(3): 1-303.

—— (1948). The free-swimming planktonic copepoda: geographical distribution. *John Murray Expedition, Scientific Reports*, 3(3): 317-592.

Sharma, R. and Rao, A. S. (1992). Geological factors associated with megabenthic activity in the central Indian Basin. *Deep-Sea Res.*, 39(3): 705-715.

Shepard, F. P. (1973). *Submarine Geology*. (3rd edition). Harper and Row Publishers, New York, London. 517 pp.

Sherman, K., Alexander, L.M. and Gold, B.D. (1990). Large marine ecosystems: patterns, processes and yields. *AAAS Symposium*. AAAS Publishers, Washington DC.

Siddiquie, H. N., Gujar, A. R., Hashmi, N. H. and Valsangkar, A. B. (1984). Superficial mineral resources of the Indian Ocean. Proceedings of the *Mabahiss*/John Murray International Symposium. *Deep-Sea Res.*, 31(68A): 763-813.

Simpson, E. S. and Schlich, W. R. (1979). *Initial Reports of the Deep Sea Drilling Project – Under the Joint Oceanographic Institutions for Deep Earth Sampling (JOIDES)*. Vol. XXV. US Government Printing Office, Washington, DC. 884 pp.

Smith, S. L. (1982). The northwestern Indian Ocean during the monsoons of 1979: distribution, abundance and feeding of zooplankton. *Deep-Sea Res.*, 29(11A): 1331-1363.

—— (1988). Is the Somali current a 'biological river' in the northwestern Indian Ocean? In: Thomson, M. F. and Tirmizi, N. M. (eds.), *Marine Science of the Arabian Sea – Proceedings of an International Conference 1988.* pp. 37-45.

—— (1989). Biological indications of active upwelling in the northwestern Indian Ocean in 1964 and 1979 and a comparison with Peru and northwest Africa. *Deep-Sea Res.*, 31: 951-967.

—— (1992). Secondary production in waters influenced by the upwelling off the coast of Somalia. In: Desai, B. N. (ed.), *Oceanography of the Indian Ocean.* pp. 191-199.

Smith, S. L. and Codispoti, L. A. (1980). Southwest monsoon of 1979: chemical and biological response of Somali coastal waters. *Science*, 209: 597-600.

Sokolova, M. N. and Pasternak, F. A. (1962). Quantitative distribution of bottom fauna in the northern parts of the Arabian Sea and Bay of Bengal. *Dokl. Akad. Nauk. SSSR*, 144: 645-648.

—— (1964). Quantitative distribution and trophic zonation of the bottom fauna in the Bay of Bengal and Andaman Sea. *Trudy Inst. Okeanol. Akad. Nauk, SSSR.*, 64: 271-296.

Spencer, D., Broecker, W. S., Craig, H. and Weis, R. F. (1982). *GEOSECS Indian Ocean Expedition*, Vol. 6, Section and Profiles. US Government Printing Office, Washington, DC. 140 pp.

Stüiver, M., Quay, P. D. and Ostlund, H. G. (1983). Abyssal C^{14} and the age of the world oceans. *Science*, 219: 849-851.

Subrahmanyan, R. (1946). A systematic account of the marine plankton diatoms of the Madras coast. *Proc. Ind. Acad. Sci.*, 24: 85-197.

—— (1959). Studies on the phytoplankton of the west coast of India. Parts I & II. *Proc. Ind. Acad. Sci.*, 50: 113-252.

Subrahmanyan, R. and Sarma, A. H. V. (1960). Studies on the phytoplankton of the west coast of India. Part III. Seasonal variation of phytoplankton and environmental factors. *Indian J. Fish.*, 7: 307-336.

Subrahmanyan, R. and Sen Gupta, R. (1963). Studies on the plankton of the east coast of India. *Proc. Ind. Acad. Sci.*, Part 1, 57B: 1-14.

—— (1965). Studies on the plankton of the east coast of India. Part II. Seasonal cycle of plankton and factors affecting the marine plankton production with special reference to the iron content of water. *Proc. Ind. Acad. Sci.*, 41: 12-25.

Suda, A. (1973). Tuna fisheries and their resources in the Indian Ocean. In: Zeitschel, B. (ed.), *The Biology of the Indian Ocean.* pp. 431-449. Springer-Verlag, Berlin, Heidelberg, New York.

Sukhanova, G. N. (1962). On the tropical phytoplankton of the Indian Ocean. *Dokl. Akad. Nauk. SSSR.*, 142: 1162-1164.

—— (1964). The phytoplankton of the north eastern part of the Indian Ocean in the season of the SW monsoon. *Trudy. Inst. Okeanol. Akad. Nauk. SSSR*, 65: 24-31.

Sverdrup, H. U., Johnson, M. W. and Fleming, R. H. (1942). *The Oceans, Their Physics, Chemistry and General Biology.* Prentice Hall, Englewood Cliffs. 1087 pp.

Swallow, J. C. (1984). Some aspects of physical oceanography of the Indian Ocean. *Deep-Sea Res.*, A, 31(6-8A): 639-650.

Swallow, J. C. and Bruce, J. G. (1966). Current measurements off the Somali coast during the southwest monsoon of 1964. *Deep-Sea Res.*, 13: 861-888.

Swallow, J. C. and Fieux, M. (1982). Historical evidence for two gyres in the Somali Current. *J. Mar. Res.*, 40 (supplement): 747-755.

Swallow, J. C., Molinari, R. L., Bruce, J. C., Brown, O. B. and Evans, R. H. (1983). Development of near surface flow pattern and water mass distribution in the Somali basin, in response to the southwest monsoon of 1979. *J. Phys. Oceanogr.*, 3: 1398-1415.

Swallow, J.C., Schott, F. and Fieux, M. (1991). Structure and transport of the East Africa Coastal Current. *J. Geophys. Res.*, 96(C12): 22, 245-22, 257.

Taylor, F. J. R. (1973). General features of dinoflagellate material collected by the *Anton Bruun* during the International Indian Ocean Expedition.

Tchernia, P. (1949). Compte-rendu succinct des observations océanographiques faites par l'Aviso polaire *Commandant Charcot* pendant la campagne 1948-49. *Bulletin du Comité d'Océanographie et d'Études des Côtes,* I(8): 10-21.

—— (1951a). Compte-rendu préliminaire des observations océanographiques faites par l'Aviso polaire *Commandant Charcot* (1949-50) – 1ère partie. *Bulletin du Comité d'Océanographie et d'Études des Côtes,* III: 13-22.

—— (1951b). Compte-rendu préliminaire des observations océanographiques faites par l'Aviso polaire *Commandant Charcot* (1949-50) – 2ème partie. *Bulletin du Comité d'Océanographie et d'Études des Côtes,* III: 40-57.

—— (1951c). Contributions à l'étude de l'Océan indien et du secteur adjacent de l'Océan antarctique, faite à l'occasion de l'exploitation des résultats des stations hydrographiques du *Commandant Charcot*. I. Introduction (French and English text), II. La structure hydrologique de l'Océan indien. *Bulletin du Comité d'Océanographie et d'Études des Côtes,* III: 414-432.

—— (1957a). Observations hydrographiques faites par le Navire hydrographique *La Pérouse* dans l'Océan indien. Trajet Madagascar-Iles Kerguelen-Madagascar. *Bulletin du Comité d'Études d'Océanographie et des Côtes,* IX: 580-582.

—— (1957b). Sur l'origine des eaux profondes du nord-ouest de l'Océan indien. *Bulletin du Comité d'Océanographie et d'Études des Côtes,* IX: 545-548.

—— (1980). *Descriptive Regional Oceanography.* Pergamon Press Inc., NY 10523, USA. 253 pp.

Tchernia, P. and Lizeray, J. C. (1960). Océan indien: Observations rélatives à l'hydrologie du bassin Nord-Australien. *Bulletin du Comité d'Études d'Océanographie et des Côtes,* XII: 371-388. (IGY: 1957-58)

Tchernia, P., Lacombe, H. and Guibout, P. (1958). Sur quelques observations hydrologiques rélatives à la région équatoriale de l'Océan indien. *Bulletin du Comité d'Études d'Océanographie et des Côtes,* X: 115-143.

Teas, H. J. (1983). *Biology and Ecology of Mangroves.* Dr. W. Junk Publishers, The Hague, Boston, Lancaster.

Thangavelu, R. and Sanjeevaraj, P. J. (1985). Exploration of shell deposits in Publicat Lake. *J. Mar. Biol. Assoc. India*, 27 (172): 124-128.

Thomas, P. A. (1976). The history of spongiology of the Indian Ocean. *J. Mar. Biol. Assoc. India*, 18(3): 610-625.

—— (1983). Distribution and affinity of the sponge fauna of the Indian region. *J. Mar. Biol. Assoc. India*, 25(1&2): 7-16.

Thomas, P. A. and George, R. M. (1986). A systematic appraisal of the commercially important gorgonids of the Indian Sea. *J. Mar. Biol. Assoc. India*, 28(1&2): 96-112.

Thompson, R. O. Y. R. (1984). Observations of the Leeuwin Current off western Australia. *J. Phys. Oceanogr.*, 14: 623-628.

Thomson, M. F. (ed.) (1988). *Marine Biodeterioration.* Oxford and IBH Publishing Co., Delhi, 826 pp.

Thomson, M. F., Sarojini, R. and Nagabhushanan, R. (eds.) (1986). *Biology of Benthic Marine Organisms.* Oxford and IBH Publishing Co., New Delhi, Bombay, Calcutta. 608 pp.

—— (1991). *Bioactive Compounds from Marine Organisms.* Oxford and IBH Publishing Co, New Delhi, Bombay, Calcutta. 410 pp.

Thorrington-Smith, M. (1971). West Indian Ocean phytoplankton: a numerical investigation of phytohydrographic regions and their characteristic phytoplankton associations. *Mar. Biol.*, 9: 115-137.

Tomczak, M. and Godfrey, J. S. (1994). *Regional Oceanography: an Introduction.* Pergamon Press, Oxford, UK. 422 pp.

Tranter, D. J. (1973). Seasonal studies of a pelagic ecosystem (meridian 110°E). In: Zeitschel, B. (ed.), *The Biology of the Indian Ocean.* pp. 487-520. Springer Verlag, New York, Berlin, Heidelberg.

Uda, M. (1966). Upwelling of subpolar intermediate water in the tropical zone of the world oceans and its relation with location of regions of high productivity, including regions favorable for tuna fishing. IInd International Oceanographic Congress, Moscow. Abstract 415.

Udintsev, G. B. (ed.) (1975). *Geological and Geophysical Atlas of the Indian Ocean.* Academy of Science, Moscow, USSR. 152 pp.

UNESCO (1963). *International Marine Science*, Vol. 1, 2 and 3. UNESCO, Paris.

Untawale, A. G., Reddy, C. R. K. and Deshmuke, G. V. (1989). Ecology of the intertidal benthic algae of the northern Karnataka coast. *Indian J. Mar. Sci.*, 18(2): 73-84.

Van der Spoel, S. and Heyman, R. P. (1983). *A Comparative Atlas of Zooplankton: Biological Patterns in the Oceans.* Springer-Verlag, Berlin. 169 pp.

Van der Spoel, S. and Pierrot-Bults, A. C. (1979). *Zoogeography and Diversity of Plankton.* John Wiley & Sons. 410 pp.

Vannucci, M. (1989). *The Mangroves and Us – A Synthesis of Insights.* Indian Association for the Advancement of Science, New Delhi. 203 pp.

Vannucci, M. and Navas, D. (1973). Distribution of hydromedusae in the Indian Ocean. In: Zeitschel, B. (ed.), *The Biology of the Indian Ocean.* pp. 273-282. Springer Verlag, Berlin, Heidelberg, New York.

Vannucci, M., Santhakumari, V. and Dos Santos, E. P. (1970). The ecology of hydromedusae from Cochin area. *Mar. Biol.*, 7: 49-58.

Varadachari, V. V. R. (1961). On the process of upwelling and sinking on the east coast of India. Mahadwan Volume, pp. 159-162. Osmania University Press, Hyderabad.

Vaughan, T. W. *et al.* (1937). *International Aspects of Oceanography.* National Academy of Science, Washington, DC. 221 pp.

Vine, F. J. (1966). Spreading of the ocean floor – new evidence. *Science*, 154: 1405-1415.

Vine, F. J. and Hess, H. H. (1968). Sea floor spreading. In: Maxwell, A. E. (ed.), *The Sea.* pp. 587-622. Wiley-Interscience.

Vinogradov, M. E. and Veronina, M. M. (1961). On the influence of oxygen deficiency on the distribution of plankton in the Arabian Sea. *Okeanologia*, 1: 670-678.

Vinogradova, N. G. (1959). The zoogeographical distribution of the deep water fauna in the abyssal zone of the ocean. *Deep-Sea Res.*, 5: 205-213.

Warren, B. A. (1978). Bottom water transport through the southwest Indian ridge. *Deep-Sea Res*, 25: 315-322.

—— (1981). Deep circulation in the world ocean. In: Warren, B. A. and Wunsch, C. (eds.), *Evolution of Physical Oceanography.* Ch. 1, pp. 6-41. MIT Press.

Warren, B. A. and Johnson, G. C. (1992). Deep currents in the Arabian Sea in 1987. *Mar. Geol.*, 104: 279-288.

Warren, B., Stommel, H. and Swallow, J. C. (1966). Water masses and patterns of flow in the Somali Basin during the southwest monsoon of 1964. *Deep-Sea Res.*, 13: 825-860.

Wegener, A. (1924). *The Origin of the Continents and Oceans.* Dover, New York.

Weiss, R. F., Broecker, W. S., Craig, H. and Spencer, D. (1983). *GEOSECS Indian Ocean Expedition.* Vol. 5., *Hydrographic Data, 1977-78.* National Science Foundation, Washington, DC.

Wickstead, J. H. (1963). Estimates of total zooplankton in the Zanzibar area of the Indian Ocean with a comparison of the results with two different nets. *Proc. Zool. Soc. Lond.*, 141(3): 577-608.

Wolff, T. (1967). *Danish Expeditions on the Seven Seas.* Rhodos International Science and Arts Publishers. 325 pp. plus bibliography.

—— (1968). The Danish Expedition to 'Arabia Felix' (1761-1767). *Bulletin de l'Institut Océanographique de Monaco*, Numero Special 2: 581-601.

Wyrtki, K. (1962). The upwelling in the region between Java and Australia during the southwest monsoon. *Austral. J. Mar. Freshw. Res.*, 13(3): 217-225.

—— (1971). *Oceanographic Atlas of the International Indian Ocean Expedition*. National Science Foundation. US Government Printing Office, Washington, DC. 531 pp.

—— (1973). Physical oceanography of the Indian Ocean. In: Zeitschel, B. (ed.), *The Biology of the Indian Ocean*. pp. 18-36. Springer-Verlag, New York, Berlin, Heidelberg.

—— (1987). Indonesian through flow and the associated pressure gradient. *J. Geophys. Res.*, 92(C12): 12941-12946.

Yonge, M. (1972). The inception and significance of the *Challenger* Expedition. *Proc. Roy. Soc. Edinburgh*, (B)72: 1-13.

York, D. (1993). The earliest history of the earth. *Sci. Amer.*, January 1993, pp. 90-96.

You, Y. and Tomczak, M. (1993). Thermocline circulation and ventilation in the Indian Ocean derived from water mass analysis. *Deep-Sea Res.*, 40(1): 13-57.

Zeitzschel, B. (ed.) (1973). *The Biology of the Indian Ocean*. Springer Verlag, New York, Heidelberg, Berlin. 549 pp.

Above:
Australia's ORV *Franklin* has conducted Indian Ocean research.
Source: CSIRO.

Back cover (left to right):
Marine living resources
Angel fish, a coral denizen. (Photo: Jordi Camp)
Diver with porcupine fish. (Photo: Serbal)
Food from the sea finds far-distant markets. (Photo: F. Botts)
(Photos by courtesy of Ediciones del Serbal, Barcelona)
Modern oceanographic platforms
Shinkai 6500 submersible. (Photo by courtesy of JAMSTEC, Japan)
ORV *Franklin.* (Photo by courtesy of CSIRO)
Oceanographic end product example
Extract from *Carte du Fond des Océans*, by M. Tanguy de Rémur. (Illustration by courtesy of Hachette-Guides Bleu)